Preface

This book contains contributions to an international symposium entitled 'The Scientific and Technological Revolution and Social Sciences'. The symposium, held 6–10 September 1976 in Prague, was organized jointly by Unesco and the Czechoslovak Academy of Sciences in co-operation with the Czechoslovak Commission for Unesco.

Philip Maxwell's essay on certain aspects of recent Latin American experience was added to the symposium papers, as well as the 'Editor's Foreword', a brief bibliographical note and the text of the final report.

We regret the death, prior to the symposium, of Professor A. N. J. Hollander of the University of Amsterdam. His paper had been communicated to Unesco and it is included in this volume.

We are grateful to Professor Robert S. Cohen of Boston University who acted as editor for the present publication.

Contents

Editor's foreword

Robert S. Cohen

Boston University,
United States of America

In the past thirty years an enormous amount of literature has been devoted to the impact of science and technology on society. That very phrase resounds with our problem, for the effect of science and technology is rightly said to be an 'impact'. Indeed, it was Unesco's journal *Impact of Science on Society*, founded in 1950, which early recognized how far-reaching the matter would become. In research literature, mostly specialist approaches through the social science disciplines have been followed, not infrequently combined with interdisciplinary pathways. In fact, among the endless filiations of narrowing disciplines and subdisciplines of the sciences, one specific subject of study has demanded an integrated understanding, and resisted the development of the usual Babel of mutually incomprehensible specialist jargons. That object of scientific study is the human enterprise of science.

It is not surprising that the scholarly study of science and technology, a field which, for a while, was neatly named the 'science of science', demands an intellectual strategy that is socially based. To any informed and sympathetic observer, science evidently has had as many aspects and qualities as human culture at large. The sociology, history, politics and economics of science and technology should be integrated, and these should be joined with the psychology, aesthetics, logic and methodology, anthropology, and certainly the philosophy of science and technology. Whether all these systems of knowledge can be brought into an orderly integration is yet to be seen, but the goal remains: to grasp science and technology as a whole, in their full nature and in their ramifying effects upon the social practices of mankind as well as upon the evolving qualities of human cognition.

Abstractly considered, we might expect a similar requirement for the integration of specialist disciplines in other studies of human practice, perhaps the first candidate being the science and technology of medicine.

But while great physicians since ancient times have spoken of the 'whole human being', and medical students are often told to treat the patient as a whole rather than treat either the disease or the diseased organ, nevertheless, widespread technical specialism dominates the medical profession. The goal of integration in the treatment and prevention of illness has receded, although it frequently is championed by anti-specialist and even anti-scientific Luddite forces among the critics of medicine as it exists in the modern industrial society of the scientific and technological revolution (STR). Moreover, great medical advances, marked by an improved understanding of a great variety of problems of human illness, were accomplished at a time when the political economy, historical sociology, and philosophy of medicine were the concerns of only a few creative investigators working at the margins of professional attention. Although acknowledged as desirable, a fully integrated medical understanding is rarely achieved even by general practitioners in this time of scientific and technological revolution. In truth, it seems that the practice of medicine does not require wholeness in order to achieve the goals our medical men and women generally have adopted—wholeness either in their epistemology or in the social application of their skills and learning. Nor have the planning and development of medicine required deliberate social-scientific attention to medicine as science, technology, or institution, however enlightening the history, sociology and philosophy of medicine have separately been from time to time.

In contrast with studies of medicine, the professional scientific studies that scientists and engineers have made of their sciences and technology, and of their lives in societies, have required integrated understanding; otherwise they flounder and fail. The American philosopher, Emerson, saw the scholar as just one aspect of humanity. 'Man thinking', he said, was only to be comprehended as a part of the entire man who acts in all the other ways that men act. Thinking is acting too, and science and technology are historically evolved stages of human practice. Therefore, the cognitive network for understanding science and technology as historically novel ways of cognition and of practical work plainly would be incomplete if we were to understand only the cognitive achievement itself, along with its empirical association with observations, experiments, testing and constructing; and just as plainly would be incomplete, uninformed, even uncomprehending, if we were to understand only the social history, political economy, and cultural psychology of the sciences and technologies, omitting the technical content.

First, we need to understand the cognitive material of the scientific and technological disciplines, in order to know what their impact upon human life has been in our time, and why it is both so strong and so problematic. Second, we need to understand the social conditions of the rise and subsequent vicissitudes of science and technology in order to comprehend the array of personal and social factors brought into human life in the wake of the several stages of scientific and technological revolution in the past five centuries.

Both of these substantial tasks of investigation—of cognitive content and social context, in other words the 'internal' and 'external' modes of understanding the development of science and technology—contribute to a scientific study of science and technology. With such an informed, rational, critical understanding of the existing problems and achievements of science and technology, and of their different possible developments, practical options and social choices will begin to become sensible, to become themselves scientific and to some degree 'technical'. If 'technical' seems too cold a term for procedures of social change, too fraught with inhumane images of inflexible maximization strategies, let us say that the complete study of science and technology, the broadest interdisciplinary study of modern cognitive practice, can contribute in an essential way to rational social choices for the use and development of science and technology. Evaluative judgement and practical reasoning, then, will be supported and enhanced by this new 'science of science'.

But practical reason, at least since Aristotle and certainly in our time of game-theoretic policy science, requires practical premisses, i.e. decisions to be made about ambiguous and often conflicting alternatives. These alternatives are social values, taken as ultimate or proximate, or as merely hypothetical—even Utopian—but in any case defining the instrumental logic for dealing with social processes. Therefore, the social implications of science and technology have come to be understood as the joint results of historically situated practical decisions, however impersonal or unconscious they may be, together with the available technology of the scientific enterprise as it is. There is no escape from analysing social realities if we are to understand the full sweeping nature of science, any more than from understanding scientific concepts and technological options.

As a result, science, which has so easily accepted a powerful role in all the social conflicts of our time, will itself be comprehended best by those social sciences which most successfully clarify those conflicts. In the development of scientific and philosophical understanding of science and technology, all our sophisticated means of investigation will be useful: logic, conceptual and methodological analysis, comparative and systematic investigation of values, and the social-historical sciences of social structures, political economy, and social conflicts.

No doubt the turmoil of the twentieth century has provided the stimulus, irritant and reciprocal context for the scientific revolution; no doubt the accelerating pace of political events has forced a new pace upon scientific thought, experiment and application; no doubt the explosively expanding industrial and military economies have vastly expanded the scale, and diverted the motivations, of science; no doubt wars mobilized the sciences to do their utmost in an imaginative and inventive effort; there is no doubt, then, that since the Russian October Revolution and the America of Ford, General Electric and General Motors, and since the Fascist counter-revolutions, science and technology have reflected and supported both the progressive and reactionary forces of the century.

However, while such mutual resonances of science and society are overwhelming, science and even technology remained distinctly autonomous to some extent. From the sciences of this century there have come intrinsically valued revolutionary movements of cognition and methodology: scientific revolutions in physics and mathematics first, soon technological revolutions in power and energy, in the production and distribution of goods, and in communications; and later, scientific revolutions in the biology and chemistry of life, and the revolutionary technology of psychology in a mass culture. Situated among the political crises and opportunities of the twentieth century, and affecting their courses and outcomes, is the general and partly autonomous scientific and technological revolution of our age.

But there is also a second scientific and technological revolution: the practical revolution in the political economy. As in the previous two centuries, science has revolutionized commodity production again in our time by a new industrial revolution consisting of automated planning and control of the productive process, a revolution that is inextricably bound up with a revolution imposed on the consciousness that controls productive forces. In short, science and production are now fused.

The Prague symposium of September 1976 was organized in order to elucidate various modes of meeting the challenges posed for mankind by science and technology. These social relations of science, which have been the central concern of Unesco from its founding, affect Member States and all human beings in many ways. As a first step in elucidation, a principal topic was set: that contemporary science has undergone a revolution, and technology with it.

In this book, several themes are explored and developed from quite differing viewpoints.

First, in Part One, 'The Scientific and Technological Revolution as Social Process'. Mendelsohn investigates the social process whereby science, an international but European development with recognizable national forms, has been radically internationalized once again.

Mshvenieradze expounds the major characteristics of the scientific and technological revolution as a social process and calls for a unified approach in order to understand the phenomenon better. He stresses in this regard the importance of the social sciences which, in his view, should not only analyse the social reality but also influence social dynamics in desired directions.

Richta situates the essential place of the scientific and technological revolution in both the establishment and the solution of many current problems now recognized within the social sciences. He explores in depth the philosophical and socio-historical nature of the tasks placed before social scientists by the distinctive recent history of science and technology, and by the array of open prospects science seems to offer.

In Part Two, 'Philosophical Problems of Technical Advancement', Fedoseyev, in a broadly conceived essay, sets forth a Soviet interpretation of the scientific and technological revolution: science and technology are sources of human developments and problems, but also the recipients and objects of social forces, a dialectic of autonomy and mastery.

Ganovski writes of the differing states and qualities of the scientific and technological revolution, matched and correlated as they have been with differing economic and political structures, and he particularly notes the contrast between capitalist and socialist economic orders with respect to the threats and the promises of the scientific revolution. Ganovski has developed a theme common to the symposium by stressing the goal of reaching or preserving normal functioning of the natural environment as an intrinsic but nevertheless serious problematic requirement of all spheres of production henceforth; but he adds at once that success depends not upon a system of vetoes over science and technology but rather upon the development of social systems with conscious controls over their social processes, including a full degree of consciously planned control over their sciences.

Peccei assesses the danger of unbalanced, or even uncontrolled, developments of science, and the need for concentrated multidisciplinary investigations into the problems of the social sciences, which confront both the unavoidable technical demands of the increasingly interlinked world economic order, and the too-frequent omission of the quality of the culture of the human individual from policy discussions and even from scientific inquiries.

Taylor draws up a balance sheet for technology, and then one for science, and outlines the agenda for social-scientific research tasks, the solution to which might help in realistic social cost accounting as contrasted with currently accepted input–output commodity accounting. Taylor urges that the total costs of economic and technical progress be accounted for by what he terms 'indices of new national welfare'.

In Part Three, 'Social Functions of Science and Science Policy', Gvishiani gives a detailed critical and optimistic survey of the technical, as well as the humanistic, choices and possibilities before mankind as a result of the increasingly close, often combined, developments of political and scientific structures in recent decades.

Kröber develops the theme of the social functioning of science under the overall new situation that has occurred, namely within the society of the scientific and technological revolution itself. He sees the widely noted dysfunctional dangers of technology as may be aided in differing degrees by the differing social systems.

Piganiol proposes specific enabling and limiting conditions for research on the appropriate uses of science and technology within a scientifically reasonable and humanly fulfilling political process.

Stroetmann argues for systematic study of technology assessment, including the assessment of alternative technologies, and for careful research and evaluation of technology policy choices and practices at the various levels of productive enterprises as well as within governmental bodies.

In Part Four, 'Science, Technology and Development', Cissé emphasizes the breakdown, often even the degradation, of traditional societies, particularly the countries of western Africa, subsequent to the external imposition of technology and of associated portions of the modernization process as these changes occurred through colonial or neo-colonial ties with the world capitalist economy. Cissé continues by pointing to the serious inadequacy of Western social science today with respect to analysis of the contradictions within Third World societies. In particular, social scientists fail to understand the crisis provoked in Third World societies by the scientific and technological revolution, a crisis whose resolution demands preservation of the human personality while a non-traditional powerful productive base is being installed, a crisis which, in his trenchant phrase, demands mobilization of new 'subjective social forces'.

Hegazy sketches a sociological inquiry into the comparative failure of Third World countries to participate in the scientific and technological revolution.

Maxwell, in an article requested for this book, describes recent Latin American experience in thinking about the role of science and technology in socio-economic development, under four headings: (a) technological dependence and technology transfer; (b) technological autonomy and the support of local science and technology; (c) micro-economic historical studies of local or indigenous technological strengths in Latin American enterprises, and (d) the 'appropriate technology' approach to development.

Rahman outlines the several roles of science and technology in Third World States. He notes that there is a political use of 'a-political' language and of value-neutral descriptions of science in the literature of the capitalist world, and he stresses the need, in developed and underdeveloped countries, for consciously political socio-technical policy judgements. Rahman also looks forward to the liberation of Third World people from lingering medieval and feudal beliefs and practices by a new scientific-cultural revolution in the recently colonial and semi-colonial milieux of the Third World. Thereby he raises the problem of the opportunity offered by the social, cultural and political roles that may be chosen by scientists and engineers in Third World countries.

In Part Five, 'Social Sciences and Social Change', Filipec devotes his article to theoretical understanding and methodological issues concerning the concept of the quality of life, and he stresses the need for research on social indicators of the conditions and modes of life. Filipec describes a hypothetical model, drawn from his own research, and sketches the (static) taxonomic features of a modelled social order as well as the criteria for expanding the model to include a fuller dynamic theory.

Friedrichs describes several German reactions to the impact of technological change upon working conditions and the quality of life, and focuses upon the initiative of the large Metal Workers Union in promoting systematic research by social scientists on these questions. He emphasizes the continuity, rather than the novel or surprising aspects, of recent technological change, such as the continuing impact of mainly labour-saving changes upon workers' lives.

Gatovski describes the Soviet experience with the increasingly close association between science and industrial production, and carefully notes that one striking and reflexive implication of the scientific and technological revolution has been that science itself, now fully a force of production, can no longer be characterized simply as a creator of new knowledge. Gatovski presents a detailed methodological investigation of the ways and means to measure, plan and control the novel determining role of science in what he calls the 'science-production cycle.'

Hollander discusses the revolutionary role of knowledge and the 'knowledge industry' today, the psychic shock imposed upon tradition-centred persons by a new and rapidly established scientific-technical life style, and the grave inadequacies in our knowledge of human nature, whether seen from the viewpoint of either the natural or the social sciences. He calls for greater support of research on the broadest range of questions associated with understanding the social functions of science, and propounds a dozen incisive questions in that regard, including understanding of the nature, motivations and social functions of anti-science and anti-technology in the contemporary world.

Kansu proposes that the historically developed direct link between science and technology deliberately be dissolved so as to recognize both the ideological quality of technology and the ideologically neutral and decidedly international objectivity characterizing science; and he further argues for the mediating role that might then be played between natural science and technology by the social sciences, in order to bring rational planning and conscious awareness to the technological instrumentalities of newly modernizing societies.

Okamoto, in a case study, presents the proposal of the Japanese

Information Development Association to foster an 'informational society', and he concludes by noting the distinctive epistemological contrast between the goal-oriented (or mission-oriented) 'soft sciences' and the traditional scholarly social sciences, whether these be avowedly value-free, or critical, or otherwise motivated.

Platt explores the 'great waterfall' of global transformations and reversals of social attitudes over the past thirty years—years of political crises and also of the new scientific-technological revolution. In the course of his paper, Platt provides several agenda items: first, urgent research topics; second, imaginable potential crises; third, conceptually novel characteristics of current social problems; fourth, classification of research studies required for dealing with world-wide problems in the remainder of this century. In Platt's lists, human needs for the social sciences are at least equal to those for the physical and biological technologies.

Thring catalogues the threatening world disasters of the last quarter of the twentieth century, the harmful effects of technology, and the requisite qualities for a satisfying 'creative society'. Thring sharply distinguishes 'standard of living' from 'quality of life', and he concludes by offering a new Hippocratic Oath for engineers and applied scientists.

The volume contains the final report of the Prague symposium. In the context of affirming the primary tasks for the social sciences in theoretical understanding and their essential roles in practical action during the modern technological era, the report identifies important areas for international and interdisciplinary research, and then addresses specific recommendations to Unesco for the organization of research and the establishment of scientific communications.

Part One

The scientific and technological revolution as a social process

The internationalization of science

Everett Mendelsohn

Harvard University, United States of America

Setting the problem

One of the commonplaces in contemporary discussions of science and technology is the extent to which science and technology have served to shape modern societies and to transform traditional societies. Technological advances have dramatically altered the human relationship to nature as well as the interaction among human beings. The long-dreamed-of voyage to the moon served as a symbolic indicator of the ability to transcend natural boundaries previously believed inviolable and perhaps fulfil the ancient dreams of Daedalus and Icarus. (But this time all those involved survived the flight!) Modern medicine gives daily evidence of its ability to challenge death and promote health and longevity. Today's science probes the very boundaries of matter and the most intimate conditions of life. An aura of rationality and promise seems to surround these achievements (and potentialities) of the new worlds of science, technology and medicine. Nations and peoples of the developing world are exhorted, from both socialist and capitalist sides, to speed up their attainment of the capabilities of the new science and technology with full expectation that this will set them on the road to enriched and more secure lives.

Yet in the face of achievements undreamed of even by Utopians of the past, increasingly serious questions have come to be asked about the place of science and technology in our societies; and there has been a persistence to the questions which prevents their being ignored. Some of the questions have been present from the very beginnings of the seventeenth-century scientific revolution, while other doubts are more recent and more clearly tied to explicit modes of use or misuse of modern knowledge and techniques.

During the grim days of 1946, as Europe slowly emerged from the devastation of the Second World War, Albert Camus projected something of the attitude that was to develop during the succeeding thirty years:

> The seventeenth century was the century of mathematics, the eighteenth that of the physical sciences, and the nineteenth that of biology. Our twentieth century is the century of fear, I will be told that fear is not a science. But science must be somewhat involved since its latest theoretical advances have brought it to the point of negating itself while its perfected technology threatens the globe itself with destruction. Moreover, although fear itself cannot be considered a science, it is certainly a technique.[1]

The immediate reaction was, of course, to the atomic bomb. The deeper response was to the potentialities that were witnessed so vividly not only in the bomb but in aircraft, cannon and those myriad other weapons of war brought forth by human ingenuity during the 1930s and 1940s. The subtle controllers and moulders of human behaviour whom Aldous Huxley had anticipated in *Brave New World* seemed literally to crash over the countryside of the Second World War.[2] Even though the victory in that war was over the 'immoral' forces of Nazism, some among the victors feared the weapons of victory almost as much as they would have feared defeat. Power and violence were two elements that appeared inextricably linked to science and technology!

One might argue in response that in the weapons of war we have witnessed scientific knowledge and technique led astray to become tools for human destruction. But lest it seem that the argument may now be put aside with the advent of efforts to use science in only beneficial ways, it turns out that even efforts of this positive sort could produce severely negative consequences. Rachel Carson's *Silent Spring*, published in 1962, showed that the effects of production for abundance could lead to pollution and degradation of the environment and turn even the fields of plenty into potential wastelands.[3]

The first level argument that claimed neutrality for the knowledge and technique and urged focus instead on the *uses* and *misuses* of science and technology seemed too simple. During the mid—and late—1960s, the examination was pushed to the very nature of the knowledge and techniques themselves.[4] Probing deeply into the relations of knowledge and action, Herbert Marcuse in his provocative *One Dimensional Man* sharpened the focus:

> Scientific technical rationality and manipulation are welded together into new forms of social control. Can one rest content with the assumption that this unscientific outcome is the result of a specific societal application of science? I think that the general direction in which it came to be applied was inherent in

pure science even where no practical purposes were intended and that the point can be identified with the theoretical reason turned into social practice.[5]

It was no longer enough to be critical of the social choices of good or evil goals for the use of science but rather it is implied that in the epistemology (the way of knowing) of science itself there had been created a separate reality, justifiable to itself and seemingly independent of the ends to be achieved, seemingly separated from all normative influences. Marcuse sees a tension existing between the very basis of science and the needs and wants of society and he is critical of technological rationality as a means of dealing with the societal needs. Technology is not neutral, he argues, and it cannot be isolated from the uses to which it is put. Rather it has within it what he labels a totalitarian tendency to pull things together under its control. Even its more pleasant manifestations are elements of a technological society which has as its basis a system of domination founded upon a way of knowing and acting and which affects concepts and techniques.[6]

There were members of the scientific community who seemed able, if at times reluctant, to take up the first level of criticism of science and come to terms with the need for a conscious expression of social responsibility within science for its use or misuse.[7] From the early post-war efforts of the atomic physicists, who in Oppenheimer's dramatic words had 'known sin',[8] to the international statements of Nobel laureates and the numerous national and local committees for responsibility in science, there came a steady stream of corrective efforts. Indeed quite recently one group of molecular biologists, working on recombinant DNA, took the almost unprecedented step of calling for a temporary moratorium on certain aspects of research while the potential dangers to the public were assayed and protective procedures developed.[9] But there has been almost complete rejection of the idea that the problems which appear to be the results of science and modern technology might have their roots deep within the epistemological and conceptual structure of the knowledge itself. Scientific knowledge has been conceived of as having very special qualities—it has been characterized as objective knowledge, fundamentally neutral and enjoying self-corrective features.[10] Even the recent assaults on the positivist philosophical conceptions of science, coming from the philosophy of science (e.g. T. S. Kuhn and P. Feyerabend),[11] have succeeded in doing little more than convince science and its staunchest defenders that philosophy is indeed irrelevant. The impact of the several sorts of critiques on the broader world of learning has been quite sharp and has even been seen to raise the question of whether the nature of truth in science (and through extrapolation, other disciplines including the social sciences) is little more than the currently adopted consensus among the scientists.[12]

The conclusion I would propose is a strong one: the characteristics of the epistemology, and the concept and technique forming apparatus—and thus the cognitive structures of science—are socially constructed; science and technology are not neutral or non-normative but, like all other ways of ordering reality and understanding data, they are generated in social and historical contexts which have values and social interests embedded in their structures.[13] An examination of the history of the social construction of science will provide a fuller understanding of modes by which science and technology come to embody social and political values. But before turning to the historical evidence, there are two additional themes or problematics which are important to note and which also have their roots deep within the history of modern science.

In a recent editorial, the *Boston Globe*, a large-circulation American daily newspaper, identified one of the problems directly. Since the 1960s, there has developed an increased demand for public participation as an important functional tenet of democracy. In the case that the editorial writers were reviewing, the City Council of Cambridge, Massachusetts, had intervened to limit basic scientific research because it had found in the research a potential threat to public safety. The assumption made by the council members was that the lay public had a direct interest in research when its conduct might affect the public, and that they, therefore, could step in and at least alter the pace of scientific endeavours if not their ultimate direction. Needless to say, the researchers involved were indignant at the thought of public intervention.[14]

There is an irony in this seeming conflict between the interests of science and the interests of the public. In history, the rise of modern science is bracketed with the decline of monarchic states and was intimately linked by many among its early adherents with the rise of democratic traditions. Thomas Jefferson in the new United States and Antoine-Nicholas de Condorcet in France, for example, both saw in the concepts and practices of science strong features of support for liberating tendencies and democratization in society. The justly famous words spoken at the graveside of Karl Marx by Engels, who saw in science a 'historically moving and revolutionary force', support the observation that the proponents of science were convinced not only of its ability to understand nature, but equally of its role in constructing progressive societal institutions and traditions.

Yet, at the present, when the traditions of science are most widely spread through the societies of the European culture area, the public believes itself excluded from the most important and intimate decisions that science makes. Certainly the apparatus for decision-making in the sciences in both the capitalist and socialist nations appears very distant from the influences of the lay public, although through financing and other bureaucratic controls, the upper reaches of the State bureaucracies and political structures can provide strong influence. While the scientific community has grown in size (both absolutely and proportionately), it has also become more élitist and out of reach. The threshold of entry to the sciences and many of their projects seems to become increasingly higher and the continued specialization which has been a hallmark of successful science makes appreciation and knowledgeable understanding strikingly more difficult. While the initial aims of science might have been coupled with democratization, to some important critics, the outcome has seen scientists emerge as an 'exclusive priesthood'. Indeed, the question might be asked whether science in its present institutional structures is compatible with democracy.

The second of the additional themes requiring examination is suggested by the remark of the American comic-strip character, Pogo: 'We have met the enemy and he is us!' Or, put more directly, in the very procedures and techniques that science develops for problem solving there are the roots of new problems. The implication is that this is not an accidental feature, but instead reflects elements fundamental to science and the nature of its involvement in problems at the interface with society. While this form of the idea of 'science-generated' problems may have some novelty, a similar concept has recently been developed for medicine in the very provocative study by Ivan Illich, *Medical Nemesis*.[15] By bringing into focus 'iatrogenic' diseases, Illich forces a re-examination of the processes and structures of medicine with particular sensitivity to their role in actually creating new disease categories. Those elements of medicine itself that Illich identifies as disease-generating have some marked similarities to parts of science and serve to reinforce an interest in scrutinizing science anew.

Numerous examples flow to mind of the efforts of science bringing forth problems. The marked successes in microbiology which led to antibiotics useful in treating communicable diseases, proved to be one of the prime sources for striking population growth. We think of the development of computer capacity for storing and sorting large amounts of information, which also emerges as a threat to privacy and individual rights.

Perhaps underlying these problems, created by science (or high technology), is the sense of availability in science of direct and unequivocal solutions. Alvin Weinberg might have been the first to use the term 'technological fix', but the implicit assumptions have much earlier roots in attempts and projected efforts to transform complex social problems into much simpler situations amenable to scientific or technical solutions.[16] Deep in the core of the scientific outlook is the 'belief' that nature will ultimately yield to the concepts and techniques of science. The conscious attempts directly to extend this attitude have led to proposing and using solutions drawn from the natural sciences with a view to solving social problems. The outcome has often been the generation of new problems caused by these efforts to avoid or play down the normative components of the social situation and reach instead for the technically oriented solution.

Historical and social construction of the scientific way of knowing

Let me now turn to what I know how to do best, that is, a bit of history. The question I want to begin with is: what did science and technology look like in European society before it was 'modern'? By this I mean the period before the transformation of the high Renaissance or late sixteenth and early seventeenth centuries, prior to the time when knowledge and technique were linked in the way now characterized as modern and scientific. Many of the attributes ascribed to non-Western cultures are in some large measure applicable to the relations of society, science, and technology in the West, prior to the period of the scientific revolution, the technological revolution, the introduction of capitalism and the Reformation. In any event, before these cataclysmic transformations, the place of technology and science was not what it became subsequently. Had a visitor from Mars dropped down then, roughly any time from the fifth century A.D., Europe would have seemed an unlikely place for the scientific and technological revolutions to occur—or for technique to be introduced as the rationale of human activity. China, I would guess, would have seemed a much likelier locale. Its technology was far more developed; it had a more rationalized commerce and a more sophisticated bureaucracy. The mandarins made their counterparts in the West look like amateurs, in terms of the use of knowledge, of written language, of symbolism and in terms of their understanding of the position of technique in human life.[17]

But then the events of the late sixteenth and seventeenth centuries changed the European picture so drastically, as to make the earlier picture almost unrecognizable.[18] What happened? During the period of transformation, those whom we now identify as scientists, but who then were often skilled artisans or engineers or literate craftsmen or alienated intellectuals, emerged as revolutionaries, in the social sense of the word. They were fighting for their share of the wealth and power, and they represented a variety of people within their society. Due to a complex set of reasons, the old order was under attack, and all authority, both ecclesiastical and secular, was under challenge.

Among the attackers were those who wanted to use their knowledge and control of nature as weapons against existing authority and wealth;

some of them called themselves alchemists, hermetic philosophers or experimental philosophers—the range was extremely wide. They engaged in a series of common activities which were critical of the established institutional forms. They debated the old texts; they hounded the scholastics; they challenged the old universities with their monopoly of teaching and learning; they questioned the traditional relations of knowledge to wealth and power (extant both in the feudal order and in that of the urban merchants and guilds). Through a number of compromises, they ultimately cut out social space for themselves and they helped in the establishment of a new social class—the emerging bourgeoisie. It turned out that these literate artisans and the intellectual contesters were useful to others who were also at odds with the old social order. In point of fact, they were often drawn from the same families of the rising bourgeoisie. Michael Walzer's fascinating study of the Protestant revolutionaries in England points out elements that permit us to see the close connections that existed among the several levels of social, religious, and intellectual change occurring in England.[19] The detailed studies of Christopher Hill provide additional evidence of the coincidence of belief and action among the dissenting religionists, the emergent capitalists, and the new scientists.[20]

One of the many things to emerge along with the changing social picture was a new epistemology. This new epistemology was the result of linking the empirical modes of the artisans and inventors to the new forms of rationality being developed by Francis Bacon and others. This linking of the rational and the empirical gave a new view of what humans could do. According to Bacon, Descartes and Galileo, humans could not only understand nature, but could control it as well.[21]

Rationalizations for technology had always existed, but this new epistemology spurred technology to unprecedented achievements and gave it a social importance it had never had before. As technology, and with it science, developed, the epistemology began to compete with and eventually overtook the other, once orthodox modes of existing thought. Certainly over the succeeding three centuries it replaced the other modes through which humans ordered reality as the established epistemology of every society in Western Europe. (Though, of course, the others continued to exist in less and less orthodox form: there were still alchemists, astrologers and religious mystics, but by the time Europe emerged into the twentieth century, they had been forced to the fringes.)

The idea of control over nature, introduced during the scientific revolution, is now not only a core value of science, but a part of the general understanding of our society. Francis Bacon's formulation that to understand nature is to be able to control it (and vice versa) was among the most important of the formulations developed by him and other ideologists of the 'new science'. In fact, control of nature for human benefit was given religious sanction in the writings of Robert Boyle, Bacon, and other seventeenth-century authors.[22] This underlines the point that new science and technology, as transferred to developing societies, carry a basic epistemology as well as a value structure. This epistemology is a potent one, and while we may ask other real questions about it, there is probably not one of us who would question its strength or its ability to give its users continued success in their confrontations with nature.

A further shaping feature of the scientific revolution was the notion that the scientist could know and control nature, but was not responsible for that knowledge or control. In 1660, for example, at the founding of the Royal Society of London, the authors of its charter openly proclaimed:

> This society shall eschew all discussion of religion, rhetoric, metaphysics, morality, and politics; we shall assume no dogma; we shall feign no hypothesis; we shall only know nature through experience.

The image that emerges is of a sharply restricted realm of inquiry.[23]

There are historical reasons for this rejection of all normative and subjective elements in science. In 1660, Charles II was restored to the throne and the Church of England was once again the established religion. The scientists and technologists who wanted to find their place and establish social space within which to conduct their activities had to proceed cautiously. Similarly on the Continent, as the Counter-Reformation drew its boundaries with Galileo in 1633 and Descartes in 1637, it became clear that if science was to survive through the currents of the Counter-Reformation, the restoration of the Crown and the re-establishment of the Church, it would have to make peace with the secular and religious authorities surrounding it. For science did pose a very real threat, as the Pope well understood when he condemned Galileo. Its new method of ascertaining truth represented a significant challenge to traditional authority. In addition, it seemed to pass intellectual authority into the hands of new classes in the society. Thus, the Accademia del Cimento in Italy had reasons for saying: 'We will deal only with experience. We won't deal with theory, or at most, theory used only to link one set of experiments to another.' What developed in this period is the self-proclaimed notion of normative neutrality—of eschewing the subjective, eschewing the social, political, ethical and the religious questions, while at the same time gaining knowledge and power. As Bacon put it so forcefully, *scientia est potentia* (knowledge is power). Just as there were deep societal roots to the search for knowledge and power, so there were compelling societal reasons to adopt an epistemologic value neutrality, which in reality, of course, left the normative elements of science to the control of the dominant social and political forces of the society. It was not that normative neutrality was actually gained by science, but rather that science ceded any right to challenge the norms of power and wealth in its host societies.

Every human culture from the ancient agricultural or the later Iron Age has possessed a technology—the ability to alter nature to some degree. But to deal with nature in a fully ordered and integrated manner as the basis of the stystem of control was something new. The addition of the new rationality of science to technology changed both the nature and the uses of technology itself, while at the same time the 'positivist compromise', which made science disclaim any normative concerns, gave the nascent industrial and capitalist movements a perfect tool. In so far as science and technology were neutral, the operators could command things and gain control over nature, without themselves taking responsibility. The implications of the extension of both the epistemology and technique to the human sciences are striking. A social science constructed on the model of the natural sciences would deny interest in the very human subjective and normative phenomena, while at the same time giving practitioners the power to alter and direct human actions.

It turns out, of course, upon examination, that neutrality does not exist. Embodied in all science and technology is a series of values—values issuing from the very human experience which created it. If technology and science were created by a newly emerging class, and adopted by practitioners who were all male, all white, and almost exclusively European, then inherent in the very nature of the knowledge and technique is a class, culture and sex-oriented value structure, bounded by the European experience and probably blind to all others. Indeed, one of the enormous contemporary arrogances is to talk about Europe as the scientific and technological society—as if no other society in human history had had a technology of its own with which tasks were carried out. Europe has developed a culture—which for our argument is shared by the United States and the Soviet Union—based on a strong epistemology (a way of knowledge), which ignores certain questions and whose aim is to gain control over things and, by extension, over human beings. It is no accident that in the nineteenth century the extension was exploited and a conscious invention called the social sciences made. As generally formulated, a definite attempt was made to link the normative indifference of the earlier natural sciences with a desire to extend understanding and control to the human realm. Now in the twentieth century we have inherited a vision that allows technique to be developed with its inbuilt value structures unrecognized. For instance, the very concept of efficiency is one based upon an economic construction not universally found in other societies. There are elements which other peoples consider in deciding whether or not a given science or technology should be used, which, in the European framework, lie outside the parameters of efficiency.

I have tried to lay out a perspective from which the epistemological bases of science and technology can be viewed. While I shall not go through its development during the nineteenth-century industrial revolution in detail,[24] we can say that the transformations which occurred then only reinforced the steps taken during the two previous centuries and literally wiped out all opposition, expunging any sense that nature ought to be treated as something from which one asked permission before creating change—for example, when one considers the popular images of St Francis on the one hand, who would address a tree before he cut it down and apologize to the spirit within it, and on the other hand the image of industrial armies in the nineteenth century moving a mountain to strip the coal out of it, without any question of the assumption that humans had a right to do this. Even more striking is the moving of people to the site of new machines and factories and putting humans on the schedule of the machine. For the onlooker, the industrial revolution represented the true fruits of the earlier efforts to control nature and provide a rational mode of ordering the physical (and the social?) reality by which humans are surrounded. The epistemology was socially shaped; the boundaries of the new scientific disciplines responded to the interaction of the new knowledge and the institutions within which it was becoming embedded; and the very concepts themselves responded not only to the challenges of cognitive imagination, but more fundamentally to the human perceptions of the immediate social and historical realities.

From transfer to transformation

Late in the nineteenth century, when the Chinese confronted the existence of European (Western) science and technology, they enunciated what they hoped would be the guiding principle of their response: 'To become Western in technique but remain Eastern in essence'. The Chinese intellectuals attempted to deal with a problem that their European counterparts didn't realize existed—that science and technology were deeply culture laden and that, unless great prudence were shown, the adoption of Western knowledge and techniques would be accompanied by a decline in the culture of the East. Their initial efforts succeeded only in postponing the large-scale introduction of Western science and technology. They clearly had miscalculated the strength of the cultural ties of science and technology; science and technology, it turns out, could not travel without bringing their cultural baggage. The second round of attempts which came, thoroughly mixed with the widespread modernizing efforts of the Sun Yat-sen revolution of the early twentieth century, had great success from the protagonists' point of view, but also succeeded in largely submerging or supplanting the considerable indigenous science and technology of traditional China.[25] The current Chinese response of rediscovering and revaluing their earlier traditions and integrating them into the modern or Western format of science and technology provides explicit recognition of a problem widely sensed today in developing nations.[26]

The movement which links the interest in intermediate technologies in Ghana, Ayurvedic medicine in India, and traditional agricultural practices in parts of South-East Asia represents a conscious attempt to maintain cultural continuities and avoid the oppressive effects of a total commitment to knowledge and technique that does not emerge from the lives and experiences of the indigenous populations.[27] Few societies have taken the path of Burma and attempted to deny the outcome of the scientific and the industrial revolutions, isolating themselves from exchanges, but an increasing number in recent years have begun to ask penetrating questions about the social content and the political and economic implications of dependence on science and technology from the industrial world—both capitalist and socialist. These new attitudes are particularly visible in societies where resources are limited and foreign exchange difficult to obtain. There has been a growing recognition that the technologies of the industrial world are resource wasteful and capital intensive, while the needs of most developing societies are for production modes which are resource conserving and labour utilizing.[28]

Both the historical case studies which are now becoming available and many contemporary examples suggest new perspectives for the study and understanding of the sharing of science and technology. First, and perhaps most difficult for those of us trained in the traditions of Western science and technology, is the necessity of recognizing the culture-boundness of our knowledge and technique and the need to explore fully the implications of sharing culture-carrying knowledge and artifacts. If science and technology are socially and culturally constructed forms, reflecting the social experiences and social structures of the societies in which they were generated, what is it that is shared with the societies to which the knowledge and technique are carried? Certainly it is much more than the instrumental knowledge previously imagined. Implied in this knowledge and its mode of production and use are myriad values reflecting the social relationships of the initial stages of discovery and innovation. Does it matter that science, its way of knowing and application were developed and institutionalized by a group who were almost uniformly white, male, and drawn from the middle classes of European and North American societies? Further, the very process of sharing has occurred within a context of use or of goal to be achieved. Generally a message of domination and control of nature were implicit (sometimes explicit). In addition, in the hands of both European and indigenous 'modernizers', science and technology have been seen as a means of transforming traditional knowledge and belief as well as transforming indigenous social structures, value systems and modes of production. Is it accidental that this process has led so often to external domination or, at minimum, to the creation of crippling dependencies? The science and technology which originally developed as liberating forces in the European context (although they later degenerated even there to create oppressive relationships) are viewed by many in the Third World as containing the threats of domination and oppression.

Among the modes of response which have been adopted towards Western science and technology has been a conscious attempt to create hybrid forms utilizing elements of both modern and traditional knowledge and practice. This seems particularly apparent in those areas of science, technology, and medicine where significant segments of the public come into contact with the practices and concepts. It turns out, on examination, that the hybridization which had occurred for many years and been perceived as incomplete copying or local misunderstanding of the Western idea or technique, is probably more correctly understood as an adaptation to a local social or cultural milieu and a partial transformation of the original. Indeed the process has been referred to as 'indigenization' of Western concepts and practices. This is viewed as distinct from the procedure by which only the local social structures and cultural traditions have been adapted to the intrusion.

This conscious and unconscious adaptation of Western science and technology suggests that the process of sharing is much more nearly two-way rather than one-way and that it represents an exchange rather than a transfer, an interaction rather than an impact. The exchange may be viewed in terms of the transformation of Western science and technology as they become embedded in new cultures and reflect new social experiences. Thus it is necessary that we study the transfer of Western science and technology from a position of appreciation of indigenous principles and practices and particularly of the social relations and modes of production found in traditional forms of scientific knowledge and technique.

Among the most cherished aspects of modern science has been its international character. The ability of science, both in method and in content, to cross national boundaries and still to be understood and applicable, has been acclaimed. Science has been international, but only in the European culture area—that is, in those nations that shared the transformations of the rise of capitalism, the scientific revolution and the industrial revolution. In other words, the internationalism of science is limited to those societies which share their cultural traditions and which have undergone the changes in modes of production and social organization that created modern Europe and North America. As the community of nations and societies with regular exchanges of a political, economic and social nature increases, a new meaning to internationalism in science begins to make itself felt. Joseph Needham has referred to a similar concept as 'scientific ecumenism', the implied bringing together and interaction of the multiple traditions of knowledge and technique.

It is apparent that science and technology reflect the values of the context of their development. In the exchange of knowledge and technique, the nature of the implicit values often becomes evident. At this point there is a chance not only for a conscious acknowledgement of the carried values, but even more important for their scrutiny for congruence with the goals for nation and society building. There is a real opportunity for choice so that those values supportive of basic human needs and responsive to fundamental freedom and justice can be advanced both in the new societies undergoing their own scientific and technological revolutions and reciprocally in the industrialized nations where science and technology have too often seemed beyond the bounds of true social control.

Notes

1. Albert Camus, *Neither Victims nor Executioners*, edited and reprinted in English by Robert Pickus, 1968.
2. Aldous Huxley, *Brave New World*, London, New York, 1932.
3. Rachel Carson, *Silent Spring*, Boston, 1962.
4. See the type of criticisms advanced by the most articulate spokesperson for the American counter-culture, Theodore Roszak, *The Making of a Counter Culture; Reflections on the Technocratic Society and its Youthful Opposition*, New York, 1968; and *Where the Wasteland Ends*, New York, 1972.
5. Herbert Marcuse, *One Dimensional Man*, p. 146, Boston, Beacon, 1964.
6. This concept is pursued from the perspective of 'critical theory', in William Leiss, *The Domination of Nature*, New York, 1972.
7. See the interesting collection of papers edited by Martin Brown, *The Social Responsibility of the Scientist*, New York, Free Press, 1971.
8. J. Robert Oppenheimer, 'Physics in the Contemporary World' (1947) reprinted in *The Open Mind*, New York, 1955. The reactions of the scientists have been carefully studied: Alice K. Smith, *A Peril and a Hope, the Scientists' Movement in America, 1945–47*, Chicago, 1965.
9. The initial call for the moratorium was issued by the Committee on Recombinant DNA Molecules, of the National Academy of Sciences (United States) in a letter signed by the committee members and published in *Science*, Vol. 185, (26 July 1974) p. 303. See the report in the same issue of *Science*, p. 332–4. The subsequent history of the moratorium and the growing public involvement in local decisions concerning the seriousness of the hazards and whether work should be pursued is worthy of much deeper study.
10. The *locus classicus* for this view is to be found in the recently collected sociological studies of Robert K. Merton, *The Sociology of Science*, Chicago, 1973.
11. Thomas S. Kuhn, *The Structure of Scientific Revolutions*, Chicago, 1962; Paul Feyerabend, *Against Method, Outline of an Anarchistic Theory of Knowledge*, London, 1975.

12. See the critical argument proposed by Hilary Rose and Steven Rose, 'The Radicalisation of Science', *Socialist Register*, 1972.
13. I have explored a number of these ideas at greater length in two earlier papers: 'Should Science Survive its Success?' in R. S. Cohen *et al.* (eds.) *For Dirk Struik*, (Boston Studies in the Philosophy of Science 15), Dordrecht, Holland, 1974, and 'A Human Reconstruction of Science', *Boston University Journal*, Spring 1973.
14. The case involved the desire of a group of molecular biologists at Harvard University to conduct research on recombinant DNA in a new facility to be constructed at the Biological Laboratories in the City of Cambridge. Another group of scientists, including Nobel Prize-winner George Wald, objected on the grounds of potential biohazards and urged the City Council to intervene using its authority to act under the local public health and safety ordinances.
15. Ivan Illich, *Medical Nemesis, the Expropriation of Health*, New York, 1976.
16. Weinberg's 'dream' for what science might do is spelled out succinctly in his paper 'National Laboratories and Missions' (1955), reprinted in *Reflections on Big Science*, p. 126–44, Cambridge, M.I.T. Press, 1967. See especially p. 141: 'There is a possibility that the technologically oriented research institutions may contribute to an unexpected degree to the resolution of problems that now seem to be primarily social. I refer to the possibility of devising "cheap technological fixes" that afford shortcuts to resolution of social problems.'
17. The magisterial studies by Joseph Needham and his associates bear ample witness to this claim: *Science and Civilization in China*, 1954– , already running to some eight or nine bound volumes.
18. The critical literature on the scientific revolution has grown enormously in the past few years. Still, among the most valuable contributions were the series of papers published in the 1930s and 1940s by Edgar Zilsel. They examine various elements of the sociological roots of modern science. A collection, including those papers originally published in English and translated into German, has recently been edited by Wolfgang Krohn, *Die sozialen Ursprunge der neuzeitlichen Wissenschaft*, Frankfurt am Main, Suhrkamp, 1976. An English-language edition, edited and with an introduction by Robert S. Cohen and Everett Mendelsohn, is planned for publication next year. See also several important additions to the literature: M. L. Righini Bonelli and William R. Shea (eds.), *Reason, Experiment and Mysticism in the Scientific Revolution*, New York, Science History Publications, 1975; Charles Webster (ed.), *The Intellectual Revolution of the Seventeenth Century*, London, Routledge, 1974, as well as *The Great Instauration: Science, Medicine and Reform, 1626–1660*, London, Duckworth, 1976.
19. Michael Walzer, *The Revolution of the Saints*, Cambridge, Mass., Harvard University Press, 1965.
20. Christopher Hill, *Intellectual Origins of the English Revolution*, Oxford, Clarendon Press, 1965.
21. Paolo Rossi, *Francis Bacon, from Magic to Science*, translated by S. Rabinovitch, London, Routledge, 1968 (1957).
22. See the recent study by Eugene Klaaren, *Belief in Creation of the Rise of Modern Natural Science*, New York, Erdman Publishers (in press).
23. See the apologia for the early Royal Society by Thomas Sprat, *History of the Royal Society* (1667) edited by J. I. Cope and H. W. Jones, St Louis, Washington University Press, 1959.
24. The nineteenth-century social transformations of science have not been given as much attention as the early period of the scientific revolution. See my own examination of the institutional status of science, 'The Emergence of Science as a Profession in Nineteenth Century Europe', in K. Hill (ed.), *Management of Scientists*, Boston,

Beacon, 1964. See also the suggestive study by J. D. Bernal, *Science and Industry in the Nineteenth Century*, London, Routledge, 1953.

25. See the recent studies by Peter Buck, 'Western Science in Republican China : Ideology and Institution Building', in E. Mendelsohn and A. Thackray (eds.), *Science and Values*. New York, Humanities Press, 1974; 'Order and Control: The Scientific Method in China and the United States', *Social Studies of Science*, Vol. 5, No. 3, 1975.
26. This process was explored by Ralph Croizier, *Traditional Medicine in Modern China*, Cambridge, Harvard University Press, 1968. Recent visitors to China have brought back reports of some interest; Science for the People (eds.), *China: Science Walks on Two Legs*, New York, Avon, 1974.
27. See for example the perceptive study by Charles Leslie, 'The Modernization of Asian Medical Systems', in J. J. Poggie and R. N. Lynch (eds.), *Rethinking Modernization*, p. 69–108, Westport, Conn., Greenwood Press, 1974.
28. E. F. Schumacher, *Small is Beautiful, Economics as if People Mattered*, New York, Harper & Row, 1973; see also the critical study from an anthropological perspective by Richard N. Adams, 'Harnessing Technological Development', in Poggie and Lynch (eds.), op. cit., p. 37–68.

Towards a unified and multidisciplinary approach

V. Mshvenieradze*

The borders of the various branches of science overlap and new theories emerge at the interface of the established disciplines. Differentiation and integration are two aspects of a single dialectical process of development in contemporary science, and constitute its very essence. Hence the significance of a multidisciplinary approach to current scientific problems which are themselves extremely complex and require the well co-ordinated and combined intellectual efforts of specialists in various branches of knowledge.

A profound awareness of the social aspects of the scientific and technological revolution (STR) cannot help but lead to the conclusion that it is inseparably linked with all spheres of life. One could hardly name any domain of social activity that is not touched and influenced directly or indirectly by the latest applications of science and technology. Such major problems as the relationship between man and biosphere, man and his environment, etc., not to mention patterns of production, industry and agriculture, all bear strong marks of the impact of developing science and technology. Social sciences, culture, education and information must become more dynamic and flexible, and develop to a considerable extent in order to meet the challenge of the STR. Take the problem of population: it seems hardly possible to conduct profound research, for example, in fertility or family planning, urbanization or rural development, and yet ignore the social implications of the latest findings in chemistry, medicine, psychology, biology, and particularly in genetics, genetic engineering and molecular biology. All this gives rise to a number of new problems in sociology, law, ethics, economic, political studies, etc.,

* The author was Director of the Division for the International Development of Social Sciences, Unesco, when he wrote this paper. His analysis was intended as a personal and substantive contribution to the work of the Prague symposium. The views expressed therein however are his own and do not necessarily reflect the opinions of the Unesco Secretariat.

and emphasizes the social responsibility of scientists in a rapidly changing world: are they observers or participants? In one sense, perhaps, they are the creators of a new world.

Social nature of the scientific and technological revolution

The complexity of the problems to be solved needs the combined efforts of various professionals, as the STR becomes more and more internationalized. However, it is not a goal in itself but a creation of the very mankind which it should serve: thus it has to satisfy the growing needs of human societies, and the broadest possible discussion of its social aspects would seem to be of the first priority. The development of the contemporary STR has caused many more problems than it can solve by itself, thereby emphasizing, however paradoxical this may sound, both the growing significance particularly of the social sciences (these problems being of a social nature) and the need for close collaboration between social, natural and technological sciences. In a certain sense it has produced new dimensions of human problems. If one does not stay on the surface of events but digs into the deeper strata of social, scientific and technological realities, one has to admit that the contemporary STR is, in essence, a social phenomenon. Therefore, the social and human dimensions of this multifaceted and multidimensional phenomenon should be taken into consideration in the fullest possible way, and the wide social programme of the STR, based on the unified interdisciplinary and multidisciplinary approach, should be elaborated in order to exercise, experiment and monitor it, and adjust it to the needs of mankind, to humanize nature and provide civilization with scientifically rigorous tools for further advancement. It needs to be approached not simply emotionally but rationally and scientifically, which means that a genuine multidisciplinarity ought to be exercised.

While not wishing to detract from the significance of the natural and technological sciences, I would like nevertheless to point out that the contemporary STR might be considered first and foremost as a social process.

First, it is social because basically it has emerged from social needs in order to satisfy them; its main purpose can be described in terms of the improvement of social life.

The main thesis of so-called 'technological pessimism', according to which technology and science implicitly bear certain irrational motivations, neglect humanistic principles of the development of personality, and the spiritual needs of human beings, and are fatally bound to detract from the role of utmost importance that ethics and morals have to play in social development, appears to me groundless.

This kind of 'technological pessimism' logically leads to social pessimism which very often results in the pseudo-criticism of those 'consequences' of technological progress for which technology itself simply cannot be blamed. The problem here seems to be not to strengthen and deepen arguments for negative criticism but rather to find rationally justified, constructive and positive ways of using technological progress for the benefit of mankind, for social progress. Neither science itself, nor technology, can be blamed for the misuse and abuse of their results.

The fundamental pattern of a given social structure essentially determines the specificities of its relationship to technological progress and nature. In the last analysis, it is for mankind to understand the potential danger of the STR. It should not dominate peoples and nations, and create and constantly reproduce an alienation of man's social essence. People and nations can use the potentials of scientific and technological progress and channel them to the service of a chosen social, economic and political progress. Individuals can reach and enhance their identities and the real conditions for their all-round development in human communities.

Although the STR touches the widest range of spheres and problems from sea beds and ocean floors to cosmic flights and moon probes, any of these problems may be given a meaning in its social or human dimension. Certainly scientific research should not and cannot be reduced to merely direct and utilitarian or pragmatic needs. The cause-effect relation here is not that simple. However, it is clear that even such socially remote objects as warm rocks may acquire a social dimension when the problem of the need for geothermal energy arises. It is not human society that exists for technological change, it is science and technology that exist for human society.

Second, the STR is a process because it represents dynamism, and permanent change. Undoubtedly it has its phases of development, each of which has its peculiarities and specific features. Research geared towards investigating the interrelationship between the general and specific patterns of the STR and their actual manifestations in various societies seems to be very important. Are these, the general and the specific features, two different frames of reference or can they be considered as belonging to a single paradigm? How can one dovetail social progress with scientific and technological advancement?

Here the importance of a unified approach, and the elaboration of some fundamental methodological principles for it, appears to be of primary significance. Should technological progress be taken as a criterion of societal development or should social indicators be applied to scientific and technological advancement to measure the extent of their fruitfulness? In other words, does technological progress have an independent value or should it be assessed within the frames of societal and human reference? Finally, to what extent should technological advancements be used for defining the adequate level of development of a given society?

No methodology should be reduced simply to measurement. It has qualitatively new implications, much broader and much deeper, and concerns first of all the concrete modes of scientific thinking (as opposed to common sense). These are based on the awareness of objective interdependence between the internal logic of the advancement of the STR on the one hand, and the social, economic and political motivations for its development on the other. One important feature of the internal logic of scientific and technological development consists in the fact that it not only engenders new discoveries and achievements, but by doing so permanently reproduces conditions for further development, each time rising to a higher level with rich potentials and perspectives. This process of development is unending. The methodological aspect here shows itself not as an ability to measure this process, as can be done by any outsider using traditional statistical methods slightly changed and adjusted to the circumstances. Decisive patterns of scientific methodology, among other things, provide any branch of scientific knowledge with its sufficient means of self-awareness. The self-consciousness of science (which is always a product of human activities) is a fundamental assumption on which the social responsibility of scientists can feed. It is also for methodology to specify a concrete combination of approaches to be used in a certain particular branch of research. At the same time, it cannot develop 'recipes' from within and voluntarily, like a spider. Methods of research depend mostly on the object of research. Thus methodology plays an active role in programming and monitoring research but the very efficiency of this activity, in turn, depends upon degrees of correct understanding of the objective nature of the subject-matter of reality, objectivity of methods, and how these methods fit with the objective constraints of a given segment of reality.

Compared with nature, social reality is much more complex and complicated. This is why social research as well as its methodology need highly elaborated tools and instruments of analysis, which are difficult to develop and apply with the same precision as in technology or natural sciences. It is even more complicated scientifically to analyse problems which emerge on the borders between the various domains and which are of a multidimensional nature, like scientific and technological progress.

If I emphasize the important role of the social sciences in the understanding of the essence of scientific and technological progress, which is indeed social, this is because it has long been ignored or at least not given proper prominence. However, one should not exaggerate this role and fall into the opposite extreme of considering social sciences as a kind of panacea. What is really needed is to elaborate a new unified approach which must be multidisciplinary because it has to deal with such a complex, multifaceted and multidimensional phenomenon as the contemporary STR. An active approach needs to be worked out based on a profound knowledge of its general patterns and specific features.

Activity versus veto

I willingly endorse the idea that scientific and technological progress makes human labour more productive and more active. The question arises, however, whether this always means that our attitude towards it becomes more constructive. How shall we activate our position in governing scientific and technological progress? For example, to deal with the interrelationship of man and the biosphere, almost all societies have now issued a number of vetoes and prohibitory laws to protect the natural environment. This is certainly correct: We have to preserve stability of the ecosphere and keep the balance of ecosystems. Do these laws and vetoes, however, show an active position? Or do they manifest a passive and defensive attitude towards the offensive of scientific and technological progress? Even when certain production processes, in order to meet prohibitory requirements, are transformed into exclusive production cycles: does this constitute an active position? Finally, are all these vetoes and taboos the only ways to protect the ecosystem? Examples and questions like these can easily be multiplied.

The modelling, or theoretical reproduction, of the processes of the STR, of its trends, internal logic and perspectives in different social structures may become a fruitful source of incentives for the international exchange of regional and national experiences.

New parameters of contemporary technological-scientific and cultural progress very often challenge traditional points of view, call for a new approach and a new scientific orientation to help to distinguish between real and illusory social changes. This also needs a profound knowledge of social dynamics, social contradictions and the ways in which they can effectively be resolved. The true knowledge of these problems embraces not merely scientific values but also the social values which in the present reality should be assessed on the international scale.

It seems important not only to discern the links between the STR and social structures, but also to investigate the internal contexts of these linkages. These contexts can reveal themselves principally in a threefold interaction between social structure and scientific and technological progress. They can accelerate the mutual development, slow it down, or impede it. The situation can change from time to time.

Reciprocal acceleration is a normal (and most profitable)

interrelationship. The necessary prerequisites of scientific and technological progress are naturally conceived in the depth of a given social structure and are engendered by social needs themselves. The social structure, although it is open to a large scale of options, necessitates scientific and technological progress and creates unlimited conditions for its development thereby favouring its own advancement. The process of self-perfection does not always go smoothly. Moreover, it means systematically overcoming various obstacles and solving contradictions. Correct solutions can be found when the priorities are well defined. The first priority belongs to social needs which are inseparable from the needs of the individual. One can see here how explicitly the interrelationships between the STR and the social revolution are made.

A given social structure slows down the development of scientific and technological progress when the latter outstrips the possibilities of the former. The social structure is not ready to bear and 'digest' new and further achievements of science and technology and especially their social consequences. The boundaries of the social structure are very narrow in this case. However paradoxical, a good many scientific and technological discoveries which are really needed by society can be frozen because their introduction under a given social structure leads to overproduction and economic crises, increased unemployment, etc. The objective tendency of science and technology towards socialization is restrained by a social structure which is not sufficiently flexible. The pattern of social stratification and hierarchy of economically determined ranks prevents the social interest from emerging. These contradictions are insoluble within a given social structure. In this case, the very development of science and technology may not only be retarded but also distorted through the advancement of particular branches of knowledge and technology at the expense of harmonious overall movement.

There may equally be situations in which a social structure impedes scientific and technological progress by not giving it significant support; e.g., where technology is imported. This transfer of knowledge and technology is followed as a rule by an attempt to combine a certain kind of wholesale import of technology with maintenance of the traditional social structure of the country in question.

The transplantation of science and technology from developed and highly industrialized countries to developing ones could be quite an artificial undertaking if the social structure of the recipient countries were not being adjusted accordingly, i.e. if there were no facilities for its endogenous development backed by a realistic planning of the establishment of the whole infrastructure, skilled labour, etc. In other words, an insufficiently developed country can simply become a dependency of a developed country, a mere external market. Its social structure shows strong features of dependency and adapts itself to the interests of developed countries rather than finding its own optimal path. In brief, a social structure hinders scientific and technological development when it is incapable of creating the social prerequisites and internal needs and facilities for its natural advancement. The imported scientific and technological products are alien to this society; it can certainly swallow them but cannot organically 'digest' them, in a manner which could lead to their reproduction on an expanded scale.

Otherwise put, the 'transfer of technology' even when it is followed by the 'transfer of knowledge' cannot automatically ensure fitness to a given structure. Paradoxically, there will be effects but no internal direct causes of these effects: there will be needs but no causes. The real challenge here seems to be not to restrict oneself to transfer but to combine it with the creation of appropriate social mechanisms and capabilities of internal production of both adequate 'intellectual technology' and 'technological intelligence'.

The contemporary STR has its own intrinsic logic. It needs, it uses, and it reproduces the finest and most powerful intellectual technology. Therefore, it should concentrate not only on external resources but also, and with growing emphasis, on the inner world of societies: social structures, values, ethical norms, psychological and cultural resources, human aspirations and consolidated human energy.

Taking all these things into account, one cannot ignore that the interpretation placed upon the social aspects of the STR, as well as other problems of social knowledge, is often an area of ideological controversy. This reality should be readily accepted.

The world at present is not homogeneous. A strictly scientific analysis of various social systems is supposed to establish both the interconnections between the components of a particular social system, and those between the systems.

Social science, in my opinion, should have predictive capacities so as to influence social dynamics in desired directions. On the other hand, social structures should be flexible enough to absorb new and progressive phenomena and to give birth to the needs for further social progress. The question of bringing the levels of social development up to the level of technological and scientific development cannot be reduced only to the transfer of knowledge or the transfer or technology. This is a necessary but not a sufficient condition to harmonize the development process.

Scientific and technological revolution and environment

Among the vital problems which attract the attention of human communities today, the problems of the environment occupy a special place.

They are real because they need to be solved. A great many books, articles, documents, films, etc., have been produced to date concerning various aspects of environmental problems. One can even say that we face a certain literary 'pollution' on this subject! True, great progress has been made. However, there are many important questions awaiting solutions. One fundamental achievement which deserves to be strongly emphasized is that overall attention has been drawn to the environmental dimension. This, in turn, has meant the start of all kinds of activities—political decisions, research, social movements, environmental education, including the United Nations' designation of 5 June as World Environment Day, each of which gave birth to a deeper understanding of the environment calling for both theoretical understanding and practical action.

The United Nations Declaration on the Human Environment in 1972 proclaimed that:

> In the long and tortuous evolution of the human race on this planet a stage has been reached when, through the rapid acceleration of science and technology, man has acquired the power to transform his environment in countless ways and on an unprecedented scale. Both aspects of man's environment, the natural and man-made, are essential to his well-being and to the enjoyment of basic human rights—even the right to life itself.[1]

The 1975 Helsinki conference on security and co-operation in Europe strongly emphasized the necessity for international co-operation in the protection of the environment. Some crucial questions emerged: How is the environment to be understood? What is the interrelationship between development and environment? How does one reconcile national protective laws with nuclear weapon tests, marine pollution, transport facilities, etc.? Which kind of environmental problems should be considered as global, which should be solved at national or regional levels, and what are the interrelationships between them? What are the most efficient ways for citizens to participate in the defence of the environment? And many others.

Here we shall briefly discuss two questions. One concerns certain tendencies which have emerged from the recent evolution of the most typical views on the environment. The second, which is narrower, is related to some specific aspects of the STR and its impact upon the environment.

To argue that environmental problems have to be considered in close connection with development or education, in cultural contexts or within the frames of legal sciences, or that they need to be treated as global and/or specifically determined, have become rather banal approaches today. No doubt, any approach should not only consist of negative and prohibitive statements but also imply positive and constructive deliberations: the idea of protection, therefore, should be linked with the idea of improvement. An active and rational approach to environment has become an imperative goal for mankind, since one of the basic and inalienable human rights is the right to normal living conditions, to life itself. Therefore, in speaking of present societies and individuals one should bear in mind that besides social, economic, cultural, political, scientific and technological achievements, environmental benefits must be taken fully into account.

If one attempts to discern the fundamental tendency in the literature on environmental problems, it is the growing comprehension of these as social problems.

A purely practical way to cope pragmatically even with local environmental problems (a natural approach for any social community) inevitably leads to a deeper understanding of the internal links between social and environmental variables as well as of the necessity to consider local problems as integral parts of the global ones, which, in turn, leads to the recognition of a general environmental theory and practice, which, again, in its turn, is conducive to multi- and interdisciplinary approaches to such a complex phenomenon as environment.

One can take at random one of the latest publications, for example an article by Peter H. Gore and Mark B. Lapping 'Environmental Quality and Social Equality'.[2] The authors examine the Adirondacks region of New York, which is a famous and unique environmental resource. Since the 120,000 permanent residents of the region are among the poorest in New York State, the authors argue that the preservation, protection and enhancement of the region's delicate ecology will jeopardize the chances of the residents for economic development. In their own words, the authors discuss 'the problems and contradictions which arise when an economically "poor" region is mandated to remain environmentally "pure"'. They come to the conclusion that 'trade-offs between development and environment have far-reaching implications' and that 'the need for economic and social variables to be examined in addition to the more 'environmental' factors of forests, wildlife, and streams, should be underlined as well as the necessity of comprehensive (rather than 'fragmented') planning initiatives. In other words, there is an urgent need 'to include social quality of life measures' in the concerns of environmentalists.

Even this example, which was selected quite haphazardly and which deals with a local situtation, shows, and this is typical, that no environmental problem can be solved in isolation from social dimensions. Any 'partial' solution can be no more than partial, becoming a pseudo-solution if regarded not simply as a partial one. What is meant here by 'partial' is a fragmented approach to the complex problem in the sense of quality. It is also 'partial' in a quantitative sense if we take the environmental problems as global ones, necessitating the combined efforts of the international community.

Here another problem arises: whether or not specifically national solutions can be used for wholesale export to other countries. If each nation has to develop its own system of environmental measures best suited to its specific and particular needs, natural and social circumstances, traditions, requirements, goals, etc., does this not contradict the very idea of the global sense of environmental problems and the possibility of elaborating a general approach which can fit the requirements of other countries? Here lies one important methodological principle which should deal precisely with the problem of the specific (national) and the general (international, global). This problem is a solvable one.

The elaboration of a general outlook should certainly be fed by the achievements of national approaches, the former not being understood, however, in a primitive way as mechanical summations of the latter. It should be considered as a new synthesis, not reducible to any number of separate approaches, but a qualitatively new approach which, in being formulated scientifically, will no doubt feed back and enrich the specific approaches. As, in reality, basic environmental problems are simultaneously of a global nature and of a particular character, indivisible and still divided in the variety and differences of the planet's natural and social contexts, so also the method to approach them should essentially reflect, or be a reflection of, the same kind of dialectical unity, and use both the general and the specific approaches.

A certain number of components should be taken into account, for instance on a country level; certain new factors should be added if an approach is carried out at the regional level (for instance, differences in social structure which are not easily to be detected in a country) and some previous factors eliminated (either as already absorbed in the more general approach or because they are of a very particular character, such as the organization of a local transport system or the building of a new factory, etc.).

A similar thing happens as we change the scale of analysis ascending from the regional to the international level. The 'ascent' of analysis means its deepening, since it disengages itself from details, which are characteristic of specific phenomena and concentrate on the most general and essential points which play decisive roles in the understanding of things and processes, the details very often being particular manifestations of the essentials. All this gives the opportunity to consider specific and particular matters not as self-sufficient and isolated but as integral parts of the whole, as elements of a system, thereby providing an approach to understand the complexity of the subject-matter and its global nature, and to use systems analysis. Certainly, the efficiency of the more general approaches depends on how and in what way the particularities of the previous levels have been taken into consideration and absorbed.

Understanding of the social character of environmental problems gives rise to the idea that systems analysis should be closely tied to multi- and interdisciplinary research, which, in a theoretical sphere, implies combined efforts and excellent co-operation between specialists of social, natural and technological sciences.

In the light of this general argument it may seem unnecessary to point out that, in industrialized countries, environmental problems are related to industrialization and technology while in developing or underdeveloped countries they are related to the absence of industrialization and technology. The contemporary STR has already affected all countries, although to differing extents. So long as there is only one biosphere, common to everybody, radio-active fall-out does not need an entry visa.

The impact of the STR on the natural environment should be measured socially, not only in the sense that today's biosphere is both humanized and human, but also in relation to the 'input-output' character of the STR. Occurring in the sphere of social production, it not only has positive outputs which are often followed by negative consequences, but also absorbs natural and human resources which, if not rationally planned, can lead to fundamental difficulties in the very near future, not to mention posterity.

In speaking of the STR, one tends as a rule to put more emphasis on technology than on science which, if mentioned at all, is only so in connection with technology or social production. Such sciences as biology, logic, psychology, medicine, economics or sociology, are rarely mentioned. A fundamental error is made when the whole group of human sciences are either ignored or shown as victims of technological progress. The basic idea which merits strong emphasis is that man is an integral part of his environment and the active creator of the latter.

In analysing the problem of man and his environment, it is supposed that man is the centre of the environment, but the very meaning of 'centre' can easily be lost if man is singled out as a consumer and not considered as a producer, and not only of his environment but also of himself. Man exists only in his environment, and environment makes sense if it is understood as a unity which includes man. In other words, the environment should be understood not simply as a fact but as a process in which man plays the decisive role. That is why the environment can and should be measured by man's activity, i.e. with a social dimension. To paraphrase an old saying, human beings at every stage of their development have the environment which they deserve.

The social impacts of the contemporary STR should be measured and valued by mankind's intelligence and self-understanding, which today has reached the highest level in history. This inspires a great deal of optimism and hope that the social impacts of science and technology will finally be put completely under the control of socially oriented policies to serve vital popular interests. Thus a scientific approach to the development of technology, and particularly to scientific progress itself, becomes one of the imperatives of our time.

Some aspects of contemporary scientific progress

RIGIDITY AND FLEXIBILITY

There have been several recent attempts to elaborate a scientifically justifiable approach to rationalizing priorities in research policy. For a start, a number of international meetings were held to consider criteria of research quality so as to lay the theoretical foundation for a rational research policy. The tacit objectives were supposed to apply to contemporary sciences and methodology.

I must begin with a few critical remarks on one of the most widespread theories which claims to have elaborated the application of the methodology of scientific research programmes to the appraisal of ongoing scientific research, leading to the evaluation of certain ongoing research projects, and to the assessment of what is known as the Popper-Lakatos approach.

Without going into too much detail, one can always express some dissatisfaction with the general orientation of this kind of theory if it looks to the past rather than to the present and, even less, to the future of contemporary scientific knowledge. Besides, science is understood by it only as a product of the individual efforts of scientists. Too little attention is paid to the socially conditioned and internally governed dynamism of scientific knowledge. Social scientific practice as well as social needs for the development of the science are almost entirely ignored. One also cannot be completely satisfied if such theories indicate certain difficulties rather than attempt to overcome and resolve them. The two aspects, critical and constructive, should ideally be complementary.

Critical analysis should not be reduced to mere negation and rejection. However, any basic statement needs to be carefully examined. Let us take, for example, the elements of a theory of 'ideal types'. These are described as: (a) consisting of truths and giving us 'ultimate explanations'; (b) having true self-reflection (to know to be true); (c) being precise at all levels; (d) meeting the requirements of a deductive system of knowledge.

This description needs to be further elaborated and adjusted. What do the 'ultimate explanations' mean? If they are not 'ultimate' in a concrete sense for this particular place and time, and if they are not conceived to be a type of Bible, then 'ultimate explanations' make no sense. Having in mind scientific systems of knowledge, it seems difficult to assume that there can be any ultimate explanation of anything (with the probable exception of certain trivial tautologies) which can be conceived as non-corrigible and non-revisable. I readily admit the existence of stable scientific principles but only on the assumption that they show a great variety and diversity of manifestations. The openness of these principles, which is especially needed to comprehend science today, must also imply a certain rigidity. Equally, one cannot entirely agree with (c): one can hardly expect the same precision (provided the precision is strictly determined) at all levels; more realistically, precision varies from level to level. Finally, the question arises why an 'ideal type' of theory should limit its requirements to so easy a requirement as only to possess strictly logical or deductive connections between its statements. This seems quite usual for any reasonable theory, and no exception should be made for an 'ideal type'. Apart from this, there are different systems of knowledge, among them non-deductive ones. Rigidity should not prevail over flexibility.

It is sometimes claimed that to obtain a true theory, no criterion is needed. This is difficult to accept. Any scientific theory should possess a criterion for testing its veracity. The absence of criteria generally speaking is very 'convenient'. It gives many 'advantages' and endless 'freedom', in particular that one can do anything under the label of scientific research.

If a theory is scientific, then it has criteria. Certainly there are no eternal and static criteria. The concrete character of criteria manifests itself in two ways: (a) for the different levels of one theory; and (b) for the same level of the same theory which develops and, therefore, needs new criteria.

For the most general criterion for any scientific theory one can always appeal to practice. However, the interrelationship is not that simple. It creates some difficulty when criteria are opposed to scientific practice. One can agree that there exists no extra-logical certainty, but one cannot agree with the idea that the criteria of truth, or of certainty in any system of knowledge, should always be found within this system.

Here we can only say that there is no rigorous criterion in any science to distinguish true from false. There is no eternal magical formula according to which we can once and for all determine truth. Truth changes, and criteria change with it. However, despite changes, this does not mean that truth does not exist. A rigorous dichotomy between criteria and scientific practice also seems dubious. Should one not suppose that scientific practice itself can serve as a criterion? This makes it easier to connect the analysis of science with technological applications.

Today one can in effect hardly dissociate pure science from technology which is now a necessary basis for scientific research. There is no empirical certainty at any level of science. Here another problem arises: the bi-polarization of depth and certainty. Is this opposition invulnerable? If one posits ultimate certainty as superficial and reduces it to scientific data, then this certainty cannot be applied to scientific theories. Is scientific data more reliable than theoretical findings? It might be more logical to assume that each level has its own criterion of certainty. The notion of certainty also needs concrete comprehension and application. Again, it should not be considered as something eternal and unchangeable. Any attempt to apply empirical certainty directly at a theoretical level can introduce incommensurability, owing to a change at the levels of reasoning.

A scientific methodology always needs to follow strict logical reasoning and must not exaggerate facts. So-called 'facts' cannot be used twice in one and the same relationship: once in the construction of a theory which often is highly hypothetical, and then again in its support. This is a misuse of 'facts' and does not even meet the requirement of common sense.

A scientific theory is dynamic. One should certainly avoid clashes between theory and experiment, which is supposed to be easily resolved on the assumption that the 'hard core' theory will be surrounded by a 'protective belt' of auxiliary theories. One can argue: the more auxiliary the assumptions, the looser the 'protective belt', the latter being subject to the risk of failing to fulfil its protective functions and leading to the need for a further protective belt. No doubt the higher the level of generalization, the more plausible the theory and the less directly applicable.

CONCEPTIONS AND FACTS

The construction of any general and unified theory, whether of the STR or of the environment, should in the last analysis be based on facts and their interpretation. Modes of interpretation concern first and foremost methodology. Scientific methodology applied to correctly chosen facts builds up a scientific conception.

What are 'facts'? The interrelation between conception and interpretation of facts is complicated even if we give the term 'fact' a strictly one-dimensional interpretation. 'Mere facts', 'pure facts' or 'naked facts' hardly exist. By this we do not challenge the existence of the earth or that of the Atlantic Ocean, which are objective facts. Any established fact becomes a subject for science and scientific methodology when it is recognized, properly selected and becomes an integral part of conception. One must assume that the existence of a certain conceptual framework determines the very way 'facts' come into existence. One can call them 'facts of science' or 'scientific facts'.

Objective facts are surely independent of human consciousness. However, this may create the illusion that there are pure scientific facts which exist in the absence of human interpretation. If only recognized by so-called common sense and absorbed in a usually non-critical way, facts cannot play an important role in science. They obtain significance by becoming a part of the conceptual framework, their objective character being a basis for one-dimensional interpretation. The superficiality of common sense thereby easily transforms it into 'common nonsense'.

There were indeed many facts (everything on earth and even beyond it) which pointed to the law of gravity but only Newton was able to shape them into a scientific modality. In this sense one could claim that the 'facts' which indicated the existence of the law of gravity came into existence only after Newton's discovery. Discovery is not the invention or creation of facts; it is the correct interpretation of facts which continue to exist after they are given a specific interpretation. The conceptual framework therefore becomes a certain mode of ordering the facts. It always has its intrinsic logic and easily resorts to intuition.

In other words, and in a wider sense, a conceptual framework can be understood as a wider apprehension of the world. Concrete understanding of the world, systematic and continuing enrichment of the world outlook, become essential features for the construction of theory. It is this conceptual framework which stipulates the very selection of facts from empirically fixed things and events.

To avoid possible objections, it should be emphasized that our description of the active role of the conceptual framework (human consciousness) does not at all diminish the importance of individual scientists and their creativity. However, the main idea stressed here is that scientific discoveries, whatever their importance, should not be attributed only to individual geniuses. Other matters should also be taken into account, especially the connection between social needs and scientific discoveries which do not always have a clear-cut and immediately obvious cause-effect relation. If something does not exist at the surface or is conceived with difficulty, that does not mean that it does not exist at all. Two main factors, social needs and the internal logic of development of prior scientific knowledge, should be taken into account in an understanding of scientific facts and discoveries. The history of science shows that when there is a real social need for scientific development, especially an economic (and, unfortunately, military) one, certain branches of scientific knowledge develop faster than could be achieved by a dozen universities. It seems quite logical to assume that in contemporary scientific thought and world outlooks, in the specific model of comprehension of empirical material, the whole history of human culture is 'encoded' and fixed. If we continue to take the law of gravity as an example, it is difficult to assume that if it had not been for Newton we would still be without an understanding of it (there is always another Newton). Contingency is an unavoidable form in which necessity exists and manifests itself.

Thus, any theoretical construction expresses certain trends of thought which are in the last analysis socially oriented. So the understanding of the social needs and of the internal logic of the development of social thought (which is by its very nature contradictory and essentially dialectical) can add food for thought and shed light on how the critique of the 'individualistic' approach to history and scientific discoveries can be conducted. While not wishing to diminish the contribution of individual scientists to the history of science, it is not very serious to reduce the history of science only to individuals, ignoring the fact that they were expressing directly or indirectly, certain social needs.

The social nature of scientific knowledge

There is a difference between criteria of truth and criteria concerning scientific quality. What kind of projects in scientific research should be given priority? And what is the main criterion to recognize progressive lines of scientific research? I think that today we still do not possess sufficiently convincing criteria to determine both the absolute and the relative values of scientific research results, or to compare the significance of various research lines.

There are certain 'criteria' which one could take into account. Among them, special emphasis can be given to: (a) the practical effectiveness of scientific research for welfare societies; (b) new theoretical generalizations; and (c) significance for other sciences.

There is one idea on which I would like to place special emphasis, namely the human dimension of scientific progress. This aspect is often ignored. It is, however, this very aspect that recently encouraged the development of new branches of knowledge connected with the STR (environment, the biosphere, ecology, the quality of life, etc.) and that underlined the considerable importance of the social responsibility of scientists.

Today, any scientific system of knowledge, especially in the natural sciences, is becoming industrialized and vice versa; industry is more and more based on science, which means that in trying to analyse perspectives of fundamental scientific trends we should not merely be guided by historical examples but should take into account the conditions of the development of scientific knowledge today and, especially, tomorrow. Like any other phenomenon, science develops spontaneously. Among other things, we must acknowledge that a new sphere of human activity is emerging which considerably differs from science and technology as they are understood today.

Theories should be considered dynamically and not simply as the ready-made knowledge fixed in them, and their verifiability should not be over-simplified but treated differently at various dimensions.

In order to induce real growth of the gross intellectual product (GIP), one needs to take fully into account that scientific knowledge always was and is now more than ever determined by social conditions. The potentials and perspectives of GIP should be measured first and foremost by the width of research which is not reduced to everyday needs, although the most general social needs should certainly be reflected. One may claim that the linkages between social needs and social sciences are obvious, but what about natural sciences? To make this idea clearer, one could give a very short account of the history of science in order to show certain objective connections between social needs and the leading role of selected branches of the natural sciences.

The question arises: which of the natural sciences plays a leading role? A simple answer would be that the leading natural science is the one that seeks to solve, and really does solve, the most vital problems of a society and its development: economic, cultural, scientific, technological. The position of leadership in natural sciences is not at all invariable: it shifts in the course of history. In the seventeenth and eighteenth centuries this role was occupied by mechanics; in the nineteenth century, an interconnected complex of various natural sciences—chemistry, physics and biology—could be considered as leaders. In the first half of the twentieth century, one could see subatomic physics in the foreground; and in the most recent decades leadership has shifted to those branches which are closely linked with new technology: biology, physics, chemistry, electronics, cybernetics, and astronomy. Although there are groups of sciences, one can nevertheless single out biology as a leader among leading natural sciences. We may consider ourselves as living on the eve of a 'biological era' which may, however, not last very long; the next leader may be a science which borders on both the natural and social branches of knowledge, for instance, psychology.

The latest findings of genetic engineering which today are at an exciting phase can be taken as a conspicuous example. Biological and biochemical research now involves separating and recombining DNA. 'Gene therapy' has resulted in gene-splicing, becoming thereby the most promising branch of scientific knowledge with extremely refined procedures and methodology. It has also raised a great number of crucial problems in the ethics of science.

The idea of a 'leading science' (or a 'leader-science') should be understood in a rather flexible way. It can belong to a group of sciences—or in the future even to no science proper but to technology or scientific methodology or rational management of sciences; this reflects multidisciplinarity and meets the requirements of the emergence of new branches of scientific knowledge with multidimensional approaches, responding thereby to challenges of social practice and of the development of other sciences. In other words, traditional approaches should not act as a strait-jacket upon creative efforts at innovation.

In the twentieth century, the development of physics brought it 'down' from atoms to their particles. Chemistry, on the other hand, went 'up' from atoms to the macro-molecular. The development of new branches of chemistry was stimulated by the practical needs of social production which showed interest in the development of a number of synthetic materials. Here technology played an extremely active role and did not fail to orient the intellectual activity of chemists to analyse the chemical structures of materials and produce them artificially. On its way chemistry encountered the process of bio-synthesis.

We can say that many fundamental natural sciences are so closely connected with technology that it is often hardly possible to make clear-cut distinctions between them (nuclear physics, atomic energy, cybernetics, automation, bionics and the wide range of cosmic research, etc.). The same can be said about methodology. The most important practical implications have not yet been discovered in cosmic research; those that have (e.g. telecommunications, meteorology) are not be considered as most essential. Through interconnections between sciences, astronomy is now being transformed through cosmonautics from a purely observational science to an experimental one. Should we expect that future development of natural sciences will proceed 'independently' and from 'within', or that this line will be stipulated by technological and, in the last analysis, social needs? In attempting to approach any new general theory, whether it is an elaboration of a unified conception of the STR or of the environment, a number of things should be taken into account.

SOME FEATURES

There is much evidence to show that: (a) contemporary natural sciences and a growing number of social sciences are now closely linked to technology; (b) the needs of social practice usually stimulate the development of scientific thought; (c) one of the specific features of contemporary scientific knowledge is a complex and frontal approach to research; (d) the most fundamental scientific problems today emerge on the borders between existing sciences; (e) natural sciences as well as social sciences are not only studying and exploring nature and society; more than ever before they are engaged in the perfection of their own tools of research; (f) scientific research is increasingly characterized by introspection and self-consciousness aimed at an understanding of itself; it gradually acquires the ability for deeper self-reflection; and (g) science develops in geometrical progression or, to be more precise, exponentially, with accelerating speed of accumulation and multiplication of knowledge: the more knowledge is accumulated, the more knowledge it is necessary to obtain.

Today, science has become an integral part of human activity, a certain and quite specific form of it. There is a striking contrast between sciences past and present. A good half of all the scientific data which we use today in various spheres has been acquired during the last twenty years. More than 90 per cent of scientists involved in research over the whole history of mankind are our living contemporaries.

While stressing that scientific progress is indispensable to technological progress and vice versa, one should certainly point out that contemporary scientific and technological progress is closely linked to social and economic structures as well as to levels of societal development, the nature of this connection being complex and often contradictory.

One should not expect immediate practical results from scientific discoveries. This is especially true today when a complex and frontal approach is imperative. Studies in depth of a number of processes, especially in pure theoretical sciences, may not have immediate application at present or even in the near future. Moreover, if scientific studies are always oriented merely to solve everyday problems, then the transition to more general, complex problems may become impossible. Any rigid limitation can seriously inhibit the perspectives of scientific progress, which should remain more open-ended. Such progress provides new possibilities to formulate strategies of scientific research which may have considerable influence on the further development of scientific thought. The difficulty is that this idea could become clear only in the future. There is a paradox: to evaluate ongoing research projects before obtaining results, when it is precisely the results that should be evaluated. Scientific searching is not always research, and in this sense it is better to pay the cost of scientific and intellectual production than not to attempt to use all the potentials for future scientific and technological progress. This all-embracing approach to scientific research may sometimes quite unexpectedly open a number of avenues to new discoveries. Again, contingency here should be taken not in itself (which is a common-sense approach) but as a manifestation of necessity.

However strange it may sound, the criteria for quality of research results are something very concrete, and although notions of an international standard have emerged, these depend in many respects upon the level of development of a given society. Concrete social practice is an indicator which should be taken into consideration in identifying the most appropriate, acceptable and useful ongoing programmes of scientific research in a given society. Scientific progress, although it should be approached objectively, can hardly be measured only by science itself.

ON HYPOTHESES AND TWO LEVELS OF KNOWLEDGE

Hypotheses can be considered to be one of the forms in which scientific and technological thought moves and develops. In other words, any new knowledge appears and is transformed into new theory through hypothesis. The comprehension of the hypothesis should not be restricted only to a deductive approach and new generalizations. A merely formal approach very often does not even touch on the problem, although it may create an illusion of its resolution. There is a much greater difference between new empirical data and new theoretical knowledge than is understood by a common-sense or a purely 'logical' approach. To comprehend the very nature of a hypothesis, it is not sufficient simply to begin with new empirical data and hypotheses themselves. One should start with the analysis of the real process of knowledge, the latter being conceived (as having some necessary stages, phases or levels) as a system. It also seems necessary to determine the point where the hypothesis becomes an absolutely necessary step for the further development of knowledge and where the natural path of such development proves to be a hypothesis. Certainly, this level should embrace the conditions prepared by previous development and which make it impossible to move along in the same mode, and simultaneously determine the content and directions of further development.

One of the ways to construct a theory, which concerns primarily those sciences where observation and experiment play dominant roles, may be the following.

First of all, one should distinguish between two levels of knowledge.[3] Both include scientific research and are not rigidly divided by a strict dichotomy of empirical sensualism versus rationalism. The first level includes all the tools of scientific research (observation, selection, experiment, hypothesis, simulation, etc.) and deals with that reality which is beyond knowledge and consciousness. The second level is the construction of a theory as a definite system of knowledge; it includes modes of theory-building, their development, etc., and, as a rule, deals with knowledge itself, with the apparatus and functioning of scientific knowledge. These two levels can be roughly compared (if one can introduce notions from economics) with the production of consumer goods and that of producer goods. Both levels are closely interconnected.

The first level is to be considered as a basis and fundamental content of science; the overwhelming majority of scientific discoveries are made there. The second level is founded on the first and should be considered as a higher stage with its own specificity: its 'empirical' basis is the first stage. It is therefore connected with reality indirectly, through the first stage. It deals mostly with the problem of how to formulate certain typical modes of obtaining knowledge. Both levels have their advantages. Today when science has become part of social production (production of scientific knowledge), large-scale experiments are very costly and time-consuming. Here one can see the practical utility of the second level: without previous theoretical work none of the costly practical undertakings or scientific experiments could be made.

Both levels deal with the problem of truth which it is necessary to obtain not only for a deeper comprehension of reality but also for the creation of realistic grounds for making decisions. If, at the first level, truth can be determined by reference to the reality which it reflects (truth being understood as a correct, adequate reflection of reality), at the second level this criterion does not apply in the same sense. However, the truth or error of the theoretical statements at the second level can be tested against their conformity with the first level. If the criteria of truth for the first level can be understood as practice in a wide sense (practice here being conceived as the domain of interrelation between man and nature, man and man, individual and society), then for the second level we can use as auxiliary criteria propositions of the first level which have been immediately proved to be either true or false. These criteria also are not rigid and unchangeable. They are quite concrete and vary according to the levels of knowledge, although in the long run only practice should be considered as the most reliable criterion.

Notes

1. United Nations document, Conf. 48/14/Rev. 1, p. 3.
2. *American Journal of Economics and Sociology*, Vol. 35, No. 4, October 1976.
3. Principally, it seems that for any science including those axiomatic and in a wider sense deductive (and non-deductive) systems of knowledge, it is always possible to point directly or indirectly to an appropriate empirical basis. In the various sciences these bases are obviously different.

The role of the social sciences

Radovan Richta

Academician, Czechoslovak Academy of Sciences

With every passing day our epoch is richer in revolutionary changes, which have an impact on the fundamental dimensions of the contemporary life of society and which, with ever-increasing urgency, call for scientific insight into the nature of the processes going on.[1] Never in its age-long development in the most diverse forms, ranging from the primary reflections about social ties in Oriental mythology and heroic epics to the complex, present-day system of social sciences, has social thinking played a role in man's life comparable to what it plays today; nor has it been confronted with such challenging demands on the range and depth of problem-solving, on temporal span, and on the precision of its conclusions.

Not so long ago, it was customary to assess findings concerning the development of society against the restricted background of a single country, continent or cultural tradition. Today, no mode of social thinking can be considered relevant unless it entails potential answers to problems affecting all countries, all continents, the entire world.

We can well remember studies which, dealing with human and social problems on the one hand, and the problems of nature and the environment on the other, programmatically referred to a completely different conceptual world. Today, however, it is evident that without a comprehensive grasp and control of the inner relationship between a people's active approach to nature and their interrelations in society neither natural nor social sciences can function successfully.

Only recently, if it was suggested that matters of social theory and practice be assessed and solved from a long-term perspective, quite a few specialists were regarding it as a rather risky novelty, as 'prophetism', or as a quest for 'Daniel's kingdom'. Nowadays, this frequently involves more than the lifetime of a single generation, i.e. a dimension once taken into account, according to Balzac, only by a few dreamers and Parisian families who had paid for their tombs. And yet there is growing evidence that today conceptions and projects concerned with the solution of grave social problems—if they ignore long-term perspectives in the formation of social and natural-technological conditions of life and if they bypass the entire chain of secondary, tertiary and further consequences of human activity—easily turn into anachronisms and hazardous adventures.

Even today, the development of society is frequently interpreted as resistant to comprehensive intellectual analysis and theoretical reproduction, either with regard to the complexity of ties and the contingency of interactions in society, which are supposed to preclude cognition and only allow for a narrow specialized description, or with reference to the value basis of man's life and conduct, which allegedly does not admit of a rational explanation but confines us to mere 'intuition', or 'empathy' (*Erlebnis*), etc. Nevertheless, even this *reservatio mentis* of both empirical social sciences and the traditional speculative '*Geisteswissenschaften*' (human sciences) is clearly becoming untenable under present-day conditions, when the entire complex of human social action, including any kind of scientific activity, is pervaded by the omnipresent and fundamental question of the actual preservation of life of man and society; when the truthfulness and precision of social-scientific knowledge is becoming a significant social value, which under specific conditions and in the hands of specific historic forces can affect the solution of this fundamental issue.

Today, natural sciences and technology are penetrating deeply into all areas of social life; they have become a powerful force in modern production systems, in the transformation of the environment, in the organization of human co-operation; they have begun to lay bare new and grave, burning problems facing man and society. At such a time, social sciences all over the world are confronted with the great task of bringing into a sharp and timely focus the nature of the scientific and technological revolution (STR), its position in the complex of social processes and its relation to current revolutionary changes as well as to the nature of the prospects it can open up for society.

True recognition of the nature, manifestations and laws of the STR in different and intrinsically opposed social systems is inherent in what today is one of the most crucial orientation points indispensable for the functioning of the entire system of the social sciences. At the same time, this is an area where the answer poses the reverse question: What are the social sciences to us today?

Do we understand what we know?

On the surface of contemporary social life, the STR manifests itself in the dramatic advance and mounting tide of scientific discoveries and industrial technology. From the point of view of a method which confines its approach to a mere description and generalization of empirical data, the question of the nature of the entire process is reduced to the specification of the explosive spread of a quantum of scientific information, the accelerating advance of technological innovations, and the increase of the growth-rate of labour productivity in production.[2] However paradoxical this may appear, the fact remains that the process of the STR, which displays so many striking symptoms and whose course was witnessed by the scientists themselves, remained in the consciousness of a considerable section of the social sciences and of science in general only as a familiar though largely unspecified background of life; as a kind of general spontaneous motion, which due to its contradictory external manifestations can be easily but non-committally quoted as a reference; as a kind of modern fate which we gradually encounter, just as in Diderot's story, *Jacques le Fataliste*, in everything we lay our hands on.

Wherever theory had responded to the newly arising material conditions of the life of man and society in this restricted, pragmatic way, in the belief that what is known is *eo ipso* understood, it soon found itself skating on thin ice. On the one hand, unregistered and unanalysed essential changes, occurring under the surface of the commerce between society and nature, exploded right under its touch.[3] On the other hand, theory was predictably cornered by the social dimension of the STR, which it failed to substantiate theoretically, not to mention preparing the ground for the intervention of social practice.[4]

The development of living conditions in society, already so difficult for this type of thinking to understand and predict, has gradually assumed a 'shocking' form. Uncertainty as to what further advances of science and technology may bring, the groping in the double darkness of social and technological dimensions of the future, the upheavals in nearly all the traditionally established values and conditions of human life, have led some of the social sciences in the technologically most advanced capitalist countries to ever new attempts to cope with these issues. A section of Western literature continues to react through a self-deafening avalanche of technocratic publications that emphasize the immense power of contemporary science and technology, appealing to feelings reminiscent of *hubris*, the pride of the condemned, in an attempt to encourage further advances along the same road, and not infrequently arguing that this is the one and only option left.[5]

On the contrary, the other part of the social sciences in the West, confronted with the STR, rebels, on bended knees as it were: while denouncing the Promethean motif and all the ideological forms and practical activities which it symbolizes, it has recourse to the deepest repentance that human reason has ever been able to engage in, repentance not before man and society or their creations, but before a mysticized and long-since non-existent 'unspoilt' nature. This approach returns man's intellectual universe way back to the very threshold of social thinking, to the hymns of 'rhapsodic intellect', to the *Kabbala* and the *Yoga-samadhi*, and to the gnostic *apocatastasis*, the illumination of the fallen soul, returned, in fact, to the abyss.[6]

It is noteworthy how fatal is the fact that for all these approaches, whatever the difference in conclusions, the STR has remained no more than a hazy image, a sum of examples which, though frequently quoted, lack both conceptual elaboration and definition of relevant social ties. An enumeration of scientific and technological innovations as examples, however exhaustive or interesting; the extrapolation of general tendencies on the basis of fragments of reality of the most varied provenance—all this is hardly sufficient to shed light on the nature of the STR as a social process of profound social significance for the entire current epoch.[7]

The sum total of the analyses carried out in the past few decades reveals that under the surface of the changes in science and technology, changes within their system, in their mutual interaction and functions, representing the external phenomena of the STR, there are profound *revolutionary transformations in the structure and dynamics of social productive forces* that proceed on the basis of a comprehensive application of science as an immediate productive force gradually pervading all the other components of productive forces—both technological and human;[8] thus calling forth fundamental changes in the position of man within their system 'in the position and function of man in the process of labour'.[9] Hence, new dimensions for the development of social productive forces are opened up. The STR is not reduced to an intensive, one-dimensional advance of technology. First and most important, in analysing the mounting potential of technological factors, 'we should not bypass the social aspect of technology'.[10] Second, each technicalization of production is tied up with a complex of broader changes in the structure of productive forces.

Out of the variform process constituting the essence of the STR and reflecting its essentially social nature, the current superficial approach singles out, as the one and only exclusive component which overlaps and obscures all other elements, the powerful, vigorous advance of science and technology viewed in terms of their mutual linkage,[11] in the form of an autonomous, fatal, unfettered 'self-feeding' process[12] that transcends the power of man and society and defies their control. It is on this point that all the strongly allergic currents of technocratic and 'cultural-critical' thought, incapable of penetrating to the social essence of the scientific and technological revolution, surprisingly coincide.[13]

It is evident that this conception has its background in a certain social reality, in the concrete conditions of the capitalist world in the West under which, during the past few decades, the STR has been taking place, or more precisely in the conditions surrounding its initial emergence on the surface of the life of this society. Transformations in the structure of productive forces, tied up with the STR, have indeed so far materialized predominantly in proportions revealing the marked prevalence of immediate technological components. Science as a productive force exercises and continues to exercise its revolutionary functions in the production and reproduction of human life primarily through its application in technology; the other components of this process have so far asserted themselves on a far more restricted scale or in forms directly subordinate to the technological reconstruction of production. Thus, for instance, the profound qualitative changes in the organization of the social process of production, linked up with the application of new scientific conceptions in the field of cybernetics and information theory, proceed as a transformation of the process of production 'from a simple process of work into a scientific process',[14] as anticipated by Marx. So far, however, in the given social conditions, they have mostly had the character of a preparatory stage preceding the introduction of new automated technology. Only to a relatively limited extent (only under certain social conditions or as an indispensable counterpart to new technological procedures), has this area also seen major changes in the field of workers' qualifications and of general factors associated with the application of science as a productive force which is directly linked to the activity of working people, and hence with the 'development of the productive force of individuals',[15] in which Marx saw the most viable form of social wealth.

Incontestably, in the one-sided course of changes in the structure of productive forces, a certain role was played by the fact that the STR itself has reached only its initial stage; that, so far, it has not pervaded the entire production basis, not even in the technologically advanced countries; that on the social scale it has not yet created the preconditions for a mass release of the potentialities of the working people from immediate production operations, for the full tapping of all the potentialities inherent in their activity during the process of adapting and reshaping the entire complex of the production and reproduction of human life. It is only natural that, in so far as the cause of difficulties stemmed from the inherited, historically restricted level of productive forces, further progress of the STR will be paralleled by a gradual reduction of difficulties and a general acceleration of development.[16]

Nevertheless, the open crisis phenomena clearly observable in the 1970s in the life of some of the technologically most advanced countries in the West, which have brought about what is perhaps the most drastic 'disillusionment' with scientific and technological progress, the most severe frustration of rational thinking and transformation of reality that modern history has ever witnessed,[17] have shown that what is at issue under the existing conditions are not merely the difficulties attendant upon the early stages in the shaping of the new structure of productive forces. Characteristically, the more countries afflicted by crisis phenomena make use of the achievements of the STR, the higher the inner tension in their life, and the greater the difficulties they encounter in this particular area.

Obviously, the causes of these contradictory phenomena lie deep in the given social conditions which affect the current mode and rate of scientific-technological development.

Social dimensions of the scientific and technological revolution

The symptoms of disproportions between the tide of material technology, on the one hand, subordinating all other components of production, and the retarded development of the human components of productive forces, on the other, resulting from the suppression of 'a world of productive capabilities',[18] of millions of people whose lives are in some of the technologically most advanced countries constrained by the limits of the simple reproduction of the labour force—all this is neither an eternal, natural condition of the production and reproduction of human life, nor a fatal attribute of the STR. What is at stake here is the law-like consequence of the relations of capital, which, for the past few centuries, has been shaping and at the same time limiting the basic parameters of the development of productive forces and civilization in general.

The social nature of capital consists in the separation of the objective conditions of the process of production, accumulated in the course of the entire development of social labour, from the subjective, human components of the production process; in its constitution as an independent power, dominating living labour as a 'superseding subject',[19] governing the course of production and determining the general material conditions of people's lives; and at the same time, in the reckless augmentation of production conditions, which had acquired a social nature and yet remained in private hands at the expense of the living components of social labour, at the cost of their stagnation. It was precisely these social production relations, these forms of motion, that formed the framework for the development and operation of modern productive forces (including science and technology), that provided the pattern for adaptation in their basic structural and developmental rhythms. Here we arrive at the social sources of the acute imbalance between the contemporary explosive rise of material powers created by society and the growing defencelessness of the human components of productive forces, compelled to service this civilization apparatus: the disparity between technological progress and the social failure of the entire preceding epoch.[20]

Interests directly or indirectly arising from the requirements of the self-expansion of capital, have for centuries, since the industrial revolution of the eighteenth and nineteenth centuries, determined the mode of development and application of science as a special production potential, separated from labour and counterposed to it[21] in the shape of the power of past materialized labour over the living labour of the workers and hence over their destinies.[22] Until recently, these interests impressed their own orientation and conditions upon the development of science.[23]

Specific goals allied to the drive for capital profits once mediated the outbreak of a modern 'revolution in the means of production',[24] which created the technological basis of the industrial revolution of capitalism. Gradually, they have converted the 'domination of past labour over living labour' not only into a general social relation but also into 'technological reality'[25] within machine-based factory production[26] and finally also within the modern system of consumption.[27] They delimited the basic developmental trends of the industrial technology of the nineteenth and the early twentieth centuries,[28] and predetermined its structure and dynamics.[29] In the long-term perspective, they have based the development of productive forces and civilization in general on the continuous overpressure of their technological apparatus and the reckless exploitation of their social, human components. They have put the stamp of an extraneous power, existing in its own right, uncontrolled by man and society, on the artificial modern, material conditions of life.[30]

The STR exacerbates the contradictions underlying the development of capitalist society since the Industrial Revolution.[31] It demonstrates them on the surface of life on a scale which gives rise to fear and apprehension. This is precisely the framework in which we today are witnessing the outbreak of the 'crisis of science and technology',[32] which, with mounting vigour, is engulfing the industrially advanced capitalist world.

The deepening contradiction between the rapidly expanding scope of possibilities opened up by contemporary science and technology and the stagnation of social reality in the advanced capitalist countries was bound to call attention to the crisis-ridden situation facing society in the West. It is no mere chance that the first signs of disillusionment and sharp criticism levelled against science and technology coincide remarkably in time with the date of one of the greatest achievements of science and technology: the lunar landing.[33] That demonstration of the immense power of science and technology in space must have made people in the technologically advanced countries aware of the weakness in the utilization of the achievements of the STR to the benefit of man's life on earth.

The vision of a constant deepening of the crisis of scientific and technological development, worsening with every new scientific and technological achievement, has now become so haunting that not infrequently it has led to conclusions that were unimaginable in the world of science thirty years ago, when V. Bush announced the advent of 'the endless frontier of science',[34] i.e. that science and technology have exhausted all their possibilities, that, just at its very outset, the entire process of the STR must be halted.

In actual fact, the crisis phenomena encountered in this sphere today are on no account inherent in science and technology themselves. The scope of new possibilities for scientific discoveries broadens with each new scientific discovery or, as Leibniz predicted, with each new contact with the unknown. And time and again arising from them are further astounding prospects of their practical utilization for man's benefit. The frontiers we are facing here are not specific to the STR, to the cognition and transformation of the world in general. They constitute one of the symptoms of the general crisis of the society in which this science and technology operate.[35] They only express the contradictions and constraints of the one-sided, deformed course of the STR in conditions governed by the profit of capital, and this is the ultimate frontier beyond which neither science nor technology can develop without disaster, if and in so far as it is not matched by corresponding development on the part of man and society.[36]

The existence and development of science in the restricted form of a closed system separated from the working majority of society and opposed to it in the shape of technological application[37] does not represent the one and only, eternal and true form pooling man's efforts and longing for rational cognition. What we are confronted with here is no more than a historically transient, conflict-ridden framework within which science develops as 'universal social labour', 'general intellect', as accumulated wealth of society in its most solid form.[38]

Similarly, the existence and advances of technology, in the form present in the material incarnation of capital, serve as a brutal means towards controlling the labour force,[39] towards reducing its value and securing capital profit and, finally, towards petrification of the rule of capital even at the expense of its own transformation into an open force of mass destruction. All these are not eternal, general means of transformation of the world by man, but only one, contradictory, transient form of technology. In its essence and in its general development, technology represents the materialized forces of social knowledge, the substantial forces of social man.

Of course, if the nature of science and technology as well as the prospects of the STR are deduced not from their entire role in the development of society but from their empirical forms and tendencies to be observed in the advanced capitalist countries, then we are reasoning in a vicious circle: here then only the specific historical forms of the spontaneous operation of capital figure as basic, permanent attributes and tendencies of the STR, and of science and technology in general. Thus, science is interpreted as a matter of the élite,[40] relying on the system of 'value-free' rationality,[41] focusing on technological domination and 'justification' of this domination,[42] etc.

Seen in terms of this approach, the STR appears to be no more than a series of distressing ventures that degenerate into 'the cancerous growth of technology-conditioned economic production'.[43] The general reversal of subject and object, dominating the capitalist system and reflected in the fact that 'we are becoming the servants in thought, as in action, of the machine we have created to serve us'[44] is then accounted for as the inseparable and insurmountable play of fate between man and his work.

The social sciences that are not capable of discerning the specific social roots of this reversed state of affairs generalize the limits confining the operation of science and technology within a social system controlled by capital, present them as the very nature of science and technology in general.[45] They ascribe to contemporary science and technology the character of a kind of new 'absolute' power in society; they refer to an era of 'absolute technology' creating 'absolute power'.[46] They regard the fulfilment of Bacon's dream of the New Atlantis, which foreshadowed the road of the Industrial Revolution, '. . . the enlarging of the bounds of Human Empire, to the effecting of all things possible',[47] as the fatal moment in mankind's history.

It is in this modern myth of an 'absolute technology', which reflects the deformed, one-sided channels and blind alleys of the STR under social conditions controlled by capital that, once again, the most varied, mutually adverse currents of thought which abound in the Western literature on the social sciences, converge: those describing as the 'triumph' of science and technology a situation, under which, according to J. von Neumann, all that is technically feasible will become deed,[48] and those, which on the other hand, see these new conditions as offering evidence that the development of science and technology has gone beyond man's control, has followed a course of immanent uncontrollable development, and is 'dominating' man.[49]

In fact, the basic assumption of these and similar arguments presents a reversed picture of the reality in the capitalist countries: it is not the technological conditions and feasibilities that are in themselves decisive for the implementation of the projects; on the contrary, the interests of capitalist profit are crucial here, asserting themselves in a society of this type. Profit selects out of the wide range of technological feasibilities a relatively restricted number of projects for implementation; and where technology is not available, all means are employed to enforce its provision.[50]

According to the evidence advanced by technologists, the gradual elimination of dangerous, monotonous, simple labour operations, not infrequently detrimental to health, in other words, procedures characterizing 'slave labour',[51] has long since, at least in countries displaying the highest standard in productive forces, become technologically feasible. At present, this issue represents one of the primary requirements in the endeavour aimed at a purposeful shaping of human life and human activity. In the socialist countries, preparations are under way for its comprehensive mastery with participation of an entire set of social and natural sciences.[52] And yet this process proceeds in the capitalist countries at a snail's pace, unevenly and, in view of a cheap labour force, even with relapses, simply because it is incompatible with the interests of an economic system that continues to be based on profit.[53]

Equally feasible and within the reach of technology in the capitalist countries is the possibility of contributing substantially to speeding up the raising of living standards in developing countries. Here too the point at issue involves a fundamental question of contemporary social development. And yet, as a result of specific socio-economic relations and motives attendant upon imperialism, the gap between the richest and the poorest countries has generally widened.[54]

The STR has rapidly expanded the range of technological feasibilities; a never-ending series of new achievements has now become available to man and his transforming activity; new productive forces are entering the process of reproduction of human life. The number of the variants of the development of productive forces is clearly mounting, raising the question of choice.

All this, however, does not imply that in the period of the STR, technology has become an absolute, independent, autonomous power, which can achieve anything, and has seized the control over the future of man and society. On the contrary, powerful changes affecting the technological components of productive forces are placing unprecedented demands on the development of the social and human components of productive forces. Only where the scientific and technological revolution itself is being narrowed down to a restricted shape; where the development of its social and human components has reached a dead end; where the creative activity of social labour has not been given the scope indispensable for its development; only under these specific social conditions may the sudden rise of objective technology appear as its mounting overpressure and ascendancy. The point here is not the excessively rapid or multi-faceted development of technology, but the rather slow and restricted development of man and society in a considerable part of the present-day world.

The high-flown statements about the omnipotence of technology only provide evidence that in such a society man is powerless.

The type of development of productive forces based on still mightier technology as opposed to the increasingly powerless labour force may appear wide-spread. But today it can no longer be regarded as the one and only or the optimal variant of the development of the material basis of human life.

What the STR really brings in its wake is the gradual transition of a whole series of a specific functions, from their former exercise by man in the production process, to the functions of technology.[55] This is the first general and highly significant aspect of the current structural changes in productive forces. The second important aspect of this process reflects the ensuing changes in the position of man or, to be precise, of the workers' collective, which is thus gradually released from simple executive functions within the immediate process of production. This is where a whole area, mediating the social reproduction of human life intervenes substantially. It is here that the dialectics of productive forces and social production relations prove to be the crucial dimension.

What changes in the position of man in the structure and development of productive forces do social systems mediate, or are capable of mediating, in connection with the application of science as a productive force?

With regard to the resolution of this issue, the process of the STR under way proceeds, beneath the surface of the current changes, along two different and substantially contradictory lines. Where the entire advance is limited by the separation of science as a special production potential, subsumed under capital and opposed to labour,[56] the changes in productive forces are confined to the narrow circle of the increasing technological application of science, while the human components of the development of productive forces are only adapted to the means of production and do not enter the development of productive forces as a new factor in its own right. With the exclusion of simple labour functions from the immediate production process, the broad masses of the working people are relegated to the status of an element more and more subordinate to the operation of the productive forces of society, not actively participating in the process of shaping the material conditions of their own lives. Where the system of social relations is not capable of mediating development on the part of man, the positive potentialities of the development of technology will sooner or later be exhausted.

A completely different line of development of the STR is followed where science as 'universal social labour' is not separated from the labour of all individuals by the barrier of capitalist private ownership, not opposed to labour as an extraneous force. Under such conditions, science pervades the entire process of production and reproduction of human life, being mastered both in its materialized and in its ideal form by the entire collective of the working people acting as the social subject. Such a social system not only allows for but actually mediates a gradual change in the position of man in the structure and development of productive forces. Hence a far wider course for the stream of the STR is opened up: science penetrates as a general social productive force into all other components of productive forces,[57] both technological and social.[58] Along with the continued technological application of science, the utilization of science mediated through the social organization of the production process, workers' qualification, the creative activity and initiative of all participants in the production and reproduction of human life, in a word, those productive forces that are linked with the development of the abilities and activities specific to the social individual, acquire the role of a component steadily gaining in significance.

The process of releasing man from simple executive operations within immediate production is paralleled by a corresponding development of his functions and activities in the preparation, control and perfection of the social production process, the introduction of new social productive forces arising from the mastery of science by the working masses. Only such conditions do, in fact, open up full scope for motion in all components of the productive forces, for structural changes set in motion by the STR. Along with the actualization of all these possibilities, the STR begins to face the prospect of a far-reaching historical process, working towards an organic linkage of production and research, labour, science and education.

This line of development, of course, presupposes a radical change in the entire position and character of science in the life of society, and a corresponding change in the position and character of production. The most important prerequisite is the removal of the barrier of capital as an antagonistic social power, based on anti-social, particular interests, subordinating science and separating it from labour, boosting its technological application irrespective of the consequences for man and society. On the contrary, it calls for a well-elaborated system of social ownership of the means of production, a social utilization of science and all its applications on the basis of socialist relations of universal mutual co-operation among men.[59] This social substance can neither be avoided, nor replaced by specific forms of intervention, assessment and stimulation. For it is only under such social conditions that science emerges in its true essence as 'the most reliable form of wealth, its product and producer alike'. On the other hand, 'the evolution of science, that ideal and practical wealth', proves to be a form 'of the development of human productive force'.[60] For only here does technology acquire its substantial nature in the form of objectified essential powers of man,[61] which in the transformed nature accumulate resources for the self-transformation of man and society. Under such circumstances, production ceases to be the mere production of things or goods and the extended reproduction of capital; it acquires the meaning of a progressive production and reproduction of human life, based on the general social development of man, of the abilities, potentialities and activities of the working people.[62]

This course of full-scale development of the STR under the conditions of socialism calls forth substantial changes, not only in the structure of productive forces but also in their dynamics. The intensive development of the productive forces of society, based on the application of science as a social productive force, is thus not a one-sided growth in the sphere of the objective, technological conditions of production; on the contrary, it increasingly acquires the nature of planned development in the sphere of human components of the productive forces. The mass development of workers' creative activities gradually set in motion, in socialist society, tendencies leading towards a steep and general growth of qualification, rationalization activities and management participation leading to initiative in the all-round application of contemporary science: all these signs unmistakably indicate that today generous scope is opening up for an unrestricted development of productive forces, based not on the reproduction or extension of the existing situation but on 'the transcendence of each point of departure'.[63] In other words, the STR has acquired corresponding social dimensions, has gained its adequate social hinterland.

The dynamics of the productive forces is then made dependent on the extent to which science is incarnated in the means of production, technique, technology and, in the long-term perspective, on the capacity of the resources of science utilized in production along with the activity of the working people and their entire life.[64] As the advances of science and technology are to a great extent at the stage attained in the general social development of the potentialities, abilities, needs and activities of the working people,[65] the groundwork is being laid here for the promotion of a new relationship in the general process of the production and reproduction of human life: from the point of view of society as a whole, full-scale development of the productive forces linked with the STR calls with increasing urgency for a corresponding universal development of the productive forces, of all individuals, of the entire working people.

At a certain stage of the process of applying science as a social productive force, preconditions are created under which the general social development of the workers, the promotion of man's abilities, needs and activities, and, in the final analysis, the general social development of man as an end in itself[66] become the most effective mode of augmenting the productive forces of society. This is the crucial turning point in the evolution of productive forces, which indicates a way out of the contradictions still besetting the entire development of the material, technological basis, as well as highlighting the historical necessity for profound social revolutions.

Where social production relations utilize these preconditions inherent in the sphere of productive forces, the full-fledged development of the social individual, in turn, affects as 'the major productive force' that of universal labour.[67] It is no longer the simple exertion of man's labour power but 'the acquisition of his own general productive force, his understanding of nature and the mastery thereof through his existence as a member of society, in other words the development of the social individual, that appear as the major pillar of production and wealth'.[68]

Within these ties of the STR arising in socialist society, social sciences—though Kant did not believe that they would ever, like accountants' books, express themselves in poetic language—are today in the position to specify the law-based lines of further development, in contrast with the social and human dimensions of all the preceding stages of civilization: creating a flexible, dynamic basis of human life, they release a wide scale of 'socially created means of development'[69] accessible to everybody. This new revolutionary development overcomes the traditionally restricted basis of productive forces, confined within the limits of the 'contradiction of development' that indicates the reduction of man's development to the privilege of the ruling minority and hence of the reproduction of class antagonisms.[70] The experience of the socialist reconstruction of society proves that social inequality in the fundamental conditions of life, the existence of special classes, monopolizing the means of the development of human abilities (and hence their social form, private ownership of the means of production) have entirely ceased to constitute the historical basis of the progress of civilization under the conditions of the scientific and technological revolution.

Present-day social productive forces, linked with the universal application of science, already cannot be controlled and governed by non-social interests extending no further than private profit of an enterprise, trust or multinational corporation. The civilization potential of capital as a form of the motion of productive forces has now been exhausted. Under the conditions of the STR, social control and mastering of the immense social powers operating in the production and reproduction of human life has become a historical necessity. This presupposes the formation of socialist production relations based on the co-operation of all members of society and opening up the prospect of communist relations which are the general mutual development of man by man.[71] Parallel to this is the gradual development of new economic forms and criteria of efficiency, based on the economy of social labour and the gaining of 'time at man's disposal'[72] as well as the acquisition of new values of human life and norms of social activity, transcending the horizon of private ownership[73] and the concomitant 'categorical imperatives'.

Where productive forces do not acquire this adequate form of motion in production relations and social patterns, and development continues to be enforced by 'external spurs' of the profit interests of a special class, the mounting technological potential fails to become a means of man's general social development.[74] On the contrary, it tends to rush forth, in its cloven structure, ever further beyond its historically substantiated boundaries, beyond the horizons of its own 'dystopias', converting an increasing amount of productive forces into forces of destruction.

Opposed social relations in the contemporary world exert mounting pressure under the surface of similar outward phenomena in production, affecting the evolution of the technological basis in the opposite direction. With each diverging step of social reality, divergent trends or tendencies in the structure and dynamics of productive forces become apparent. All available evidence supports the fact that, with the further advances of the STR, the diametrical differences in the social forms of motion between capitalism and socialism will also be reflected in the differing, divergent lines of the development of science and technology, of the entire contemporary civilization.[75]

The development of society and the transformation of nature

With the sweeping changes in the structure and dynamics of productive forces, the STR generates a qualitatively new situation in the entire process of the transformation of nature by society and man.

The 'modern' attitude towards nature, connected with scientific cognition and the technological control of a number of natural processes, has in fact been shaped in the process of the industrial revolution of the preceding centuries, the period of the introduction and use of machinery in capitalist factory production. As a result, this attitude entails the historical boundaries specific to this particular stage in the development of society and technology.

In the primitive communities, man and nature were not separated from each other in terms comparable to the new age. Nor were they opposed to one another.[76] They were conceived in a unity, unmistakably traceable in the ancient Oriental tradition, in the principles of Tao. Even in ancient communities, man's bond with the surrounding world was strong enough not to permit any abstract isolation of the conception of nature from the conception of man in the thinking of that epoch.[77] And vice versa, human community was not excluded from the conception of nature.[78] The great majority of the thinkers of the period, in Greece just as in India and China, considered social and moral contemplation superior to practical technical thinking.[79]

Life in civilizations based on slavery was not directly and deliberately oriented towards domination over nature. Ancient man knew no fear of the rule of technology. Motifs of Golem, Frankenstein, Moreau, Caligari, Faustus, etc.,[80] were alien to him. All his hopes and apprehensions centred on questions of 'fate', 'fortune', etc., which dominated men, nature and even the gods. According to Plutarch, the temples of ancient towns were more frequently consecrated to these forces of fate than to the patron gods.

What all these strands of ancient thought seem to reflect is the fact that the immediate basis of life was the *polis*, the community of the free, who 'by nature' ruled the slaves, providing through them for the reproduction of the conditions of life, thus ensuring the commerce between man and nature. Conditions prevailing in modern bourgeois society are different. It is precisely man's separation from nature, the subordination of nature to the external purpose of production in the form of transformed, civilized and industrialized nature that constitute the new (negative) formal indication of their inner unity.

Descartes's conception of man as the 'master and owner of nature', Bacon's endeavour to achieve through knowledge the *imperium hominis* over the universe, and Hobbes's counter-position of civilization (*status civilis*) and nature, were put into effect by capitalism and the industrial revolution, thus giving expression to the specifically modern approach to nature as the external object, which must be conquered, controlled, and used in production.[81]

Descartes and Bacon formulated the question of man's domination over nature in a rigorously abstract form, wherein the reality of what in actual fact constitutes the basis of this, i.e. man, as the active agent, the subject, 'master and owner of nature', is lost. This is an adequate expression of the fact that what is at issue here is no longer the sovereignty over nature of the particular man of the Middle Ages, and far less of the original natural being, but the domination of a general, abstract individual: the product of a specific social development in the bourgeois system.[82]

The historical grounds for the general practical application of the approach to nature, viewed as a separate mechanical object detached from man and brought under the control of his powers through knowledge, was provided by the capitalist production relations.

In this way it is only capital that introduces universal appropriation of nature as well as the social bond itself by the members of society . . . thus creating a level of social evolution which makes all the preceding levels appear as a mere local development of mankind, as a mere natural idolatry. It is only here that for man, nature becomes a mere object, only a matter of utility ceasing to be recognized as a power in its own right.[83]

Social power over nature, materialized in capitalist, machine-based industry, develops here concurrently with the denaturalization of social relations, which begin to constitute an autonomous, spontaneous social force separated from and dominating the individual, independent of the immediate influence of natural elements. Capital exploits all the forces of nature in the service of the production of wealth, at the same time converting the masses of human beings into a means of its own self-expansion.

Within this historically transitory zone of the mutual separation and confrontation of man and nature, the continuous transformation and exploitation of nature developed into a universal mode of man's existence and thought. And vice versa: for man, nature became a matter of mere external utility, unrelated to his own essence. These conditions of production and reproduction of human life gave rise to *anschauende Naturwissenschaft*,[84] a special type of abstract natural science, that is 'value-free', science in the 'true meaning of the term', which ignores the social conditions of the historical subject of cognition and transformation of the world, and which excludes the entire process shaping this subject from the purview of 'exact sciences'.[85] Science, relinquishing any integrated social project in its base can, understandably, be used (or misused) by anybody.

Galileo, as one of the founders of 'modern' analytic/synthetic natural science, was aware that the 'reduction of universal motion' through the conception of 'constant elements' to a single objective basis, and the exclusion of all subjective considerations as accidental, corresponds to the conception of the world as a great mechanism and of science as a rational basis of technology. Bacon offered his method as the 'new machine of the mind' suitable for the 'business of cognition', oriented to control the mechanism of nature. Descartes did not flinch from handling all nature *more geometrico*, i.e. as a substance 'denoted by geometricians in terms of magnitudes'; contrary to Plato, he stressed the usefulness of technical focusing on the transformation of nature, and its superiority over contemplative speculation as the true basis of 'modern' science.

On the other hand, these concrete historical conditions laid the ground for the isolated 'social theory' of the bourgeois epoch which eliminated man's active relation towards nature, and separated the immediate production process from social life.[86] This theory likewise attempted to construct social sciences *more geometrico*,[87] interpreting society as a sum total of individuals in the sense of Locke's social mechanics or W. Petty's 'political arithmetic' (the original designation of bourgeois political economy), as a huge mechanism composed of mutually interacting human particles, which are the private owners.[88] Consequently, the true nature of society, which, according to Marx's characterization, is an 'aggregate of these relations, in which the agents of this production stand with respect to nature and to one another',[89] remained a mysterious uncontrolled background for the social sciences of this epoch, a background reflected in the 'invisible hand' of economic life (English political economy, especially Smith and Ricardo), the *Weltgeist* of social development (German classical philosophy, especially Hegel), the 'course of civilization' (French political theory and historiography, as represented by Guizot), etc.

In the intellectual development of the entire preceding epoch, the growing awareness had to assert itself with utmost difficulty that man's approach to nature as the object of external utility is a historical relation formed by the capitalist mode of production, that the entire structure of 'modern' science (its inner structure and external limits), which developed along with the industrial revolution, inherently reflects the needs of the 'system of general exploitation of natural and human properties'.[90]

Bacon linked his idea of man's conquest of nature with the endeavour to bridge the contradictions among people, with the aim of making them 'turn with united forces against the Nature of Things, to storm and occupy her castles and strongholds'.[91] Saint-Simon expected science to replace the exploitation of man by the 'exploitation of the terrestrial globe, of external nature'.[92]

Nevertheless, the exploitation of man by man through exploiting nature has never ceased to constitute the actual social basis and objective which only continued to reproduce itself in the process of the expanding, reckless onslaught on nature attendant upon the rise of factory-based industry and the technicized environment of the epoch of capitalism. The exploitation of the working masses by capital established and predetermined the orientation and the rate of this process for the whole century to come. It was the limits and inner contradictions of this specific, restricted form of the transformation of nature and the development of society, defined by the framework of capitalist social relations, that were and, indeed, had to be pushed *ad absurdum* with the first steps of the STR.[93]

Thus, before our eyes today, the power and weakness of this historically transient approach to nature have developed to the full. On the one hand, the capitalist mode of exploiting nature replaced the original, natural conditions of life by conditions transformed by man. There is no doubt about the fact that, in so doing, capitalist production fulfilled an important mission of civilization: without the society-made technological basis of the reproduction of human life, just as without society-made social relations independent of spontaneous natural processes, the control and regulation of living conditions would not be possible. But the enthusiasm over 'dominating nature', 'the conquest of nature', 'the triumph over nature' can never obscure the fact that, in the bourgeois system 'to the same extent as mankind conquers nature . . . man becomes a serf of other men'.[94]

The more this type of 'domination over nature' is put into effect, the more its one-sided, deformed nature comes to the fore.[95] The symptoms of ecological devastation, the crisis in the way of life, and the grave social effects of this mode of applying science and technology, have underlined the irony of the fact that it is precisely the 'man-made' world (in fact, of course, only a world modified by society), wherein the goals of the capitalist society have been imprinted over the past few centuries, that may prove a fatal factor.[96]

Of course, the social theory that conceives this explicitly historical type of exploiting nature as an external and natural one, and cherishes the illusion that the limits the STR strikes against in the capitalist part of the world are exclusively coextensive with the limits of our planet (or at least with the limits of the biosphere), has become too restrictive for the contemporary advances of society, science and technology.

Romanticism, on the other hand, seized upon the negative tenor of this approach and turned euphoria over 'the subjugation of nature' into a lament over a world in which 'nature becomes one single, enormous refuelling station' where man 'approaches the complex of Being as raw material for production' and subordinates the world to the 'broom and order of production'.[97] Proceeding from this vantage point, romanticism launched a 'radical' attack against the entire 'modern' science and technology from Galileo, Bacon, Newton and Descartes, identifying them with science and technology in general and relating the causes of all present-day evil to the 'imperialism of modern science'.[98] All that it was able to counterpoise to the 'one-dimensionally oriented' science of the epoch of capitalism, rejected because of viewing the world with abstract, 'dead eyes', and to the 'dissecting' and 'deadening' technology, was the restoration of the myth, and the slogan of the 'religious renewal' of the Western world[99] through the revival of natural animism and a Westernized Oriental tradition.

A century ago, Engels pointed out that 'we by no means rule over nature like a conqueror over a foreign people, like someone standing outside nature—but that we, with flesh, blood and brain belong to nature'.[100] Hence his critique of the 'specific backwardness of the past centuries', i.e. of the Lockean 'metaphysical mode of thought' built on the foundations of the entire 'modern science', viewing the objective world as a system of machines, grasping natural process 'not in their life but in their death'[101] and simultaneously excluding man from nature and society as an *a priori* abstract subject who lost his social ties, his history, his character of the active agent of practical social activity.[102]

The restricted nature of modern science in the period of the industrial revolution of capitalism, the inadequacy of its theoretical and methodological basis, have long since become a frequent target of criticism voiced also by thinkers from different cultural traditions. One may recall the bitter complaints of L. N. Tolstoy who claimed that modern science remains indifferent to the sense of life.[103]

Of great significance for our entire epoch is the sometimes underrated fact that the specification of a positive way out of this situation, based on the fully fledged development of the dialectics of the transformation of nature and the development of man and society, is associated with the names of Marx, Engels and Lenin.

If the core of the reckless exploitation of nature (and the source of corresponding theoretical systems) was constituted by the laws governing the exploitation of man, then the road to a new, different approach to nature inevitably leads through the area where the sovereign power of capitalist profit over man's destinies is overcome.

Only on this basis, connected with both the reshaping of objective reality, nature, and the essential revolutionary transformation of man and society, can the sense and mode of the entire process of production and reproduction of human life change, along with the very foundations of man's approach to nature. And indeed they do: people are gaining control of the conditions surrounding them and 'become, for the first time, conscious, real masters of nature' and thereby 'masters of their own social integration'.[104] To transform the process of 'socialization' of people into 'an individual act of freedom', of course, presupposes the creation of conditions conducive to people's general social development, introducing the type of relations among people characteristic of advanced socialist society.

Thus, conditions of the most developed, purposefully controlled transformation of nature in the STR are combined with conditions of the revolutionary transformation of society and the general development of man. Only this can provide the basic framework for the gradual, real formation of the historical process of mastering natural forces, i.e. as Marx puts it: 'those of the so-called nature as well as his own nature'. And with them the really new, modern theoretical approach to nature originates, based on 'the grasping of man's own history as a process and the awareness of nature as his own real body'.[105]

Men cannot control their own conditions of life unless they themselves transform these conditions, unless they attain a situation under which all conditions are modified by society and not left at the mercy of natural elements. Historically, this determines the need to apply science to the fullest in the transformation of nature and social life.

Proceeding from these principles, the founders of Marxism-Leninism laid down new foundations of science. This new conception transcends the horizons of the 'modern science' of the epoch of capitalism, oriented towards external profit by means of utilizing nature. At the same time, it goes beyond the horizons of ancient thought developed, according to Aristotle, to satisfy the needs of the free. For the coincidence of the transformation of the world and self-transformation of society is inherent in its foundations. From the methodological point of view, this new type of science differs substantially from the system generalized in the epoch of the bourgeois industrial revolution.[106] It does not exclude the starting point of social cognition from the real process of the transformation of the world and the development of society and man as an *a priori* subject, an abstract individual. It does not reduce the objective world to a lifeless, passive set of mechanical *res extensa.*[107] On the contrary, it links the process of reflection of objective material reality with the process of its transformation;[108] it incorporates into the process of cognition and transformation of reality the theoretical analysis and practical transformation of the social and material preconditions of the activity of the concrete subject of cognition and transformation of reality.[109] It conceives the development of society and man as the highest form of the dialectics of the material world.

It becomes apparent that Marx's ideas of the new foundations of unified science,[110] which enable and call for the inner unity of natural and social sciences,[111] are fully adequate to the tasks of the present stage of the process of the transformation of nature, and correspond to the demands placed on science by the rapid growth of its functions in the life of society,[112] and by its responsibility for the most fundamental values in the life of social man.[113] The link between the endeavours of the natural and social sciences, a precondition for the complete mastery of the new level in the process of transforming nature and the development of society, has become one of the characteristic features of science in the socialist countries, a source of a number of successes in the settlement of the great problems of the present.[114]

With the advent of the STR, man's active approach to nature developed from originally scattered, individual attempts at locally modifying it into a complicated comprehensive global process, including the control of the development of the 'biosphere'[115] as well as man's own evolution. The main issue here is no longer man's control over this or that section of nature, but a purposeful mastery of the entire process of mastering nature by society[116] in the interests of the development of man, his relations and capacities.

This corresponds to a practice which bases its approach to nature on the purposeful social control of mastered nature as society and man's 'own body'[117] along the lines of 'ecological strategies'[118] of society as a whole, opening up scope for man's capacity to regenerate, 'reproduce' the natural conditions of life. Only this conception, systematically implemented in the process of the socialist transformation of the world, is capable of finding a way out of the reckless devastation of nature—a process called into life by the industrial revolution of capitalism—and at the same time to attain the goals of the humanistic programme,[119] namely to avoid the fashionable pose of the so-called 'critique of anthropocentrism',[120] which frequently indicates a gross idolatry of nature concealing the suppression of the endeavour of billions of working people aimed at the creation of dignified conditions of life and development for everybody.

The new stage in the process of commerce betwen society and nature has brought a new degree of inner complexity into all changes in different aspects of social productive forces. It links the establishment of networks of cybernetic technology with the formation of a new power basis, with acquisition of new materials, with intervention into the genetic code of living organisms, with the foundations of bioengineering, with the endeavour to control the functions of the human brain and the ways and means of affecting man's behaviour, etc. Interventions into the environment along with projects of the regulation of climate and with other achievements of the STR have reached a stage where the global conditions of human existence become, just as with astronauts, dependent on their continuous artificial reproduction and their preservation by man in closed cycles.

The new, higher dynamics of productive forces,[121] linked up with the STR and hence with the revolutionizing of relations towards nature, can only be mastered by a social system which views the full and unrestricted development of productive forces as constituting the basis of life and as a social need, and which is capable of regulating purposefully the development of man as the subjective force of the transformation of nature.

The crucial issue of the present is not the halting of the process of transforming nature, the taming of the STR or its reduction to the level of the possibilities of the old social order, which failed to revolutionize the subjective components of social development. On the contrary, the key issue of science today, and its most difficult practical task, is the purposeful control of social processes, connected with the building up of a socialist society based on general mutual co-operation and geared towards man's general development, so that the human community would constantly be able to make use of the released immense natural forces, to 'regulate its development in interaction with nature'.[122] Society at present is not confronted with the task of clipping the wings of science or technology; on the contrary, its vital interest is to expand science and its applications into new dimensions, capturing the entire complex of the transformation of nature and society. At the same time, the greatest and most complicated tasks facing contemporary science emerge in the social science field and are, in the final analysis, connected with the need to secure priority for questions concerning man's general social development in the unified system of science.[123]

Human society represents a law-based, higher stage of development in the life of the universe, an independent form of the motion of matter, which is not reduced to mechanical or biological changes[124] that are spontaneously recorded in genetic structures. There are no predetermined, unsurmountable limits in the current form of these processes in space and time.[125]

The revolutionary motion in forming the conditions of human life previously only dissolved old, obsolete antagonistic formations and cleared the ground for new forces. Today, however, this motion becomes the actual basis for such a society, based on mutual co-operation and mutual development of man by man. And vice versa: with the attainment of this level of general human co-operation in the socialist society, the STR becomes an indispensable basis for the life of social man.

In the past, mankind only experienced revolutionary moments; nowadays, however, it has entered a *revolutionary epoch.*

The scientific and technological revolution and the 'crisis of perspectives'

All the miracles of the Old Testament, the fabulous accounts of the Vedas, which Tagore considered as symbols of 'people's unbiased imagination', as well as the fantastic discoveries described by Leonardo da Vinci, many of which were confirmed in practice—today all this pales in comparison with the forces and processes that contemporary science and technology have set in motion. Man is rapidly approaching a state of affairs where he can not only entirely change the living conditions on our planet but also make it permanently uninhabitable.[126]

The underlying dialectic is as follows: the achievements of science and technology, which have the power to lay the basis for the general development of human forces, are *eo ipso* able to destroy man universally. This dilemma becomes the omnipresent, concomitant feature of contemporary science and technology; this is the borderland in which all further steps of the STR will be taken.

In this sense, the gravest feature of the beginnings of the STR is the fact that a disproportionate amount of world expenditure on science has been spent on the production of means of mass destruction. The constructive tasks connected with the control of the commerce between society and nature and, *a fortiori*, with the solution of burning social and human problems, often represent a mere fragment of the general potential of research. In some developed countries which, until recently, set the pace in a number of technological branches, the whole contemporary system of science came into being on the basis of programmes of so-called 'big science', oriented primarily towards the production of means of destruction.[127] In contrast, other components of science, especially basic research, became a mere appendix, a by-product or service, financed as overhead.[128]

Over the past few decades, this disproportion has evidently blocked an enormous amount of the positive potentialities of science and technology, and postponed the solution of a number of vitally important problems of human life and the development of whole continents.

Within the forthcoming decades, the development of the means of mass destruction is bound to go far beyond the limits of absurdity and, unless halted by a ban on the evolvement of new systems of weapons and new technologies for mass annihilation, will result in the emergence of a range of possibilities and gadgetry that will defy technological control.

At a time when the old technologically advanced world of capitalism and the new, nascent system of socialism coexist, when hundreds of millions of people are getting rid of colonial subjugation, the increasing potential for mass annihilation begins to open up a special dimension: the threat of a possible attempt of the forces of the Old World to regain their privileges becomes a threat endangering man's existence in general.

Mankind now disposes of power sources amounting to some 10^9 kW. Within the next few decades, it will probably be capable of generating power of the order of 10^{13} kW from plentiful natural elements, gradually nearing the power of solar radiation on earth.[129] What will be the effects of this awe-inspiring power, once it is controlled by the concerns or multinational monopolies, which are not responsible for anything to anybody? Or by the attendant aggressive 'military-industrial complexes'? Or simply by terrorist gangs?

A number of reports today list symptoms of the gradual changes affecting the climate and the living conditions of whole continents. Science and technology are faced with the challenge of reconstructing the entire process of production on the basis of closed cycles of metabolism, the reproduction of the most important renewable resources of the planet, the artificial control of the circulation of such elements as oxygen, carbon, etc. Further tasks ahead include a purposeful environment strategy, the formation of an artificial climate in a number of regions of the planet and the control of the entire biosphere. All this, however, requires the existence of corresponding social conditions. Without steady relaxation of international tension,[130] without substantial changes in the foundations of social organization and motivation of people in a number of countries,[131] no project elaborated today in ecological laboratories can successfully fulfil its function.[132]

Perhaps the greatest hopes but also the gravest fears as to the prospects of the STR arose in connection with the penetration of science into the nature of the genetic code, the problems of 'bioengineering'[133] and 'biomedical innovations'.[134] The fantastic possibilities for the control of life processes but also the hardly imaginable dangers ensuing from the application of the discoveries of microbiology and genetics,[135] the prospects of complicated 'genetic battles' in the cultivation of new strains of bacteria, viruses, etc., have even led to the demand for a moratorium in certain branches of research.

The idea that future advances in genetic control, embryonal therapy, etc., which obviously can regulate but also monstrously deform the development of human disposition, would be governed by capitalist profit interests is indeed terrifying.

Equally wonderful possibilities but also no less shocking issues arise from the rapid advances of the 'computer revolution'.[136] In particular, they evoke extensive problems in connection with the rapidly developing research into the human brain and projects for scientific control of human behaviour. What will this progress in science and technology bring in its wake once it is set in the system of commercial manipulation of man? Compared with the world experience of deafening advertising not hesitating to appeal to man's primitive instincts and desires, experiments with the artificial excitation of the brain centres of pleasure which, in rats, unmistakably deafen all other stimuli, thus sweeping their entire life into the monstrous circle of narcissism, are certainly 'spine-chilling'.[137] A system which makes the achievements of productive forces serve the formation of people's needs in compliance with the requirements of the extended reproduction of capital, can, with the contemporary revolutionizing changes in science and technology, arrive only at monstrous life-styles producing needs based on self-deafening and a 'civilization contained within the crude barbarism of need'.[138]

The standards for the transformation of man's living conditions have evidently reached a stage where each further intervention represents not only a wide-ranging technological task but also a grave social and human problem. It is therefore imperative that each major contemporary project in the sphere of natural and technical sciences be followed not only *ex post facto* but also *ex ante* by adequate social scientific research.

Under the conditions of the STR, each advance of science and technology is, in its way, of a 'global' character. This is not because it would affect different social systems in the same way and in the same measure. As a rule, the contrary is true. The point is that each of these advances has to be assessed in terms of a whole chain, or rather tide, of current and forthcoming changes it can influence. Within the world literature concerned with the contemporary development of man's living conditions, there is not one single problem that in itself would be technically insoluble; nor is there any foreseeable discovery that, under specific conditions, society would not be able to cope with.[139] The solution of some problems is, however, increasingly dependent on the solution of other problems; hence '... when considered in their global interrelationship, they ... necessitate drastic innovations in technology, but even more so in the sociopolitical and institutional fields'.[140]

Concrete analysis will always show that the key role in the solution of major problems connected with the provision of material, technological conditions of life is played by social relations. A genuinely sound prospect for contemporary civilization can only be based on a development of man and socicty that displays a purposeful orientation centring on the formation of the material conditions of man's life as the mediating link within this process; but as to the general solution, it will be hard to attain. In a considerable part of the world, production relations do not correspond, however, to the level attained in the development of productive forces: antagonisms in the structure of society are the main source of disproportions, even in the commerce between society and nature.[141]

As a result of deep-rooted conflicts between the current motion of productive forces and the inadequate nature of production relations, millions of people in the technologically advanced countries of the West[142] experience one of the greatest crises affecting the values of human life and activity in the entire history of civilization. Reflections on the perspectives of society under the conditions of the STR mirror the prevalent feeling of the *transitory* character of the period, an awareness of the general crisis-ridden situation now facing bourgeois society,[143] which reveals with increasing clarity its antagonistic foundations. As J. McHale declared in his address to the Rome Conference on Future Research in 1973: 'Our environmental, socio-economic, urban and technical crises ... are institutional crises deriving from specific modes of economic, social and industrial practice'.[144]

In discussing the diagnosis of contemporary development, social scientists in the West increasingly use labels such as 'spasm', indications of the 'decline of the industrial civilization of the planet',[145] etc. Even approaches painstakingly trying to retain an optimistic view have to admit that prospects for the future development of society under the conditions of the STR make one 'lose one's nerve'.[146] In this way, their arguments make one feel that the 'scenarios' for the future claiming that 'things are going rather well' are in actual fact just what the scenarios of the masters of *commedia dell'arte* originally were—improvisations on predetermined themes.

It is precisely at a time when the prospects of the STR and social development provide the topic of inquiries steadily gaining in intensity in the field of social sciences that the feeling of uncertainty in face of future developments (the deep '*crisis of perspectives*') reaches its climax in bourgeois society: never before has the notion of an impending decline[147] and the advent of a new, unknown epoch, been so widespread as today.

'The absence of persuasive positive images of the future'[148] is so obvious here that the spokesmen for this society begin to refer to the present as a period that is 'unique in the history of Western civilization' because 'the future speaks a foreign language to us today'.[149]

Viewed from the angle of a society losing its prospects, the current turning point appears to be an attained 'climax' of social evolution, i.e. 'not only in our ability to improve our own condition as human beings, but also in our ability to change Nature itself'.[150] Thus, all one may expect will be just the downward trend in the people's 'quality of life'.[151] Hence, the ever more frequent comparisons between the present day and the last days of the Roman Empire, the eclipse of an ancient civilization.[152] Those who not so long ago cherished great hopes for the future, today produce obituaries steeped in pathos over a world afflicted by an incurable 'civilization malaise',[153] and thus swell the ranks of those who have identified the crisis in the perspectives of bourgeois society with the exhausted history of mankind.

In actual fact, underlying the current 'crisis of perspectives' is the historical limitation of bourgeois society whose inner antagonisms were laid bare by the STR. It is clear that this bourgeois system is incapable of solving the problems concerning mankind's future development. It fails to provide a settlement even in the most technologically advanced countries, not to speak of the international sphere.

A classic example of the dead-end reached by bourgeois society is the intolerable amount of social-class inequality accumulated by it over centuries of development under the banner of formal 'equality'. The deep-rooted confirming contradictions in the general sharing of the achievements of modern civilization in the technologically advanced countries of capitalism[154] and the ominous widening gap between living standards in metropolises and the subsistence level prevailing in the developing continents increasingly lead to capitulation of bourgeois authors as regards prospects under the conditions of the STR.

In addition to the fundamental social roots of this crisis of perspectives, as well their expressions, there are limitations apparent in the system of the social sciences, a system that has gradually separated natural sciences as value-free science, as science in the 'true sense of this term', from the humanities.[155] The great social science system of the capitalist epoch, its classical conceptual system,[156] developing since the *Cinquecento*, the Enlightenment and Reformation together with the industrial revolution of capitalism, arrived at a certain idea of the laws of social evolution and ultimately proclaimed *savoir pour prévoir, afin de pouvoir* as the principle of modern 'positive science of society'. Nevertheless, it is beyond doubt today that they are incapable of providing answers to the major issues concerning the present-day and future development of society under conditions in which the powerful social revolution merges with the stream of scientific and technological changes. For the actual basic reality which these social sciences accept as eternal, natural and absolute, and towards which they are, therefore, *ex principio* uncritical, are the general conditions and abstract expressions of social relations created by capitalism.[157] A more profound historical dialectic of productive forces and production relations in the process of the transformation of the world and the self-formation of society, which is the key to the perspective of social development under present-day conditions,[158] lies outside the horizon of their entire complex.[159]

Despite the accumulation of an ever-growing amount of individual findings about society, a disintegration of this historical system of social sciences has been taking place in the course of the last century, initially into individual branches and currents of thought, increasingly resorting to superficial, vulgar elements. Today, however, the development of social reality as a whole has passed beyond the range of the bourgeois theory of society. The first intimation of the idea of objective laws governing social development was in fact gradually abandoned, and the scientific application of this idea to study of the future became the concern of Marxism.[160] What was left over from the social science system of the epoch of capitalism is alternately buttressed up by purely technological projects, which lack the sense of their own social provenance, or patched up with colourful, exotic social mysticism.

Frequently used today to evade the 'crisis of perspectives' are scientistic conceptions presenting science as the universal 'oasis of order', a panacea for all social illnesses and diseases[161] or concomitant technocratic projects of the so-called 'technological fix', a kind of universal technical mechanism for the solution of social problems. The following telling questions are then posed:[162] 'To what extent can social problems be circumvented by reducing them to technological problems?' or 'Can we identify quick technological fixes for profound and almost infinitely complicated social problems?' Or, more to the point: 'To what extent can technological remedies be found for social problems without first having to remove the causes of the problem?' This last question invalidates the whole conception, one that ignores the basic conditions of objective social processes.

Basically, the same methodology (the difference being only in reversed signs) is discernible in the romantic conceptions of the 'counter-culture'. Instead of the development of science, its liquidation is clad in 'the garment borrowed from a multitude of exotic sources'[163] and a transition to 'drug-thought, mysticism, impulses' figures as an 'oasis' for the future.[164]

Similarly, the halting of technological progress (the alleged source of disaster), the 'decline of civilization', the 'Dark Ages', 'life in an ant-hill system', etc., are presented as the specific 'technological fix'.

However understandable may be the disgust at the conditions of technologically advanced capitalism, and the escape of the overfrightened to rural communities, can we seriously accept the idea that a ban on nuclear energy, television, etc., will bring a higher quality of life to man and the settlement of the burning social problems to the world?[165]

These two concepts of the social sciences reacting to the STR are combined in the theory of the limits of growth proceeding from the tenet that 'a global collapse of human affairs is inevitable if the present course of mankind is not radically changed'[166]—primarily if 'the transition from the past dynamics of growth to a future condition of world equilibrium'[167] is not achieved, if 'conditions of ecologic and economic stability'[168] are not created or, at least, if the pace in major areas of civilization is not slackened to mere 'organic growth'.[169]

The calculations extrapolating the symptoms of the current crisis of the bourgeois system into the future reflect some of the new dimensions in the commerce between society and nature; in particular, this applies to the attempt at the specification of 'the mutual interaction between major factors on a world-wide scale'. Their predictive capacity has, however, been drastically reduced in that the development of social relations, today's social revolution[170] is not listed among the major factors; likewise excluded from the list are the qualitative revolutionary changes under way in the development of productive forces.

And yet, according to the conclusions offered by the Club of Rome, 'the real limits to growth, then, are not material but political and social, or, if more fundamentally viewed, imposed by the behavioral and motivational patterns of man . . . '[171], i.e. they are to be sought precisely in the areas which were ignored in these calculations of the future.[172] The rational content of the world models for the future is thus reduced to the statement that all calculable tendencies specific to the current development of productive forces in the capitalist countries are working towards an ultimate disaster, provided we abstract from the revolutionary changes on the part of society and technology.

Further generations of mathematical models of world development meet with difficulties since, in bourgeois society, one can hardly find a coherent theory of social development that could form the basis for the strategic conception of scientific and technological development. Under such circumstances, prognoses of the future fall victim to the compilation from different variants of economic growth in individual sectors of the present-day world, and to different petty-bourgeois and radical political slogans on the basis of subjectivist notions about the redistribution of development.[173]

Due to embarrassment over the loss of perspectives, melioristic conceptions of the social sciences today call for the transition from material growth to psychosocial growth.[174] Thus hopes for the future are in fact pinned on to the area of way of life, which, throughout the period of the STR, has witnessed the most palpable manifestations of the breakdown of the bourgeois system. In practice, this suggestion calls only for another interpretation of the existing social reality in the minds of men.

A parallel line of development is followed by ventures undertaken to compensate for the restricted and deformed nature acquired by the STR under the conditions of capitalism by a systematic taming of technology,[175] its subordination to the control of social assessment,[176] by a reorientation[177] of the deviant development of science and technology, a different channeling of research, redesigning the trajectories of scientific and technological development and even something on the lines of a design-science revolution.[178]

However, none of these trends are based on the corresponding constructed and shared conception of changes in social conditions in which the reorientation of science and technology would become feasible and hence the function of curing the evils of the 'technological epidemic, which is proposed for the social sciences, is hardly likely to be anything more than a police function'.[179]

Certain demands placed on the conception of social development are formulated by economic, sociological and philosophical trends which in face of the current scientific and technological development and its false course postulate the idea of intermediate,[180] or 'convivial' technology.[181]

Constituting the social basis of these conceptions is, of course, not the prospect of overcoming the limits of capitalism but the return to a pre-capitalist, 'Buddhist'[182] type of economy, based on use-value; to the romantic idyll of 'convivial' natural relations in groups rid of general social ties and potentialities.

Likewise, at present, the social sciences in the West are faced with grave difficulties in formulating a positive outlook for the future of the developing countries under the conditions of the STR, i.e. for a solution of one of the key issues of the development of the whole of human society in the decades ahead.

The social system, which for several centuries has been building its industrial base, *inter alia*, through the exploitation of colonies, has left these countries on the brink of disaster, and is unable today to offer any solutions. A part of the social sciences in the West seeks salvation in the eternal coexistence of completely different civilizations, occupying diametrically different levels as to their ways of life.[183] For it is, as a matter of principle, not possible, so the argument runs, for all the peoples of the world to attain the general living standards prevailing in the most technologically advanced countries, not to mention the conditions for the general development of human forces.[184] This would allegedly completely ruin the social and natural conditions of our planet. The other part of science in the West confronts the working people in the advanced countries with the gloomy prospect of a drastic reduction of their vital needs to a mere fragment of the current level as the one and only alternative way out.[185]

An increasing number of authors in the West describe the situation of the Third World as almost insoluble in view of the disproportion between the limited resources and the current population in these countries. This view, though, only reflects the defenceless position of this type of social science in relation to the major social processes of the present. For one thing, the 'lack of means' is relative at a time when the world, under the pressure of war-instigators, spends U.S. $250 billion on armament annually, i.e. a sum exceeding the total national income of the poorer half of mankind. Secondly, population trends are dependent on social conditions, on living standards; the steep upward population trend in the developing countries is, to a great extent, the logical outcome of the deplorable conditions existing in this part of the world as a result of the monopoly-based exploitation policies once practised on a world-wide scale. In the past decade, a number of useful international activities have been undertaken for the utilization of the achievements of contemporary science and technology in the interests of the developing countries. Experience shows that the STR can play a significant role in overcoming the persisting 'civilization gap', of course only under quite specific social conditions.[186] Most of the world projects and theoretical analyses concerned with this process[187] make it clear that it is precisely the socio-economic conditions, which decisively determine such an application of the achievements of the STR, that would lead to the actual obliteration of the burning civilization gap. One of the most relevant aspects of the project for a new economic order, which, of course, can only coincide with a new social order, is the provision of a real basis for a purposeful utilization of the processes of the STR conducive to the solution of social tasks, to the levelling out of substantial differences in the standards of the general development of civilization, to the speedy upward trend in the developing countries. In these terms, the elaboration of this programme is a test of the viability and operativeness of the major conceptions of the social sciences today.

Up-to-date science and technology frequently enter the life of the developing countries without continuity with their preceding history; in a certain sense, without a mediating economic potential, and hence without the usual social basis, social forces and motivations. Under such circumstances, the introduction of the latest technology may produce unexpected effects which, paradoxically, run counter to original intentions.[188] Moreover, the pitfalls emerging here are to a certain degree also related to the social deformations inherent in the pattern of the scientific and technological development, taken over from the West.

In these countries, the achievements of the STR can be used

successfully and with no great losses only following a thorough, concrete social analysis basing the technological projects in question on a comprehensive and systematically implemented scientific conception of the social development of these countries, in particular of the formation of the subjective social forces furthering this accelerated development. An integral part of such a conception is also the country's own strategy of scientific and technological development, which naturally has to utilize the findings of modern science and technology (eliminating those elements that lead to the crisis of Western life-style,[189] and at the same time it may also promote indigenous technologies (in so far as they can establish ties with further social and technological transformations).

This, of course, cannot dispense with an independent contribution being made by the indigenous research potential of the developing countries[190] and the corresponding top-level technological development in certain key areas of the economy.

If this part of the world was frequently subjected to 'recommendations' to the effect that it should follow only at a distance the scientific and technological development of the former metropolis and pay special attention only to the earlier, more accessible, technology, then it is clear that the consequent implementation of this line would transform the developing countries into a permanent periphery of world development, into an economic appendage of the monopolies of the industrially advanced world, the eternal supplier of goods produced by a cheap labour force (non-equivalent exchange relations). This would turn the developing countries into a permanent source of insoluble world problems. As long as countries exist on our planet that are condemned to live the unequal life of a cheap labour force on the verge of mere subsistence, as long as a scope conducive to general development is to remain closed for billions of people, there is no hope that the otherwise fully soluble problems of population growth, control over commerce between society and nature, etc., will be tackled successfully.

The failure of the social sciences, which were constituted in the advanced capitalist industrial countries, to cope with the conditions of the STR in the developing countries, reflects the partly deliberate, partly factual resignation of this entire intellectual system faced with the solution of a great epoch-making task: the creation of the social and technological conditions for all inhabitants of the planet, which would allow for the gradual transformation of each human life into the permanent process of the development of the forces, potentialities and activities of man.

This one and only law-based and feasible prospect for the development of contemporary society seems incomprehensible, fantastic and unrealizable to bourgeois social thought. The cause for this, so they argue, is man himself, the growing billions of human beings. Here the logical circle closes. The crisis of the prospects of a society, which for more than three centuries accumulated external wealth and ruthlessly swept away all external hindrances, is explained in words uttered by a comic-strip character: 'We have met the enemy and the enemy is us'[191] —the billions of people struggling for a truly human content to their lives. The perspective sketched by this type of social science can, therefore, be summed up as: 'The world has cancer and the cancer is man.'[192]

Unintentionally, the modern bourgeois social sciences, which once began with an apotheosis of the rights of man, express in these words their inner, theoretical, methodological and axiological collapse.

When the first symptoms of the 'crisis of perspectives' in bourgeois society became apparent, they were still accompanied by growing hopes in the social sciences,[193] which promised to find means for handling the new, breathtaking achievements of science and technology and, ultimately, to enable a purposeful shaping of the content of human life and the solution of urgent social problems.

Although frantic efforts in this direction have been exerted in the West in recent years, the overwhelming powerlessness surrounding the endeavour to elaborate social prospects under the conditions of the STR has not been countered. This fact gave rise to a wave of scepticism in regard to the social sciences,[194] boosting the popularity of cheap romanticism and, to a certain extent, inducing profound reflections on the causes of these failures.

All that is left of the original demands placed on *prévoir* and *pouvoir* is no more than a very pragmatic social theory and practice, summed up in the official slogan of the Hudson Institute:[195] a 'planned muddling through' the processes of the social and the scientific-technological revolution, irrespective of the hazards involved.[196]

Day-to-day experience confirms with increasing clarity that the solution of the major global problems of the present-day world and the creation of long-term prospects for the development of human society under the conditions of the STR are intrinsically tied up with the existence of a new, socialist type of society, which applies a conscious scientific approach to the law-governed process of its own reproduction and development. And it is one of the historic tasks of a new, Marxist-Leninist system of social sciences, to provide the adequate theoretical basis for the comprehensive scientific control of social processes shaping man's future.[197]

The role of the social sciences in the process of mastering the scientific and technological revolution

Contrary to the ancient forms of social thinking which accentuated heroic deeds and, with Cicero, interpreted social history as *magistra vitae*, the system of social sciences evolving along with the industrial revolution of capitalism was destined to specify the objective regularities of social development, those 'eternal, iron, great laws' around which, according to Goethe, all revolves, and thus to create conditions for man's activity within separated forms of the social life of this epoch,[198] an activity, subject to the 'verdict of history'.[199]

In their classical forms, the social sciences of this epoch studied—with cold detachment, frequently castigated as cynicism—the objective motion of social relations, capitalism as the substance and subject, determining the fundamental conditions of human activity, within the forms it introduced. In this way, class contradictions developed into the form of generality. They acquired the nature of general laws of social life. In its theoretical reflection bourgeois society became thus an abstract 'society *per se*', and its mode of handling the external conditions of life became the 'eternal' attribute of the relation between society and nature.

This specific mode of existence of the class element in a general form, which Rousseau already specified as the cause of the separation of the phenomenon and the substance in modern bourgeois civilization, materialized with the industrial revolution in the real forms of production. At first glance, it may of course appear that, inside capitalist industrial production where only machines and the labour force are counterposed, the social relation which moulded this production process is 'fading away'; that we encounter here the eternal, material form of interaction between society and nature. Nevertheless, this 'fading away' of the social form is just an illusion caused by the fact that the entire production process in its total range and material form (technological conditions controlling labour force servicing machinery) has been transformed into the production process of capital. This coincidence of social form and material conditions of the corresponding production was used by Saint-Simon in his conception of the 'industrial system' theory. Seen in this light, this may give the impression that, with the development of the techno-organizational apparatus of capitalist production, capitalism itself and its interests somehow got lost in the 'industrial system', gave way to 'technostructure', gave free rein to 'rationalization', etc., which, in their specific current forms, are its own expression. The forms of management originating as modes of controlling the interaction of the means of production and the labour force within capitalist factory production thus take on the semblance of 'industrial management' in general, allegedly regardless of its underlying social form.

Classical forms of social thought did not accentuate the idea of controlling society as a whole, since they conflated this principle with the functioning of bourgeois society as the 'best of all possible worlds': with the operation of the 'invisible hand' of national economy, with the spontaneous application of 'political rights and liberties', etc. Only with the aggravation of the class contradictions of the bourgeois world did the conception of control of social life push its way forward—of course, with the increasingly more evident accent on the preservation of the state of affairs, jeopardized by disintegration.[200]

When encountering the pressures of the inner contradiction, social sciences proceeding from the classical bourgeois tradition rapidly switched over to sheer 'ideology', to the positions of the 'philosophy of life', relegating laws, causal relations, etc., only to natural science, and identifying the task of the social sciences with mere *Zuschauen*, *Erleben*, *Verstehen*[201]—a relapse to the level of the original 'heroic epics', the 'religion of vision'.[202]

The idea of the purposeful control of the entire society and planned shaping of the future with the aid of scientific knowledge is thus in fact the late product of the wide-ranging crisis of the real and intellectual forms of the bourgeois system, and the concurrent rise of the new superior social reality of socialism with its completely different system of social sciences.[203] As a result, this idea was frequently accepted with unwillingness, as something at variance with the nature of society in general, as a deceptive abstraction[204] or sheer impossibility.[205] However, the development of productive forces was steadily converting this impossibility into historical necessity. Under the conditions of the present-day STR, when the entire sphere of the social productive forces of human life has been set in motion, when man's transforming activity penetrated deeply into the process of commerce between society and nature, and when, on this particular basis, all social processes began to display a new degree of acceleration, material preconditions for a purposeful control of the law-governed development of society were created, and along with them the possibilities of releasing the immense power of spontaneous processes (which at a certain stage may get beyond man's control) as well as means of mass destruction. At this particular moment, the control of social processes became—at least to a minimum extent—a law-governed, indispensable component of human history, a life-and-death issue for the generations to come. What is at stake is, first and foremost, the control of social processes geared towards peaceful coexistence and the elimination of war. Furthermore, there are processes enabling the gradual obliteration of the 'civilization gap' and the creation of means conducive to a dignified life for all the working inhabitants of this planet. Finally, there are processes attendant upon the creation of the environment and the maintenance of the required balance in the process of transforming nature. An urgent objective task facing social science at present is to build up a comprehensive theoretical basis for the control of the major contemporary, social processes and, at least, for the solution of the 'global problems' under consideration.

Under the pressure of these circumstances, the necessary material for the 'reconstruction' of the 'positive', 'nomothetic', 'value-free' social sciences in the West was expected from the theory of 'industrial management', based not on the classical tradition of the fundamental forms of life specific to this society but, primarily, on the technical-organizational, *prima facie* non-social elements, into which the antagonistic social reality was incorporated indirectly, through the forms assumed within the production process. Max Weber's special position in modern European bourgeois theory rests upon the fact that he delimited a definite space in the general structure of the social sciences for this material derived from the sphere of industrial management. In the corresponding American tradition, the same process acquired a more pronounced technocratic tenor: following F. Taylor, T. Veblen and others, the principles of 'industrial management' were gradually applied to the entire society and became the prevalent basis of the modern social sciences (economy, sociology and psychology), factually conceived as sciences of the management of society as an 'industrial complex'—whether in the form of 'piecemeal social engineering',[206] or the application of 'social technology',[207] or manipulation through 'behavioural technology'.[208]

The rushed development of ever new varieties and forms, applying the principles of 'industrial management' to society as a whole,[209] shows the steeply rising demands placed as a result of the STR on the control of the entire complex of social life and commerce between man and nature.[210] Just as the industrial revolution of the past centuries called forth, along with the new productive forces, the need for scientific control at the level of industrial enterprises, contemporary changes in the structure and dynamics of productive forces became uncontrollable without scientific management at the level of the entire society, within the whole complex of scientific and technological innovations and social development.[211] While Pascal could still raise his voice in warning: 'Woe to history determined by man!', today the problem seems to be reversed: without millions of people having asserted their law-governed objective possibilities of controlling the process of 'their own social integration' and hence the control of their own development on a general social scale, the spontaneous course of present-day civilization would necessarily end in disaster. A society which is to master the conditions of the STR can only be a 'controllable society', a society, which controls the entire complex of people's living conditions, affecting their activities, interests and stimuli; in which, therefore, the working community is capable of steadily attaining social goals, meeting the requirements of the development of all its members. These fundamental features of real 'scientific controllability' define the advanced socialist society[212] as a new type of society, for which historical development, contrary to the entire 'prehistory of mankind', is a process of a purposeful shaping of conditions for man's all-round development. Similarly, the attendant system of the social sciences acquires a completely new historical nature, developing as a theoretical hinterland for a comprehensive control of social processes. Here, of course, 'control' takes on a new sense, entirely different from the notorious managerial conceptions, in that it gradually draws in social 'self-control', 'self-formation'. The struggle for the mastery of the STR under present-day conditions necessarily becomes a struggle for the new social foundations and content of human life, and, at the same time, a struggle to decide whether, to what extent and in what temporal range the historical process in general will become controllable for society.

The accelerating alternation of the theories of 'industrial management' extended to the entire development of bourgeois society indicates that these conceptions of social sciences fall short of their task, where the programming of social processes at the level of contemporary technological possibilities is concerned.[213] Quite frequently they, just like the owl of Minerva in Hegel's well-known image, begin an active life at dusk as an *ex-post* commentary on man's plight.[214]

Despite its undoubted advances, the 'theory of management' which accumulated a number of findings about different areas of social development, and paradoxically as a result of its partial successes in the practical sphere, the social development of the advanced capitalist countries displays an unmistakably spontaneous course. It is precisely the long-term application of the 'scientific management' and 'social engineering' theories that, notwithstanding the relatively favourable conditions of the development of productive forces, precipitated a crisis in the way of life and in the social goals, which has no counterpart in modern history. Social relations, it is argued, resist 'our best efforts to guide them'.[215] This is certainly not only the result of the immense complexity of social processes, which require an advanced theory of major systems and computer technology, operating with a huge number of variables for their general quantitative specification. Far more relevant are the immanent limits of the theory of social control, inherent in the practical functioning of the 'industrial system' of capitalism.

True, modern, particularly American, trends deriving from the practical, technological realizations of 'industrial management' have long since let its social nature fade into oblivion. However, in the turbulent conditions of the STR, this social nature and the corresponding restrictions assert this force with recurrent frequency and in a variety of ways. Its outward manifestation is reflected in the absence of the general social goals of social control;[216] in the lack of general social criteria for social activity;[217] and in the fact that this system of social control pursues not the interests of man's general social development but the interests of maximum 'net income' produced by the given amount of capital, since it is solely designed for their realization.

Due to the pressure of the STR, the conceptions of a 'scientific management' of society (as an external object) proceeded from the original simple schemes of industrial hierarchization to modern system theories, employing mathematical models, operational analysis, etc. The barrier which all these conceptions run up against, the problem which they fail to solve, is the question of the subject, the active agent of social processes. From both the factual and the methodological point of view, they display the viability to cope with the problem of advancement of the subjective forces of social development along with the concurrent transformation of social conditions. For the hidden, though ever-present, domineering and invariable subject of such management activities, which supposedly will never cease to accompany 'Western civilization',[218] is and continues to be capital; on the other hand, it is society (the working masses engaged in the process of production) that constitutes the very object of this type of management. Whatever the way in which 'human values', 'inter-human relations', 'social responsibilities', 'social motivations', etc., are additionally appended to these conceptions of 'scientific management', as calculated means of 'social technology', 'social engineering' designed to attain goals pursued from outside, all this cannot change the structure of the subject and object in this type of social control, nor can it lead to the formulation of a working scientific theory of the organization and control of society.[219] This is because of the absence of the dimension essential for making the control of social processes truly operative, i.e. the dimension pertinent to a purposeful formation of the intrinsic subjectivity of social labour[220] and the intrinsic development of the working masses on the basis of the law-governed dialectics of the objective and subjective factors in history. Under the present-day conditions of the social and the scientific and technological revolution, the subjective factors of historical development are steadily gaining in significance,[221] since the range and rapidity of transformations depend both on the number of people actively participating in the process[222] and on their mastering the achievements of science and technology, i.e. in proportion to how the application of science merges with the activity of all workers. Due to this fact, the shallow managerial conceptions derived from the capitalist conditions are becoming more and more ineffectual.[223]

These requirements and limits are strikingly apparent in reflections on the control of the future development of society on a global scale. Conceptions, stressing the need of transition from the current economic growth to a 'state of equilibrium', tried to apply Wiener's idea of the 'homoeostatic function of science' to the control of this process. They elaborated projects envisaging a kind of 'world homoeostatic system'[224] or 'world sensorium'[225] which would signal well in advance the impending danger and exercise the function of a legitimate body empowered to solve certain 'global problems' of the present. The commentaries on these projects raise the justified question: Who is supposed to be the subject of this system of control, presented here as a divine power aloof from society?[226]

Owing to the mounting impact of scientific and technological innovations, calling for a more and more adequate social settlement, these unspecified restrictions on the subject of social control in terms of modern 'social technology', are in themselves developing into a serious source of danger. For they factually presuppose the existence of a narrow management 'élite' superimposed on society,[227] reviving time and again the idea that this élite might attempt through drastic interventions into the life of countries and continents and with the assistance of modern science and technology to halt, 'freeze' or contain developments within 'acceptable limits',[228] even at the cost of 'inevitable losses of freedom', under the slogan of the 'reign of terror of the good',[229] etc.

In face of these prospects of 'social technology' under the conditions of the STR, the last few decades saw the upsurge of criticism and attempts at the 'self-actualization', 'self-thematization' and 'self-reflection' of the theoretical systems and practical conceptions of social control.[230] Although giving rise to whole schools of 'reflexive', 'autoanalytical' sociology or political economy finding its 'philosophical' counterpart in the programme reducing social 'emancipation' to 'self-understanding',[231] bourgeois social science has failed to provide a satisfactory solution to the riddle of the subject of social development and control.[232] In modern systems, theories of the development of society, 'subjectivity' remained, despite all intellectual efforts, in fact confined either by the subjectivity of a special élite or to a general but powerless, subjectivity of the private individual to operate within the given pattern of social relations and hence relegated to a mere adaptation of man.[233]

The romantic currents of thought, too, confronted the aspirations of 'social technology' only with subjectivity of the private world of the individual (referred to as the 'one and only true reality')[234] deprived of his powers and unable to identify critically and to assert in practice his own social character.

European trends in the social sciences, adhering to Weber's tradition, began to seek a way out of their seriously felt 'subjective uprootedness' in the construction of certain substitutes of real social subjects presenting two 'alternatives' during the heated discussions: 'intersubjectivity' based on the 'communicative behaviour' within the given system[235] or, on the contrary, the 'intersubjectivity' of critical 'understanding' in the sense of hermeneutics, which after all 'only ensues in the self-reflection of all participants, which is successful and operative in a dialogue',[236] where only one factor of all the accumulated social forces of the present is given full rein—the power of language.[237]

It is evident that the dynamics of life under the conditions of the STR call for a much deeper and more efficient conception of the entire activity of the social sciences, commensurate with the new dimensions that the historical developments of the present gradually acquire and indeed have to acquire. An urgent need looms large for a general strategy of the control of social processes by society itself, proceeding from general social interests and aimed at the creation of conditions conducive to the development of the potentialities, abilities and activities of the entire working people for the comprehensive scientific control of all social processes,[238] within which the law-based dialectics of the transformation of the world and self-formation of soceity is being materialized. This requires general social control combining scientific, technological, economic, social and cultural aspects.[239] Science itself and its application become a controllable force, developed on a planned basis and applied[240] as an integral, indispensable component. In direct contrast to bourgeois managerial conceptions this new pattern of social control realized under the conditions of socialism proceeds from the general objectives of soceity, governing criteria and methods of enterprise management.[241] It leads to the formation of essentially new instruments of social analysis, proceeding from new, higher economic forms, from the indicators and criteria of the society-wide system of the 'economy of time', today systematically built up in socialist countries as an indispensable guideline for the comprehensive control of social processes.[242] The criteria of maximum profit of given capital are replaced here by the criteria of the maximum of 'disposable time' of the entire society, i.e. time released for the general development of man, coinciding in fact with the amount of social productive forces inroduced into the production of human life, with the growth of productivity of total social labour.

Viewed from the angle of traditional economic rationality, the substantial transformation of the content of labour might appear as a mere, largely unprofitable social measure. The same applies to the standards of education in case they transcend the immediate need for qualified manpower. The picture changes completely if we consider the sources of scientific-technological and social progress, of innovation activities of productive forces society acquires along with the purposeful transformation of the content of labour[243] or with the systematic development of education standards—as a result of the impact of these measures on the general development of man.

Secondly, there enters today as an indispensable element into the control of the social processes the purposeful and systematic formation of the actual subject of social control, i.e. the development of basic social forces of the present, and in the final analysis, of each man,[244] so as to allow for a concurrent law-based constitution at each newly arising level of the application of science and technology, of a correspondingly mature subject of social control able to master the further process of transformation of the world and the development of society.[245]

It is exactly the realization of these dimensions of social control that provides socialist society with its specific capacity to handle the subject-object dialectics and with its new higher historical dynamics.[246] The new, socialist type of social sciences do not bypass the problem of the subject of social cognition and social control as a transcendental, 'bracketed out' external circumstance. On the contrary, their fundamental principle, which means a true revolution in social sciences, a feature associated with the advent of Marxism-Leninism, is the firm alliance of social theory with social forces, which due to their general position, represent, in accord with the logic of historical development, the real subject of the control of social processes.

In the social structure of industrially advanced countries and, basically, in the contemporary world in general, there is only one force capable of overcoming the traditional foundations of the civilization process based on particular, non-social interests; only one class able to lead the entire masses of society to active participation in the shaping of history, thus establishing the society (free of class antagonism) as the subject and pushing through the legitimate general social goals into the process of social control; only one active agent that can and in fact does set in motion the subjective sphere of social development, achieving with the transformation of objective living conditions its own development as the subject. Such a force is the working class,[247] whose basic vital interest is to abolish all antagonistic social relations and their corresponding material conditions restricting the life of the workers in the capitalist society and to create for each and everybody social and technological prerequisites of life based on man's general development. The working class, therefore, is the one and only social force of the present that is able to 'extend systematically the scope of the subject of social control'[248] gradually to the entire society; to apply science fully as a 'revolutionizing, transforming' social force[249] and to link social labour consequentially with its intellectual potential into a new type of historically creative people's activity. Consequently, this is the only class able to control the STR,[250] to develop this revolution purposefully and to guide it as a process of creating material means for social relations of mutual development of man by man.[251]

The conditions of the STR intensify the general social need for the leading role of the working class in the process of fulfilling its great historic mission, the purposeful control of the entire complex of social processes, a task the implementation of which is guided by a Marxist-

Leninist party, which links this class with science, thus establishing it as the very subject of social cognition and control of the social development of our epoch.[252]

The process of fulfilling this historic task in socialist society changes the content of all economic, social-political and cultural forms, gradually reversing the established structure of the object and subject in social life and converting the process of social control into a process of shaping such a way of life, that is conducive to a gradual general development of the forces, abilities and activities of the working people.

The working class gradually, on a mass scale and with the assistance of an entire complex of socio-economic and scientific-technological measures, transforms the character of its own labour[253] as the fundamental element of people's social activity and hence of their entire way of life. Today, for the first time in history, there are being materialized in the socialist countries major scientifically grounded, society-wide projects for the linkage of labour, science and education, for the purposeful transformation of human activity and the way of life of the working people, for the all-round development of man.

It is no mere chance that the STR brought precisely this question to the fore with great urgency.[254] A socialist way of life is, on the one hand, the goal of the scientific, technological and economic progress of socialist societies; on the other hand, it is one of its important active factors[255] ensuring that the resources released by the advances of productive forces meet the goals envisaged, i.e. that they open up the scope for the general development of the forces and potentialities of the entire working people and that they avoid getting lost in activities and motivations of a purely consumer type or in areas detrimental to man's capacities. Thus, new powerful resources of the development of society's productive forces are set in motion[256]—unknown to the bourgeois society and to the respective calculations of 'scientific management' and social theory: the implemented objective changes become the source of the further development of the subject and vice versa. This cause-effect dialectic widens the circle of mutually concatenated transformations in the productive forces and production relations into a spiral, in which the full-scale development of the STR in organic linkage with the development of the social relations of the socialist society is gradually materialized.

Thus we are witness to the process of completion of the 'prehistory of mankind'—a period during which people were not masters of the process of their own social integration and remained at the mercy of spontaneous, fatal forces. The social processes unfolding here acquire the dimensions of a purposeful self-formation of the subject, society and man through the transformation of the objective world according to the laws of historical development. At the same time, as a result of the development of the social subject the great problems of the transformation of the external world become soluble.[257]

If the social sciences are to fulfil their historic tasks in face of the STR, they must proceed, in their approaches and forms, from these objective conditions of their own activity. In this way, new unprecedented horizons are opening up before them. Simultaneously, social responsibility looms large, the inevitable corollary to a purposeful control of the entire complex of the social processes at the very moment when the realization of the full-scale development of the subjective forces of the contemporary historical process, the general development on the part of society and man as well as the possibility of its complete destruction has come within its reach.

Man, who is gradually beginning to control the conditions of life on this planet and is about to mould the destinies of other celestial bodies, can no longer commend his own destiny to the stars.

Notes

1. John McHale, 'Futures Critical: A Review', *Human Futures. Needs, Societies, Technologies*, p. 13, Rome, 1974: 'There seems to be a general agreement that we live in a period of more critical transition, of revolution, and of more abrupt discontinuity in human affairs than ever previously.'
2. This was the method applied in *Technology and the American Economy* (Washington, D.C., 1966, Vol. 1, p. 1–2), a report which thus called into question the thesis on the advent of the scientific and technological revolution while preferring the conception of 'permanent technological change' parallel to the process of 'social change' in W. E. Moore's theory.
3. W. Kuhns, *The Post-Industrial Prophets*, p. 2, New York, 1971: 'But the most frequent and perhaps the most valid doubts about technology stem from the world it has shaped. As our waterways become sewage pits and the air fills with poisons, the technological utopia that has guided the American dream seems to have been grotesquely inverted.'
4. Even authors whose argument is pervaded by the spirit of eulogy sung in praise of the 'world of long-reach possibilities', 'firm pulse' of the nation at the head of the ventures in science and technology, whose 'creative virility' is 'unexcelled by any other nation in the world' (like A. B. Bromwell, *Science and Technology in the World of the Future*, p. xvi, New York, London, Sydney and Toronto, 1970) have to admit that 'our intellectual disciplines and professional societies are experiencing difficulty in coming to grips centrally and forcefully with these issues of destiny. . . . Sciences and engineering have been trying to understand the dimensions of the problem and develop organizational approaches, but they are still far from having the answers, for this is an extremely complex issue. In many respects, the greatest issues affecting the nation's destiny go unanswered, while our scholarship deals in erudite ways with the pieces.'
5. A. Burry, *The Next Thousand Years. A Vision of Man's Future in the Universe*, New York, 1974. With regard to present-day conditions, the following argument is typical: 'Our increasing ability, primarily through science, to control our own evolution creates more problems than it solves and yet is something from which we

cannot retreat.' (F. G. George, *Science and the Crisis in Society*, p. 8, London, New York, Sydney and Toronto, 1970.)

6. T. Roszak, *Where the Wasteland Ends. Politics and Transcendence in Post-Industrial Society*, London, 1973. Contrary to the 'bad, mad ontology' of contemporary science and practical transforming activity, T. Roszak finds a way out in the 'gnostic myth' in which 'the apocatastasis is the illumination in the abyss by which the lost soul, after much tribulation, learns to tell the divine light from its nether reflection' (p. 459).
7. 'None of the findings concerning the processes brought to life by the scientific-technological revolution—if considered individually—leads to an understanding of its nature, to an assessment of the prospects of its future development. . . . It is only the comprehensive and truly synthesizing approach to the evaluation of the scientific and technological revolution, an approach which pays due regard to its linkage with the fundamental social processes, that allows for an adequate specification of its nature and historical significance.' (P. N. Fedoseyev, 'Sotsial'noye Znacheniye Nauchnotekhnicheskoi Revolyutsii', *Voprosy Filosofii*, No. 7, 1974, p. 5.)
8. See *Man-Science-Technology. A Marxist-Leninist Analysis of the Scientific-Technological Revolution,* p. 366, Prague/Moscow 1973. Similarly Emil Duda, *Vedeckotechnická Revolucia*, Bratislava, 1974.
9. K. Hager, *Sozialismus und wissenschaftlich-technische Revolution*, p. 23, Berlin, 1972.
10. Y. S. Meleshchenko, *Tekhnika i Zakonomernosti yeyo Razvitiya*, p. 41, Leningrad, 1970.
11. This linkage is interpreted either as the 'technicalization of science' (E. Husserl), the rise of 'technoscience' (M. Heidegger), or as the incarnation of 'technological rationality' (M. Weber), 'scientification of technology' (J. Habermas), the constitution of 'technostructure' (Galbraith), 'megatechnology' (Mumford), etc. Underlying these and similar interpretations is, however, the same notion of creating a special, autonomous social power mediated through the liaison of science and technology.
12. A. Toffler, *Future Shock*, p. 26, New York, 1971.
13. T. Roszak sees science and science-based technology as a 'dominant force' of our times, as 'the curse and the gift we bring to history' (*Where the Wasteland Ends*, p. xxiv, London, 1973). Methodologically, this approach is analogous to, e.g., the conception of the immanent cumulative advance of science and technology as 'mixed blessings' of contemporary society specific to current technocratic conceptions—e.g. H. Kahn and A. Wiener, *The Year 2000*, p. 387 ff., New York, 1967.
14. '*Aus dem einfachen Arbeitsprozess in einem wissenschaftlichen Prozess.*' See K. Marx, *Grundrisse der Kritik der politischen Okonomie*, p. 588, Berlin, Dietz-Verlag, 1953.
15. *Die Entwicklung der vollen Produktivkräfte der Einzelnen.* ibid., p. 595.
16. See I. I. Dryakhlov, S. I. Nikishov, Y. K. Pletnikov, S. K. Shukhardin (eds), *Nauchnotekhnicheskaya Revolyutsia i Obshchestvo*, p. 208 ff., Moscow, 1973.
17. 'This anti-science goes deeper than the mere machine hatred of Luddism-simple; instead the attack on science and technology is transformed into an attack on the dynamic rationality of scientific method itself.' (H. Rose and S. Rose, *Science and Society*, p. 255, London, 1969.)
18. K. Marx, *Capital*, Vol. I, p. 340, Progress, Moscow, 1966.
19. Thus, *alle gesellschaftlichen Potenzen der Produktion sind Produktivkräfte des Kapitals und es selbst erscheint daher als das Subjekt derselben*'. (Marx, *Grundrisse*, op. cit., p. 79.)
20. As science and technology 'become more autonomous and more powerful, men experience themselves as less potent, less in control of their own destinies'.

(A. W. Gouldner, *For Sociology. Renewal and Critique in Sociology Today*, p. 76, New York, 1973.)

21. With the growth of industry, capital 'makes science a productive force distinct from labour and presses it into the service of the capital'. (Marx, *Capital*, Vol. I, op. cit., p. 341.) This separation of science as a force opposed to labour comes under the general category of separation, i.e. alienation of production conditions from labour, which is a feature characteristic of capital; of course, at the same time, under given historical conditions and for a certain transition period, this was a precondition for the development of society's scientific and technological potential.
22. 'The negation of the head (*caput*) and the will of workers, their existence as mere limbs of the factory body appears as the legitimate right of the capital which hence exists as "*caput*". It is only capital that converts the material process of production into the application of science in production . . . but only in that it subordinates the worker to capital and suppresses his own intellectual and professional development.) (K. Marx and F. Engels, *Sochineniya*, Vol. 47, p. 559, Moscow, 1973.)
23. Over the past thirty years—in the period of the advent of the scientific and technological revolution—these capital interests commanded 75 per cent of the more than $ 500 billion spent on research and development all over the world within the period under review. But contrary to the situation prevalent in the past, they were no longer the sole ultimate decisive factor.
24. K. Marx, unpublished papers on technology, quoted after *Voprosy Istorii Estestvoznaniya i Tekhniki*, Vol. 25, 1968, p. 27.
25. 'Along with the machine, the domination of past labour over living labour becomes not only a social reality but also—so to speak—a technological reality.' ibid., p. 74.
26. J. Purs, *Prumyslova Revoluce. Vyvoj Pojmu a Koncepce*, Prague, 1973.
27. '*Es ist nicht der Arbeiter, der Lebensmittel und Produktionsmittel kauft, sondern die Lebensmittel kaufen den Arbeiter, um ihn den Produktionsmitteln einzuverleiben. Lebensmittel sind eine besondere stoffliche Existenzform, worin das Kapital dem Arbeiter gegenübertritt.*' (*Arkhiv Marksa i Engelsa,* Vols. II–VII, p. 60, Moscow, 1933.
28. '*Die Entwicklung des Arbeitsmittels zur Maschinerie ist nicht zufällig dür das Kapital, sondern ist die historische Umgestaltung des traditionell überkommen Arbeitsmittels als dem Kapital adäquat umgewandelt. Die Akkumulation des Wissens und des Geschicks, der allegemeinen Produktivkräfte des gesellschaftlichen Hirns, ist so der Arbeit gegenüber absorbiert in dem Kapital.*' (Marx, *Grundrisse*, op. cit., p. 586.)
29. 'Technology has built into it characteristics of the given social relational system, especially the decision-making process on production. In that way criteria for decision-making (relations of production) are built into the means of production . . . . For 200 years, the criteria of businessmen have dominated the design and selection of technology.' (S. Melman, 'The Myth of Autonomous Technology', in N. Cross, D. Elliot and R. Roy (eds.), *Man-Made Futures,* p. 58, London, 1974.)
30. Today the fact is gaining world-wide recognition that with regard to the origin of the material and technological basis of contemporary life, there is no other way for social sciences to arrive at a correct analysis and criticism than through Marx's mode of analysis.
31. In these terms, the scientific and technological revolution was rightly described as the 'catalyser of social changes'. (T. Trendafilov, *Nauchnotekhnicheskaya i Sotsialnaya Revolyutsiya*, Sofia, 1973.)
32. The critique of the crisis phenomena affecting science and technology spread rapidly as is attested by J. M. Lévy-Leblond and A. Jaubert, [*Auto*] *critique de la Science*,

Paris, 1973. The first prediction of a 'technological crisis' dates back to the early beginnings of the scientific and technological revolution (cf. J. von Neumann, 'Can We Survive Technology?', *Fortune*, June 1955); of course real symptoms have appeared rather earlier than expected in the West; as to the 1980s, even the most optimistic prognoses acquire a gloomy tenor (see H. Gahn and B. Bruce-Briggs, *Things to Come. Thinking about the Seventies and Eighties*, New York, 1972.)

33. It was E. Ashby who, in his lecture presented to the British Royal Society on 4 March 1971, rightly pointed out this paradoxical fact.
34. V. Bush *et al.*, *Science, the Endless Frontier*, NSF, Washington, D. C. 1960. (The report was first published in 1945.) 'There is now a sudden sense that the endless frontier has come to an end,' wrote H. Brooks, commenting on this report. (J. A. Shannon (ed.), *Science and the Evolution of Public Policy*, p. 111, New York, 1973.) A. Musset issues a straightforward warning that 'in a few generations we shall run out of ideas to discover—perhaps chief of these being the discoveries of science . . . the end of scientific discovery would result in a steady drying up of invention as the possibilities are exhausted'. (R. L. Wolke (ed.), *Impact of Science on Society*, p. 215 and 219, Philadelphia, London and Toronto, 1975.)
35. 'The trouble science is in,' wrote A. Szent-György, the biochemist, 'is only part of a general crisis, and in order to understand it we must search for the cause of the wider calamity.' (*The Scientist*, Vol. 10, 1969, p. 25.)
36. N. Inozemtsev, *Sovremennyi Kapitalizm. Novyie Javleniya i Protivorechiya*, Moscow, 1972; N. Gauzuev, *Social Effects of the Scientific and Technological Revolution under Capitalism*, Moscow, 1973.
37. It is precisely in the industrial revolution, accompanying the development of capitalism, that 'the separation of science as science applied in production from immediate labour occurs, while in the earlier stages of production a restricted volume of knowledge and experience was immediately tied up with labour itself.' (Marx and Engels, *Sochineniya*, op. cit., p. 554.)
38. See Marx, *Capital*, op. cit., Vol. III. 1, p. 104; Marx, *Grundrisse*, op. cit., p. 594 ff.
39. In capitalist industrial production, science, which 'is incarnated in the machines or methods of production, in chemical processes, etc. . . . acts as an alien force, hostile to labour and dominating labour'. (Marx and Engels, *Sochineniya*, op. cit., p. 555.)
40. Max Weber's thesis that science is 'the affair of an intellectual aristocracy, and we should not hide this from ourselves' (M. Weber, *Essays in Sociology*, p. 134, London 1967), applies to the currently prevalent conception of science in the West.
41. M. Weber's conception of 'rationality' is based on the awareness that '*wir wissen, dass wir in einer Welt leben, wo nicht die Arbeit, sondern das Kapital das Übergewicht hat. Und wenn die heutige Wirtschaft rationalisiert, so betrachtet sie das vernünftige Gestalten der Wirtschaft von Standpunkt der Kapitalinteressen aus*'. (cf. *Fragen der Rationalisierung*, p. 9, Zürich.)
42. In this respect, Weber's conception of technology is developed by J. Habermas in *Technik und Politik als Ideologie*, Frankfurt am Main, 1968.
43. H. Morgenthau, *Science, Servant or Master*, p. 23, New York, 1972. Morgenthau refers directly to the danger of the perversion of science.
44. J. Galbraith, *The New Industrial State*, p. 7, Boston, Mass., 1967.
45. This constituted the actual essence of the technological fatalism of J. Ellul (*La Technique et l'Enjeu du Siècle*, Paris, 1954).
46. V. C. Ferkiss, *Technological Man. The Myth and the Reality*, p. 21, London, 1969: The definition of this new state of 'absolute power' causes considerable difficulties and is factually based on a tautology: 'To the extent that man can do all the things that he can do and knows it, we are entitled to speak of the end of the modern

world and the advent of an existential revolution' (ibid).
47. *The Works of Francis Bacon*, Vol. 3, p. 156, London, 1870–72.
48. This idea is developed by H. Ozbekhan: 'Whatever technological reality indicates we *can* do is taken as implying that we *must* do it.' (*The Triumph of Technology: 'Can' Implies 'Ought'*, p. 67, Santa Monica, SDC, 1967.)
49. E. Fromm (*The Revolution of Hope. Toward a Humanized Technology*, p. 32, New York, Evanston, London, 1968), like J. Habermas, protests against Ozbekhan's thesis, to which he refers as the scandal of contemporary science and technology. E. Fromm, however, shares Ozbekhan's and von Neumann's illusion that this thesis expresses the reality of contemporary science and technology; a truly scientific criticism, however, must admit that science and technology in the West do not do what 'they can' but first and foremost what they are paid for doing by capital.
50. It is indeed as a result of the power of capital, employing science and technology, that man feels defenceless in the face of science and technology; hence the feeling of the 'loss of control' over the entire system, the impossibility to 'direct the course' of science and technology. In his essay 'Pigs and Engineers' (*Futurist*, No. 4, August 1973), J. P. Martino debunks those who decide on the development of technology under capitalism; he argues that the production of pigs and the production of engineers is governed by the same rules of business—only the production cycle differs as to duration.
51. See M. Thring, *Man, Machines and Tomorrow*, p. 7, London, 1973.
52. See A. G. Yegorov, *The Scientific and Technological Revolution: The Personality and Culture Under Developed Socialism*, p. 34–5, Moscow, 1975.
53. Thus, the problem—V. C. Ferkiss concedes—'lies not in the autonomy of technology or in the triumph of technological values but in the subordination of technology to the values of earlier historical eras and its exploitation by those who do not understand its implication and consequences but seek only their own selfish personal or group purposes. . . . Private economic interests will decide what industries will eliminate labour and what human functions will be replaced by the computer. . . . Bourgeois man is still in the saddle. . . . Bourgeois man is incapable of coping with this revolution.' (*Technological Man. The Myth and the Reality*, p. 28, 190, 245, London, 1969.)
54. In the past decades, farmers in the United States of America have been paid special bonuses for letting 5 million acres of land lie fallow, i.e. for not exploiting the possibility of providing nourishment for a population equalling that of Pakistan. The losses incurred as a result of the non-exploitation of the potential of the industrial capacity of the United States of America due to the crisis of 1975 amounted to a sum approaching the total annual industrial production of all developing countries taken together. M. W. Thring is correct in pointing out that 'a society concerned solely with the profit motive cannot find a really effective way of helping a less-developed country to install machinery because this help is basically opposed to the profit of the richer country'. ('Machines for a Creative Society', *Futures*, March 1970, p. 47.)
55. V. I. Lenin, *Collected Works*, Vol. I, p. 85, Moscow 1963.
56. An external reflection of this objective position of science is the definition of science as an 'insulated system' (J. Ben-David), separated from labour: in the most advanced countries of the West—according to K. E. Boulding—the majority of the population regards science as something wholly outside them, as an alien force.' (*Bulletin of the Atomic Scientists*, 1970, p. 202.)
57. The comprehensive development of all components of productive forces, the utilization of the potential of the scientific and technological revolution in all its dimensions are characteristic features of an advanced socialist society. (See

G. Glezerman (ed.), *Die entwickelte sozialistische Gesellschaft*, Berlin, 1973.)

58. 'Science and scientific findings materialize through the process of labour both in the material and the personal elements of the productive forces.' (N. V. Markov, *Nauchno-tekhnicheskaya Revolyutsiya: Analyz, Perspektivy, Posledstviya*, p. 99, Moscow, 1973.) See also V. S. Marachov, *Struktura i Razvitiye Proizvodit Elnykh sil Socialisticheskogo Obshchestva*, p. 157–8, Moscow, 1970.
59. G. N. Volkov, *Chelovek i Nauchno-tekhnicheskaya Revolyutsiya*, Moscow 1972; V. N. Turchenko, *Nauchno-tekhnicheskaya Revolyutsiya i Revolyutsiya v Obrazovanii*, Moscow, 1973.
60. Marx, *Grundrisse*, op. cit., p. 439.
61. K. Marx, *Economic and Philosophic Manuscripts of 1844*, p. 97, Moscow, 1974.
62. 'The positive and negative consequences of the scientific and technological progress—all these facts bring into distinct relief the reality that production is more than the production of things, goods, services, that it is the production of man's actual necessities . . . the production of conditions of human existence, man's needs . . . the production of people's life styles and modes of their social contact.' (D. M. Gvishiani, *Nauchno-tekhnicheskaya Revolyutsiya i Sotsialnyi Progress*, p. 14, Moscow, 1974.)
63. Marx, *Grundrisse*, op. cit., p. 438.
64. The economy of social labour deriving from the activity of an innovator from among the working people in the Union of Soviet Socialist Republics, equals—on the annual average—the performance of another worker; the performance of leading innovators comes to the performance capacity of 5–20 workers. (See B. M. Kedrov (ed.), *Nauchno-tekhnicheskaya Revolyutsiya i sotsialism*, p. 94, Moscow 1973.)
65. Showing why V. I. Lenin placed such 'great emphasis on the all-round development of all members of society', A. G. Yegorov points out that 'the free and all-round development of the individual is a powerful stimulus for the growth of social production, science, technology and arts'. (*The Scientific and Technological Revolution, the Personality and Culture under Developed Socialism*, p. 9, Moscow, 1975. Similarly, S. V. Shukhardin, *Nauchno-tekhnicheskaya Revolyutsia i Obschestvo*, p. 19, Moscow, 1973.)
66. These aspects of the scientific and technological revolution are discussed in R. Richta, *Clovek a Technika v Revoluci Nasich dnu* (Man and Technology in the Revolution of our Times), p. 65, Prague, 1963.
67. Marx, *Grundrisse*, op. cit., p. 599.
68. ibid., p. 593.
69. '*Gesellschaftlich produzierte Entwicklungsmittel* . . . ' See K. Marx and F. Engels, *Werke*, Vol. 34, p. 171, Berlin, 1966.
70. Former societies have displayed a contradictory motion due to the fact that production, insufficient to the entire society, allowed for a kind of development, only if . . . one section—the minority—were . . . excluded from all development ('*für die ganze Gesellschaft unzureichende Produktion nur dem eine Entwicklung möglich mache, wenn die Einen—die Minorität—das Monopol der Entwicklung erhielten, Während die Andern—die Majorität—von aller Entwicklung ausgeschlossen wurden.*' (Marx and Engels, *Werke*, Vol. 3, P. 417, Berlin, 1962.) J. S. Mill used this 'contradiction of development' to justify the 'progress of civilization' (cf. *Elements of Political Economy*, Vol. II, p. 2, London, 1821).
71. The law-based coincidence of these processes was pointed out already by J. D. Bernal, *Science in History*, London, 1954.
72. 'The scientific and technological revolution, the control of social processes now function as the most important means of . . . the economy of time.' (V. G. Afansyev, *Nauchno-tekhnicheskaya Revolyutsiya, Upravlenie, Obrazovanie*, p. 25, Moscow, 1972.)

73. 'Science as a social productive force *par excellence*, depending—as Marx pointed out—both on the co-operation of the living, and on the utilization of the labours of those who have gone before, is incompatible with any social forms of private ownership and defies such forms continually. The communism of scientific ethos is incompatible with the definition of technology as private ownership in capitalist economy.' (R. K. Merton, *Social Theory and Social Structure*, p. 558, New York, 1965.) Attempts to revive Bacon's vision expressed in his New Atlantis and basing the social forms of the future on private 'intellectual ownership' (D. Bell, *The Coming of Post-Industrial Society*, p. 176, New York, 1973) on a 'new kind' of private ownership in relation to intellectual components (C. A. Reich, 'The New Property', *The Public Interest*, No. 3, 1966, p. 57–89) or directly on private-ownership relations towards scientific discoveries are, in the face of present-day achievements of science, no more than an expression of the absurdity of conceptions separating science from labour by the barrier of ownership.
74. 'It is only by dint of the most extravagant waste of individual development that the development of the human race is at all safeguarded and maintained in the epoch of history immediately preceding the conscious reorganization of society.' (K. Marx, *Capital*, Vol. III, p. 88, Moscow, 1966.)
75. B. M. Kedrov (ed.), op. cit., p. 362.
76. '*Das bornierte Verhalten der Menschen zur Natur ihr borniertes Verhalten zueinander, und ihr borniertes Verhältnis zur Natur bedingt, ebenweil die Natur noch kaum geschichtlich modifiziert ist . . .*'. (Marx and Engels, *Werke*, op. cit., Vol. 3, p. 31, Berlin, 1962).
77. Aristotle's definition of man as '*zoon politikon*'—a definition likewise occurring in Meng-tse—implies that man is by nature a member of the ancient *polis*. Similarly, 'the natural slave is one who is qualified to be, and therefore in fact is, the property of another'. (*Politics* I, c. 5, p. 1254 b 20 ff.)
78. For Aristotle (and, analogically, for Meng-tse) nature was only 'one of the genera of being', ruled by basically the same order as that governing a 'household' or a 'community'—i.e. the hierarchy of the free, the slaves, domestic animals, etc. (*Metaphysics* IV, c. 3, p. 1005 a 34; cf. also XII, c. 10.)
79. Where they came across technical inventors (Daedalus), they regarded them—just as Seneca—as intellectually inferior for their inability to grasp the truth about things, this being the privilege of the users of things (free men), not of the producers (slaves). This accounts for the fact that inventors such as Archimedes felt it necessary to explain that their preoccupations did not spring from practical needs but were mere pastimes.
80. What this varied pattern of tradition reveals is the modernity of such motifs and their profound link with the use of machinery in factories. (A. E. Lewis, *Of Man and Machines*, p. 31, 349, New York, Dutton, 1963.
81. It is certain that this approach was shaped in the long-term process of historical development. J. R. Forbes (*The Conquest of Nature: Technology and its Consequences*, p. 15, London, 1968), as well as V. Ferkiss, T. Roszak and number of other writers, derive man's drive to bring nature under control from the 'Judaeo-Christian tradition' while referring to the passage from *Genesis* 1, 28, cf. 1, 26, where God said: '*Replete terram et subjicite eam, et dominamini piscibus maris et volatilibus coeli, et universis animalibus . . .* ' The biblical motif of nature surrendering to man's control has, of course, an originally different meaning. It is not obtained by man's endeavour but through divine will, and it is not oriented towards man's benefit but towards the glory of the Lord. In this particular shape, the subordinate status of nature in the Judaeo-Christian tradition is only one of the forms of a tendency discernible both in some offshoots of later Graeco-Roman thought like the theoretical groundwork of Islam (especially the 'mutazili' sect) and in the ever

recurrent apostasy in sinoism (unmistakably present in Hsün-tse). Underlying the conception of man's sovereignty over nature is the acquisition and ownership of domestic animals and land, which was gaining in significance with the gradual disintegration of the ancient world, making use of the purely physical capacity of the vast masses of slaves. The modern motif of man's sovereignty over nature draws on different traditions like Leonardo da Vinci, Rustavelli, Nizami, Ibn Sina, etc.

82. The transformation of the individual into a general *a priori* subject of cognition reflects the disintegration of social orders based on direct subordination of man by man. Symptomatic in this respect is Descartes's conviction that man's reason and cognitive faculties are best distributed in that they are generally accessible to each individual. The epoch which produced and generalized 'the standpoint of the isolated individual' was of course 'an epoch of the then most advanced (and from this point of view, general) relation in bourgeois society' (cf. Marx and Engels, *Werke*, p. 61, Berlin, 1961, Vol. 13, p. 61). For the slave system and feudal society, this standpoint was neither acceptable nor comprehensible since formal equality of people was non-existent. As Vico already pointed out, abstract people (*homines*) were only plebeians in ancient times (contrary to patricians, who were '*cives*', '*viri*'); in the Middle Ages only vassals (and not '*nobiles*').
83. Marx, *Grundrisse*, op. cit., p. 313.
84. Marx designates as 'abstract' the system of natural sciences which 'excludes history' (*Capital*, op. cit., Vol. 1, p. 352).
85. Galileo Galilei regarded this historically arisen approach and the corresponding type of modern natural science as certain and natural, while declaring that the progress in natural science is independent of society, independent of any man. (cf. *Dialogue on the Great World Systems*, p. 63, Chicago, 1953.)
86. Even today this approach, while it isolates the process of work, the process of commerce between man and nature from society, is frequently revived, (cf. J. Habermas, *Erkenntnis und Interesse*, p. 61, Frankfurt am Main, 1973.)
87. 'Thus this science follows the same procedure as geometry.' (cf. G. Vico, *Die neue Wissenschaft über die gemeinschaftliche Natur der Volker,* p. 139, Munich, 1924.)
88. Locke's metaphysical theory of property and individualism, representing the counterpart to modern conceptions in natural sciences, 'in addition to its function as an instrument of class domination . . . has provided the basis for an unrelenting and remorseless assault on nature'. (V. C. Ferkiss, *The Future of Technological Civilization*, p. 30, New York, 1974.)
89. Marx, *Capital*, op. cit., Vol. III, p. 818.
90. Marx, *Grundrisse*, op. cit., p. 884.
91. F. Bacon, 'The Advancement of Learning', *The Works of Francis Bacon*, Vol. 4, p. 372–3, London, 1970–72.
92. *Doctrine Saint Simonienne: Exposition*, p. 463, Paris, 1954.
93. As soon as the spontaneous character of this onslaught on nature became apparent in its grave consequences, reversed interpretations of this process began to spread: as if the misleading goal afflicting contemporary society were man's domination over nature, while the domination of man by man were no more than its side-effect (see M. Horkheimer, *The Eclipse of Reason*, Oxford University Press, 1947; G. G. Simondon, *Du Mode d'Existence des Objets Techniques*, Paris, 1958; and others) and as if technocracy and not capital were the crucial problem of the present. This apologetic mode of putting the question characterizes, for instance, the Frankfurt school of the 'critical theory of society'.
94. Marx and Engels, *Werke*, op. cit., Vol. 12, p. 3, 4.
95. A. Vann and F. Rogers (eds.), *Human Ecology and World Development*, London, Plenum Press 1974; Michael Micklin (ed.), *Population, Environment and Social*

Organization: Current Issues in Human Ecology, p. 509, Hinsdale, Ill., Dryden Press 1973.

96. 'The artificial environment has treacherously turned into public enemy number one', says T. Rozsak (*Where the Wasteland Ends*, op. cit., p. 30): again, not without a considerable amount of apology bypassing the fact that the true sources of destruction are inherent in the social conditions, in the capital.
97. M. Heidegger, *Gelassenheit*, p. 20, Pfullingen 1959; *Holzwege*, p. 266, Frankfurt am Main, 1950.
98. A. Gehlen, *Seele im technischen Zeitalter. Sozialpsychologische Probleme in der industriellen Gesellschaft*, p. 70, Hamburg, 1957.
99. T. Rozsak as well as other authors of the 'romantic revolt' define the main objective (*Where the Wasteland Ends*, op. cit., p. xx, 463) as 'the religious revival of the Western world', 'the Great Restoration' and 'the way back' in the spirit of St Bonaventura, to whose image he adapts Buddha, Lao-tse and others. A similar situation is traceable in the melioristic, technocratic trends of contemporary thought, exemplified in D. Bell's latest publication, *The Cultural Contradictions of Capitalism* (New York, 1976), where the revival of religious creed is ascribed as one of the central motifs of the entire present-day endeavour.
100. F. Engels, *Dialectics of Nature*, p. 180, Moscow, 1966.
101. F. Engels, *Anti-Dühring*, p. 34, Moscow, 1962.
102. In reality, of course, 'it is the transformation of nature by man, not nature as such, that is the most essential and most immediate nature of human thought'. (Engels, *Dialectics of Nature*, op. cit., p. 225.)
103. 'Science gives us power, but remains impersonal and all too often ignores nature', wrote J. Nehru. (*The Discovery of India*, p. 476, London, 1946.) In comparison with the tradition of thought in which the main motif was—as in the Upanishads and, in a different philosophical form, in the Socratic tradition—the self-reflection of a particular being, and for which nature was not a mere thing but something rather like 'thou', the advance of modern science must necessarily appear as an invasion of foreign, anonymous powers converting nature into a lifeless atomized world. The point is, however, that the ancient tradition of thought—and even more the romantic currents—never could supersede the horizon of modern science and its practical efficiency.
104. '*Zum ersten Mal bewusste wirkliche Herren der Natur, weil und indem sie Herren ihrer eigenen Vergesellschaftung werden.*' (See Marx and Engels, *Werke*, op. cit., Vol. 20, p. 264.) This fact seems to have—to a certain though limited extent—filtered into the general consciousness. In his study *The Domination of Nature* (New York, 1972), W. Leiss concedes that 'the mastery of nature' can become a vehicle for liberation only if it includes as an essential component the self-mastery of men in their social relationship. Unfortunately, 'the self-mastery of men in their social relationship' need not necessarily imply real revolutionary transformation on the part of the subject.
105. Marx, *Grundrisse*, op. cit., p. 387, 440.
106. See G. Klaus, *Rationalität—Integration—Information. Entwicklungsgesetze der Wissenschaft in unserer Zeit*, Berlin, 1974.
107. More and more frequently, science derives its methodological procedures from the self-development of matter, which crosses the boundaries of the mechanical *res extensa* and displays a complicated inner structure and dynamics. It is here that Lockean metaphysics loses its theoretical hinterland. It is here that the analytical method of traditional natural science fails to operate and bourgeois social science seeks recourse in subjective *Anschauen, Erleben.* A number of observers of this crisis of the traditional foundations of science (e.g. E. Husserl, *Die Krisis der*

europäischen Wissenschaften und die transzendentale Phänomenologie, The Hague, 1954) analyse its symptoms from the standpoint of transcendental subjectivity, which, in fact, was the actual theoretical source of this crisis. Consequently, they are not capable of carrying out a true analysis of the objective social preconditions of the subject of cognition nor of providing new foundations for science that would correspond to the new dimensions in the relation between man (society) and nature.

108. See T. Pavlov, *Teoria Odrazu a Naša Súčasnost*, p. 174, 267, 274, Bratislava, 1975.
109. In this connection, V. I. Lenin referred to an approach which never loses sight of 'general preconditions of the existence of the concrete subject (= human life) under objective conditions' (V. I. Lenin, *Sochineniya*, Vol. 38, p. 203).
110. Marx's prediction of the rise of a unified science today frequently stands in the forefront of attention. (See e.g. *Sintez Sovremennogo Nauchnogo Znaniya*, Moscow, 1973.)
111. P. V. Kopnin, *Dialektika—Logika—Nauka*, p. 27, Moscow, 1973.
112. Science has reached a stage, where it necessarily turns to the investigation of itself. To quote J. D. Bernal ('after Twenty-Five Years', in M. Goldsmith and A. Mackay, (eds.), *The Science of Sciences*, p. 217, London, 1964) 'to call for the self-realization of science has become a characteristic concomitant feature of the scientific and technological revolution'.
113. S. Angelov and J. Janchev, *Eticheski Problemi na Naukata*, Sofia, 1973. The resolution adopted by the World Federation of Scientific Workers stresses the responsibility of each individual scientist for the consequences of his work and if this is within his power, for preventing its misuse for destructive and antisocial purposes.
114. Apparently different are the results specific to the capitalist countries, where neither social conditions nor the scientific basis are matched to suit this particular assignment: 'The accumulation of social problems means that much is expected from the combined contribution of natural, engineering and social sciences. The results however are disappointing', writes G. Ferné, Head of the OECD Sector of Social Sciences (*OECD Observer*, Vol. 76, July–August 1976, p. 28).
115. M. M. Kamshilov (*Filosofia i Teoria Evolutsii*, p. 235, Moscow, 1974) characterizes this change as the transition of the 'biosphere' into a qualitatively new state, the 'noosphere', thus using a term brought into vogue by Teilhard de Chardin (*The Future of Man*, London, 1964), but long ago coined by V. I. Vernadskiy.
116. In this respect, D. M. Gvishiani and J. R. Mikulinski refer rightly to the new level of commerce between society and nature.
117. 'The universality of man appears in practice precisely in the universality which makes all nature his inorganic body'. (K. Marx, *Economic and Philosophic Manuscripts of 1844*, p. 67, Moscow, 1974.)
118. G. G. Gudozhnik, *Nauchno Tekhnicheskaya Revolyutsiya i Ekologicheskiy Krizis*, Moscow, 1975.
119. This was the spirit surrounding the wide-ranging discussion on the tasks of ecology in the socialist society on the pages of *Voprosy Filosofii*. ('Chelovek i sreda ego Obitaniya', *Voprosy Filosofii*, Nos. 1–4, 1973.)
120. The thesis that the exploitation of nature has replaced today the exploitation of man is one of the cheap arguments raised in apology of the classes living on the exploitation of man. Naturally, it cannot provide a basis for the transformation of the world under the conditions of the scientific and technological revolution.
121. '*Die Wissenschaft schreitet fort im Verhältnis zu der Masse der Erkenntnis, die ihr von der vorgehenden Generation hinterlassen wurde.*' (Marx and Engels, *Werke*, op. cit., Vol. 1, p. 521.) 'Each discovery becomes the basis of a further discovery or improvement in production methods.' (Marx and Engels, *Sochineniya*, op. cit., Vol. 47, p. 553–4.)

122. J. K. Fyodorov, *Vzaimodeystviye Obschestva i Prirody*, p. 85, Leningrad, 1972.
123. P. N. Fedoseyev characterized this law-based phenomenon as follows: with each new step of the scientific and technological revolution, the role of the social increases. ('Sotsialnoye Znacheniye Nauchno-tekhnicheskoy Revolyutsii', *Voprosy Filosofii*, No. 7, 1974, p. 14.)
124. N. P. Dubinin (*Genetika in budushcheye chelovechestva*, Moscow, 1971; *Vechnoe Dvizheniye*, Moscow, 1975) proves the incorrectness of the hypothesis advanced by K. Lorenz, who reduces social motion to biological motion.
125. See I. B. Novik, *Methodologicheskiye aspekty Issledovaniya Biosfery*, Moscow, 1975. It is especially J. K. Fyodorov who develops this idea referring to the optimistic conception, highlighted by G. T. Khilmi (*Osnovy Fiziki Biosfery*, Leningrad, 1966).
126. P. Kapitza, *Eksperiment, Teoriya, Nauka*, p. 277, Moscow, 1974.
127. A. Szent-Györgyi estimates that the total sum ever spent by mankind on basic research is probably less than the total budget of the Pentagon for one fiscal year. (cf. 'On Federal Support of Science', *The Scientist*, No. 10, 1969, p. 25–8.)
128. cf. *Basic Research and National Goals*, p. 15, Washington, D. C., 1965.
129. See N. N. Semenov, *Nauka i Obschestvo*, p. 109–44, Moscow, 1973.
130. 'There is a certain gap between the already existing and rapidly mounting ability to implement measures that are global in character and the non-existence of a corresponding social mechanism not only in the regulation of such measures but also in the evaluation of their function from the point of view of the entire mankind'. (E. K. Fyodorov, *Lenin a Soucasná Přirodověda*, p. 366, Prague, 1972.)
131. 'The environmental crisis . . . reveals serious incompatibilities between the private enterprise system and the ecological base.' (B. Commoner, *The Closing Circle*, p. 276, New York, 1972.)
132. In the concluding remarks to *The Future of Technological Civilization* (p. 291) V. Ferkiss shows the brutal arrogance with which capital interests crush projects of ecological protection: 'The first flush of interest in ecology which culminated in Earth Day 1970 has waned, and environmentalists are on the defensive before a determined counterattack by big business.'
133. I. T. Frolov, *Progress nauki i Budushcheye Cheloveka*, Moscow, 1975, speaks in this connection about a new stage of the scientific and technological revolution.
134. E. I. Mendelsohn, J. P. Swazey and I. Taviss (eds.), *Human Aspects of Biomedical Innovation*, Cambridge, 1971.
135. N. P. Dubinin, *Genetika i Budushcheye Chelovechestva*, Moscow, 1971.
136. V. M. Glushkov, *Besedy ob Upravleniyi*, Moscow, 1974.
137. cf. J. M.R. Delgado, *Physical Control of the Brain: Towards Psychocivilized Society*, New York, 1969.
138. Marx and Engels, *Werke*, op. cit., Supp. Vol. I, p. 552.
139. The long-term prognoses of the scientific, technological and social development, drafted by the Academy of Sciences of the Union of Soviet Socialist Republics, prove that 'global' problems are fully soluble within the framework of socialist society on the basis of society-wide control and management. The study *New Directions for Science and Technology* elaborated under D. Gabor for the Club of Rome arrived at the conclusion that individual problems (including the questions of energy, mineral resources, foodstuffs) considered separately are not 'really critical', but that their general solution will be hard to attain.
140. M. Siebker and Y. Kaya, 'Towards a Global Vision of Human Problems', *Technological Forecasting and Social Change*, Vol. 3, 1974, p. 240.
141. Significantly, the so-called 'big problems'—world supplies of foodstuffs, the development of resources of the oceans, the evolvement of new giant power

systems—are 'very bad problems' for research conducted within the framework of capitalist entrepreneurial activities, since the solution of such problems is 'likely to be unprofitable'. (C. B. Kaufmann, 'The Future Social Climate for R and D', *Research Management*, No. 1, January 1975, p. 32.)

142. A contrary situation developed in the socialist countries, where the insufficient level of productive forces—a heritage of the past—acted for some time as a limiting factor in the development of social relations. In the advanced socialist society, however, the dialectics of the productive forces and the production relations specific to socialist society asserted itself in its fully fledged form. (cf. G. E. Glezerman, *Historicky Materialismus a Rozvoj Socialisticke Spolecnosti*, p. 295, 323, Prague, 1975.)
143. During the proceedings of the Second Congress of the World Future Society held in Washington in June 1975, almost all participants concurred, in one way or another, with Roy Amara's comment that we are faced with the necessity of rather drastic changes in a relatively short time. (cf. A. A. Specke (ed.), *The Next 25 Years, Crisis and Opportunity*, p. 2, Washington, D.C. 1975.) The world—says the *U.S. News and World Report* (23 June 1975), while commenting on the conclusions which met the widest acceptance among those present—is now condemned to 'waggle from crisis to crisis'.
144. *Human Futures—Needs, Societies, Technologies*, p. 17, London, 1974.
145. A. Toffler, *The Eco-Spasm Report*, p. 51, New York, 1975.
146. '[It] is a kind of a failure of nerve, an attitude I share. I'm scared by much current technology.' (H. Kahn, 'Things are Going Rather Well', *The Futurist*, December 1975, p. 292.)
147. In reply to those who expressed apprehension over the scientific and technological revolution, Don K. Price said: 'I do not think they are pessimistic enough. To me it seems possible that the new amount of technological power let loose in an overcrowded world may overload any system we might design for its control; the possibility of a complete and apocalyptic end of civilization cannot be dismissed as a morbid fantasy.' (*Science*, No. 163, 1969, p. 25–31.)
148. F. Polak, *The Image of the Future*, Amsterdam, London and New York, 1974, p. 231.
149. It is precisely this absence of a 'positive image of the future' that provided the starting point for theories operating with the vague, negative prefix 'post' which verbally covers up the failure of a positive solution. Speaking about the conception of the 'post-industrial society', D. Bell says: 'What these new social forms will be like is not completely clear. . . . The use of the hyphenated prefix post indicates that sense of living in interstitial time.' (D. Bell, *The Coming of Post-Industrial Society*, p. 37, 112, New York, 1973.) The fact that the concept of 'post-industrial society' is employed by nearly all the spokesmen of bourgeois society ranging from H. Kahn and D. Bell to T. Rozsak and I. Illich—although the usage reflects different interpretations—provides evidence that the real intention of this concept is only to establish a kind of 'counterweight' to the conception of socialist development.
150. R. L. Wolke, *Impact of Science on Society*, op. cit., p. xvii.
151. E. Jantsch, 'World Dynamics', *Futures*, No. 2, 1971, p. 167.
152. G. R. Taylor, *How to Avoid the Future*, London 1975, p. 105–18.
153. R. L. Heilbroner, *An Inquiry into the Human Prospect*, p. 21, London, 1975. In direct contrast to his earlier works, Heilbroner writes (p. 136): 'If, then, by the question: Is there hope for man? We ask whether it is possible to meet the challenges of the future without the payment of a fearful price, the answer must be: No, there is no such hope.'
154. A recent study by the World Bank (*Redistribution with Growth*, Oxford, 1974) which tried to specify the inequality trend in individual nations, presents a telling picture of

the differences in the incomes of social groups (differences between the incomes of 40 per cent of the poorest and 20 per cent of the richest): the socialist countries display the highest equality in the distribution of the national income (p. 7). The relevant ratio figures for Czechoslovakia, Bulgaria and Hungary are: 1 : 2.2 up to 1 : 2.8. In the United States of America, United Kingdom and Japan, they are in the range of 1:4; in Sweden, the Federal Republic of Germany, the Netherlands and India, 1:6.2 up to 1:7.2; in France, Brazil and Turkey, 1:11 up to 1:13, in South America and Rhodesia the ratios amount to 1:16 up to 1:20.

155. The disintegration of these 'two cultures' into adverse streams is described by C. P. Snow (*Public Affairs*, London 1971); C. P. Snow did not fail to notice that this gap is non-existent in socialist countries (ibid., p. 35).
156. The social sciences of this epoch, of course, did not start the moment that the proverbial 'cerebral hygiene' and the 'angelic influence' of Clotilda de Vaux led Comte to coin the term 'sociology', replacing the earlier 'social physics'. The wealth of the classical conceptual system of this epoch is represented by British political economy (from Smith to Ricardo), French political science (from Turgot to the bourgeois foundations of Saint-Simon) and German philosophy (from Kant to Hegel).
157. Smith, Ricardo and others subsumed all categories of political economy under the basic determination of the capital relation, and presented private ownership in its active form as the subject 'civilizing the world'. The same approach was adopted by the 'son of the French Revolution', Saint-Simon, who founded his theory of social order and evolution on principles governing the hierarchy of bourgeois ownership and who was fully justified in describing Guizot's attempts at defining the laws of the 'course of civilization' as the popularization of findings which he himself had published in *L'Organisateur*; the same, of course, applies to Comte's conception of positive 'order'. Likewise, Hegel's attempt to present a unified conception of the world in the form of the *curriculum vitae* of the '*Weltgeist*' was no more than a speculative expression of those material relations that mediate the circulation of capital.
158. *Nauchno-Tekhnicheskaya Revolyutsiya i Osobennosti Socialnogo Razvitiya v Sovremennuyu Epolhu*, Moscow, 1974.
159. Each individual area of the bourgeois system of social sciences assesses man and society 'using a different and opposite yardstick'. (K. Marx and F. Engels, *Collected Works*, Vol. 3, p. 310.)
160. This is how M. J. Roberts comments on the breakdown of the foundations of the social sciences in the West: 'The dominant conceptual systems in economics, sociology and political sciences all share a tendency to legitimate the status quo, which makes it difficult to see how the world could or should be other than marginally different from what it is. . . . Dissatisfaction with the state of affairs may well be the major reason for the resurgence of interest in Marxism as an alternative theoretical framework. . . .' ('Nature and Condition of Social Sciences', *Daedalus*, Summer 1974, p. 58, 59).
161. See J. Fourastie: *Les Conditions de l'Esprit Scientifique*, p. 250–1, Paris, 1966.
162. A. M. Weinberg, 'Can Technology Replace Social Engineering?' *Man-Made Futures*, p. 282, London, 1974. The social import of these questions is revealed by Weinberg's reply that although technology cannot completely replace social engineering, it can convert 'violent social revolution into acceptable social evolution' (ibid., p. 288).
163. T. Roszak, *The Making of a Counter-Culture*, p. xiii, New York, 1968.
164. C. Reich, *The Greening of America*, p. 394, New York, 1970.
165. Marx proved that bourgeois thinking has never got further than the opposite of

romanticism and hence romanticism will be a concomitant feature in the form of a 'legitimate counterpart' (*berechtigter Gegensatz*) to bourgeois conceptions right to their ultimate end. (Marx, *Grundrisse*, op. cit., p. 80.)

166. The stand of the Club of Rome on the *Limits to Growth* report in Tokyo, autumn 1973 (see *Technological Forecasting and Social Change*, Vol. 3, 1974, p. 260). This standpoint was already essentially contained in A. Peccei's study, *The Chasm Ahead.*
167. J. W. Forrester, *World Dynamics*, p. 2, Cambridge, 1971.
168. D. H. Meadows, D. L. Meadows, J. Randers and W. W. Behrens, *The Limits to Growth. A Report for the Club of Rome and Project on the Predicament of Mankind*, p. 24, New York, 1972.
169. M. Mesarovic and E. Pestel, *Mankind at the Turning Point, The Second Report to the Club of Rome*, p. 3, New York, 1974.
170. 'Neither this book nor our world model at this stage in its development can deal explicitly with these social factors. . . .' (D. H. Meadows, D. L. Meadows, J. Randers and W. W. Behrens, *The Limits to Growth*, op. cit., p. 46); the hitherto elaborated mathematical models of world development do not take into consideration the different trends operating in the development of the socialist countries.
171. M. Siebker and Y. Kaya, 'Towards a Global Vision of Human Problems', *Technological Forecasting and Social Change*, Vol. 3, 1974, p. 258.
172. A. Mileykovski, 'Novyi mif "Nulevoy Rost" i "Ekologicheskoye Obshchestvo"', *Novoye Vremia*, Vol. 49, 1972; J. Kuczynski, *Das Gleichgewicht der Null. Zu den Theorien des Null-Wachstums*, Berlin, 1973; H. S. Cole, C. Freeman, M. Jahoda and K. L. R. Pavitt (eds.), *Models of Doom. A Critique of the Limits to Growth*, Sussex, 1973. This note is likewise included in A. Peccei's comment on the two reports to the Club of Rome.
173. This, for instance, is a feature characteristic of the so-called Third Report to the Club of Rome, compiled by the 'Argentinian Team' sponsored by the Bariloche Foundation. Its scheme of slowing down the development of both the capitalist and the socialist countries in favour of the 'redistribution' of development presupposes, on the one hand, the release of resources still tied up with armament; on the other hand, this 'redistribution' approach is associated with social conceptions (towards an Alternative World Model: the Socio-Political Dimension) the realization of which would result in the defeat of the real forces capable of pushing through this release.
174. cf. e.g. G. T. Land in *The Futurist*, No. 1, 1975, p. 10; cf. also similar arguments in Land's book *Grow or Die. The Unifying Principles of Transformation*, New York, 1973.
175. 'Technology must be tamed', says A. Toffler in *Future Shock*, op. cit., p. 445. Similarly *The Futurist*, No. 6, p. 269, etc. But who, in actual fact, will be the 'tamer' under these conditions?
176. In itself, however, the evolvement of a system of 'technology assessment' will not provide a reliable social framework for this practical activity: 'Discussed in a political vacuum, technology assessment is merely in danger of being used to refine and prolong the existence of inequality, exploitation and inhumanity rather than . . . to improve the quality of human life.' (G. J. Stöber and D. Schumacher (eds.), *Technology Assessment and Quality of Life*, p. 287, Amsterdam, London and New York, 1973.)
177. J. J. Salomon, 'L'Avenir de la Science', *La Recherche*, Vol. 50, 1975, p. 922. The justified idea of reorienting science, of course, immediately induces the question: Who should reorient the scientific and technological development, how, and in what direction? (See *La Recherche*, Vol. 64, 1976, p. 158.)

178. R. Buckminster Fuller, *Utopia or Oblivion*, New York, 1969; N. Jequier. 'New Problems in Science Policy', *Management of Research and Development*, p. 166, Paris, 1972, raises the question of the application, in the relation between science and society, of something like a 'Marxist revolution'.
179. J. J. Salomon, *Science and Politics*, p. 246, London and Basingstoke, 1973.
180. E. F. Schumacher, *Small is Beautiful. A Study of Economics as if People Mattered*, p. 128, London, 1974. Adherents of the 'break-through' and 'toward stampede' are counterposed here to 'home-comers', i.e. people who are in search of their own soul, with a penitent religious undertone (ibid., p. 130).
181. I. Illich, *Tools for Conviviality*, p. xii, London, 1973. The term 'conviviality' is used here as an antonym to 'productivity' and should convey the idea that the 'accelerating crisis' of the present is caused by the failure to the machinery system; therefore, it is claimed, it should be countered by the 'retooling of society' (ibid., p. 10)
182. E. F. Schumacher admits that 'the choice of Buddhism for this purpose is purely incidental' (ibid., p. 43) and, hence, is one of the series of fashionable expressions hiding the prosaic absence of perspectives under poetic religious terms.
183. M. Guernier ('The Great Imbalance', *Human Futures*: *Needs—Societies—Technologies*, p. 40 ff., Rome, 1974) postulates the idea of the 'plurality of civilizations'; M. Guernier discerns an element of stability in the fact that Africans will have their 'own' civilization, e.g. at the 500 GNP per capita level, and the Americans and Europeans at a ten times higher level.
184. Not all people in the world can live as Westerners do today. Our earth is not rich enough to permit this. (cf. E. Laszlo, *The Systems View of the World*, p. 103, New York, 1972; similarly E. F. Schumacher, *Es geht auch anders Jenseits des Wachstums*, p. 17 ff., Munich, 1974.)
185. A. M. Weinberg, rejecting the catastrophic prognosis that the living standards of working people in the West could drop to one-tenth of the present level, nevertheless admits that a drop by one-half could become a necessity (*Science and Public Policy*, No. 7, 1974).
186. G. Skorov (ed.), *Razvivayushchiyesya Strany i Nauka, Tekhnika, Ekonomicheskiy Rost*, p. 353, Moscow, 1975; A. Y. Shpirt, R. T. Aframovich and N. I. Lukyanova (eds.), *Nauchno-Tekhnicheskaya Revolyutsiya i Razvivayushchiyesya Strany*, Moscow, Nauka, 1973.
187. See, e.g., the World Plan of Action for the Application of Science and Technology in Economic Development (of the developing countries) elaborated by the United Nations in 1971, or the conclusions of the panel of experts convened by A.-Mahtar M'Bow, Director-General of Unesco, in 1975.
188. Thus the use of fairly expensive, highly automated production technology, which absorbs a relatively small section of the labour force, may in some cases, due to spontaneous application, result in the well-known paradox: the retarded process of the transference of manpower from agriculture to industry aggravates the situation of the 'overpopulated village', while the results of the latest industry continue to form the basis of the 'consumer life' of the narrow local élite.
189. According to the Ford Foundation's Energy Policy Project, the crisis in power resources observable in certain parts of the world is rooted in the fact that over the past thirty years the major capital groups in the West have not been interested in seeking new power resources (see *Technological Forecasting and Social Change*, No. 3, 1972, p. 242). In particular, this entails a gross underestimation of the application of solar energy, which is of special significance for Asian and African countries.
190. See A. Rahman and K. D. Sharma, *Science Policy Studies*, Bombay/New Delhi,

1974. As a result of certain social circumstances, the mode of education and systematic 'brain-drain', as well as other factors, the research potential at the disposal of the developing countries may sometimes act rather as an appendix of major American and West European programmes.

191. *Human Futures: Needs—Societies—Technologies*, op. cit., p. 156.
192. This metaphorical phrase by A. Gregg, published in *Science*, No. 121, 1955, p. 681, was chosen as the motto for M. Mesarovic and E. Pestel's report to the Club of Rome. (*Mankind at the Turning Point. The Second Report to the Club of Rome*, op. cit., p. 1). With every passing year, it is quoted more and more frequently in social science literature, occurring in both sociological and psychological analyses and historical and economic reflections.
193. 'We need more social science fiction, more social science imagination as well as technological imagination', insisted D. Riesman ('Leisure and Work in Post-Industrial Society', *Mass Leisure*, p. 383, Glencoe, Ill., 1958).
194. H. Brooks points out that 'after a brief period of public belief in the promises of social sciences . . . the climate for them also has deteriorated'. ('Are Scientists Obsolete?', *Science*, 11 August 1974.) In *Science Policy and Business. The changing Relation of Europe and the United States*, Boston, 1973, Brook is convinced that the system of social sciences in the United States is relatively more advanced than its European conterparts. On the other hand, A. M. Weinberg stresses that 'unfortunately our society has relatively little experience or tradition for large-scale support of social sciences.' (A. M. Weinberg, 'In Defence of Science', *Studium Generale*, No. 9, 1970, p. 804.)
195. 'We propose the following provisional programme: the world can probably best deal with its problems and dangers by emphasizing, as much as seems feasible "planned muddling through".' (*Prospects for Mankind*, p. 59, New York, Hudson Institute, 1972.)
196. All that social sciences can do in a situation like that is to meditate upon 'how great the hazards can be in trying to handle problems in the modern world simply by trying to muddle through . . .'. (C. B. Kaufman, 'The Future Social Climate for R and D', *Research Management*, No. 1, January 1975, p. 30.)
197. S. P. Tapeznikov, *Obshchestvenniye Nauki—Moguchiy Ideyniy Potentsial Kommunizma*, Moscow, 1974.
198. In its classical form, English political economy was conceived as a science concerned with general interest; French political science as the theory of social relations controlling formally equal individuals; and German classical philosophy as the speculation on the advances of abstract bourgeois reason.
199. 'Nations and orders have never learned anything from history': this is how G. W. F. Hegel ('Vorlesungen über die Philosophie der Geschichte', *Werke*, Vol. XXVI, p. 9, Berlin, 1848) disproved the ancient view; according to Hegel, the development of society is governed by the laws of the objective spirit (*Weltgeist*) which, through their operation, pronounces its 'judgement' (*Weltgericht*) over people's lives.
200. The original slogan of 'positivist' social science '*prévoyance, d'où action*' (A. Comte, *Philosophie Positive*, Vol. 1, p. 38, Paris) gradually changed into '*prévoir, d'où régler*', but at the same time acquired a shape which J. S. Mill described as a kind of regulation obsession.
201. W. Dilthey, 'Der Aufbau der geschichtlichen Welt in den Geisteswissenschaften', *Gesammelte Schriften*, Vol. VII, p. 218, Leipzig, 1927.
202. Marx predicted that this very form would be the final product of the disintegration of bourgeois social sciences (*Theorien über den Mehrwert*, Vol. II, p. 573, Stuttgart, 1921).

203. 'Socialism represents the formative stage of a new type of historical process involving a gradual, increasingly fuller control of the conditions of the common social life by people themselves.' (See: *Sotsialisticheskoye Obshchestvo. Sotsialno-filosofskiye Problemy Sovremennogo Sotsialisticheskogo Obshchestva*, p. 128, Moscow, 1975.)
204. H. Schelsky describes the concept of the planning of the future as a 'deceptive abstraction' of those who refuse to 'participate in the contemporary life of the institutions' of bourgeois society (H. Schelsky, 'Sociologisches Plannungsdenken über die Zukunft', *Universitas*, No. 12, 1970, p. 1552). The extent to which this attitude is rooted in the economic literature of the West was revealed by the opposition Keynes's theory met with, when he departed from the idea of the spontaneous course of national economy and subscribed to the idea of economic control, comparing, though, the role of social sciences to something like dentistry.
205. 'The historical course of the atomic age cannot be curbed or controlled either by the individual or by a group of people, by a commission of statesmen, investigators, technicians, however prominent they may be, nor by conferences of economic and industrial representatives. No human organization is capable of controlling this age.' (cf. M. Heidegger, *Gelassenheit*, p. 20–1.)
206. K. R. Popper, *The Open Society and its Enemies*, London, 1945.
207. O. Helmer, *Social Technology*, New York, 1966.
208. B. F. Skinner. *Beyond Freedom and Dignity*, New York, 1971.
209. J. W. Forrester based the entire project of world models of development explicitly on this generalization of the principles of 'industrial dynamics' (*Industrial Dynamics*, p. 45, Cambridge, 1961).
210. 'Social engineering' is described in the survey of the social sciences elaborated by Unesco (*Main Trends of Research in the Social and Human Sciences*, Part 1, p. 32, Paris/The Hague, 1970) as a science which has grown immensely in scope, responsibility and status in this age characterized by an unprecedented technological progress.
211. In their survey *Main Trends of Research in the Social and Human Sciences*, the authors of chapters on economy, sociology and psychology stress the necessity of broadening the scope of inquiry of the social sciences into a comprehensive theory of social development, which twenty years ago was described by T. Parsons as 'probably' unfeasible (*The Social System*, p. 524, Glencoe, Ill., 1951). J. Piaget calls for the overcoming of the 'tragic compartmentalization' of the social sciences and of their separation from natural sciences. It is obvious that the need for the inner unity of science, stressed by Marx, is becoming one of the major topical tasks of the present.
212. 'In this sense socialism is synonymous with the commencement of the formation of a new type of historical process, the essence of which consists in the gradual and increasingly more viable mastery of the conditions of common social life by the people themselves.' (*Sotsialisticheskoye obshchestvo* . . . op. cit., p. 128, Moscow 1965.)
213. Replying to the question of the intensification of social science research, the physicist E. P. Wigner says: 'The social sciences are not yet ready to benefit from the methodology of big science. That is to say, significant material support of the social sciences would advance them little.' (*Impact of Science on Society*, No. 4, October–December 1972, p. 285.) All this formulation tells us is the simple fact that the lack of knowledge in the social sciences in the West reflects deeper, social causes.
214. As early as the 1960s, a special sub-committee of the United States House of Representatives stated that striking disproportions became apparent between the natural and the social sciences in that country (cf. *Science*, No. 3692, 1965, p. 40–2, 84). This state of affairs—perhaps except for Japan and a few other countries—still

prevails in the capitalist countries (see A. Shonfield, 'The Social Sciences in the Great Debate on Science Policy', *Minerva*, No. 3, 1972, p. 341; similarly, A. Mayne, 'Science Policies for a World of Crisis', *Science and Public Policy*, No. 1, 1975, p. 39). Quite different is the position and development of social sciences in the socialist countries, where these disciplines absorb 18–32 per cent of the total volume of research; the development of economic and philosophical sciences in the Union of Soviet Socialist Republics outdistanced as to its rate of growth the development of natural and technical sciences in the 1960s (see D. M. Gvisniani, S. R. Mikulinski, S. A. Kugel (eds.), *Nauchnotekhnicheskaya Revolyutsiya i Izmeneniye Struktury Nauchnych Kadrov SSSR*, Moscow, 1973).

215. Toffler, *Future Shock*, op. cit., p. 446.
216. See N. Fedorenko and his criticism in a Unesco publication, *The Social Sciences: Problems and Orientations*, p. 67–92, Paris/The Hague, 1968.
217. See *Technology Assessment. Understanding the Social Consequences of Technological Application*, New York, 1972.
218. P. Drucker, *The Practice of Management*, p. 4, New York, 1954.
219. cf. D. M. Gvishiani, *Organizatsiya i Upravleniye*, Moscow, 1971.
220. This entails the requirement of general participation in the advances of science and technology, in the process of innovation (see F. Valenta, *Tvurci Aktivita—Inovace—Efekty*, Prague 1969).
221. This fact was already predicted by V. I. Lenin (cf. *Polnoye Sobraniye Sochineniy* (Moscow), Vol. 45, 1964, p. 174.
222. Contrary to conceptions viewing science as a pursuit for a narrow élite, P. N. Fedosyeyev proved that under the conditions of the scientific and technological revolution, the role of the masses in this process increases (see P. N. Fedoseyev, *Dialektika Sovremennoy Epokhi*, p. 226–7, Moscow, 1975).
223. The continuing disintegration of the traditional conception of industrial management in the West, the mounting prevalence of 'muddling through' projects and 'contingency' or 'situational' theories of management have become one of the characteristic features of the day.
224. E. Lazslo, *A Strategy for the Future: The Systems Approach to World Order*, New York, 1974.
225. O. Reiser, *Cosmic Humanism and World Unity*, New York, 1975.
226. 'The separation of ourselves from the object of our concern', wrote E. Jantsch commenting on these reflections, 'will never get us there because, as members of human society, we are subjects and objects of evolution at the same time.' (*Futures*, No. 3, 1975, p. 250.)
227. D. Gabor calls for an open rehabilitation of the 'élite' idea; for only this, he contends, can implement the restrictive programme recommended by 'social technology' (*The Mature Society*, 1972).
228. As Marx and Engels pointed out, utopias, not based on objective conditions, always contain an element of inherent despotism. Understandably, attempts at materializing the biggest utopia of our age, i.e. the utopia envisaging the eternalization of monopoly capital, are associated with more and more drastic forms of violence.
229. This gloomy anticipation of the scientific and technological revolution in the West is shared by an ever growing number of specialists from different branches of the social sciences: R. Dahrendorf, K. Lorenz, J. Habermas, J. B. Douglas, K. Borchardt and others.
230. H. J. Krysmanski, *Soziales System und Wissenschaft*, Gütersloh, 1967.
231. J. Habermas, *Zur Logik der Sozialwissenschaften*, Frankfurt am Main, 1970.
232. Husserl, *Die Krisis der europäischen Wissenschaften*, op. cit., p. 5.
233. See, e.g., the Stanford System (*Changing Images of Man*, 1974) or the Michigan

Model (Kan Chen and K. F. Lagler (eds.), *Growth Policy—Population, Environment and Beyond*, 1974).

234. C. Reich, *The Greening of America*, p. 225, New York, 1970. In a society employing science as a social productive force, the isolated 'private world' must necessarily appear as paradoxical as it appeared to Pisistratus and the ancient world in general: '*privatus*' is simply he who has been deprived, excluded from sharing in the wealth of the community and public power.
235. N. Luhman in *Theorie der Gesellschaft oder Sozialtechnologie*, p. 319, Frankfurt am Main, 1971.
236. J. Habermas, *Dialektik und Hermeneutik*, Vol. I, p. 99, 1970; similarly in: *Theorie der Gesellschaft oder Sozialtechnologie*, op. cit., p. 213.
237. Contrary to this assertion, A. D. Ursul (*Problemi Informatsii v Sovremennoy Nauke*, p. 216–17, Moscow, 1975, stresses the primary significance of the basic social relations for the nature of communication processes. See also T. J. Oizerman, *Sotsialnaya Filosofia Frankfurtskoy Shkoli*, p. 360, Moscow, Nova Mysl, 1975.
238. S. Trapeznikov, 'Leninizm i Sovremennaya Nauchno-tekhnicheskaya Revolyutsiya', *Nauchno-tekhnicheskaya Revalyutsiya i Socialnyi Progress*, p. 10, Moscow, 1972.
239. See F. Kutta and M. Soukup, *Vedeckotechnicka Revoluce a Rizeni*, Prague, 1973.
240. G. M. Dobrov characterizes this situation as the transformation of science into a 'controllable system' (see 'Science Policy and Assessment in the Soviet Union', *International Social Science Journal*, No. 3, 1973, p. 322).
241. L. M. Gatovski, *Nauchno-tekhnicheskiy Progress i Ekonomika Razvitogo Socializma*, p. 289, Moscow, 1974.
242. The economy of time, as G. F. Afansyev points out while developing Marx's idea (*Nauchnoye Upravleniye Obshchestva*, p. 146–1947, Moscow, 1973), is an essential economic form of the intensive development of national economy; without its application none of the 'global' problems concerning the dialectics of goals and means in economic development can be solved.
243. cf. I. I. Changli, *Trud, Sotsiologischeskiye aspekty Teorii i Metodologii Isledovaniya*, Moscow, 1973.
244. In a socialist society, 'the scientific and technological revolution provides the material basis for the development of social relations' and, through them, the development of man's activities, faculties and abilities. (V. G. Afanasyev, *Nauchno-tekhnicheskaya Revolyutsiya, Upravlenie, Obrazovanie*, p. 187, Moscow 1972.)
245. The growing significance of the role of subjective factors of social development in socialist societies does not consist in the diminishing import of objective laws, but on the contrary in their conscious application—thus this growing significance of subjective factors is in itself an objective law. (E. Hahn in *Objektive Gesetzmässigkeit und bewusstes Handeln in der sozialistischen Gesellschaft*, p. 44–5, Berlin, 1975.)
246. V. G. Lebedev and V. I. Kushlin, (eds.) *Nauchno-tekhnicheskaya Revolyutsiya i Preimushchestva Sotsialisma*, Moscow, 1975.
247. See T. Timofeyev in *Rabotiy Klass—Glavnya Revolyutsionnaya sila Sovremennosti*, Moscow, 1977.
248. T. S. Stepanian (ed.) *Leninizm i Upravlenie Sotsialnymi Procesami pri Sotsializme*, p. 47, Moscow, 1973.
249. In this delimitation, R. S. Mikulinski integrated the entire new complex of the functions of science in a socialist society to the function of moulding a world outlook of a new type (similarly G. Kröber and H. Laitko, *Wissenschaft als soziale Kraft*, Berlin, 1976).
250. It is only the workers who are in a position to '*ihre vollständige, nicht mehr bornierte*

Selbstbetätigung, die in der Aneigung einer Totalität von Produktivkraften und der damit gesetzten Entwicklung einer Totalität von Fähigkeiten besteht, durchzusetzen'. (Marx and Engels, *Werke*, Vol. 3, op. cit., p. 68.)

251. K. Kröber and H. Laitko (eds.), *Wissenschaft. Stellung, Funktion und Organisation in der entwickelten sozialistischen Gesellschaft*, p. 49, Berlin, 1975.
252. V. G. Afanasyef, *Socialnaya Informatsiya i Upravleniye Obshchestvom*, Moscow, 1975. In this approach, social sciences under socialism diametrically differ from all the trends deriving from bourgeois tradition, which in one way or another regard the abstract individual as the only subject of social cognition and social control.
253. See V. V. Krevnevich, *Vliyaniye Nauchno-tekhnicheskogo Progressa na Izmeneniye Struktury Rabochego Klassa v SSSR*, Moscow, 1971.
254. T. Jaroszewski in *Czlowiek, Socjalizm, Rewolucja Naukowo Techniczna*, Warsaw, 1974.
255. B. Filipcova and J. Filipec, *Ruznobezky Zivota*, Prague, 1976.
256. See S. V. Shukhardin, *Nauchno-tekhnicheskaya Revolyutsiya i Obshchestvo*, p. 18, Moscow, 1973.
257. Traditional conceptions of 'scientific management' consider many of the present-day world-wide problems insoluble because they do not reckon with the development of the subject. Heilbroner's thesis that, today, the Promethean man has to relinquish the idea of his own development is, therefore, necessarily a programme of self-destruction.

Part Two

Philosophical problems of technological advancement

A philosophical interpretation

P. N. Fedoseyev

Academician, Academy of Sciences, USSR

The scientific and technological revolution (STR) has enlivened and sharpened interest in the classical problems of philosophically understanding science and practice. At the same time it has given and continues to give rise to new, specifically philosophical problems of scientific and technological progress.

First and foremost, the STR has considerably broadened the historical context in which the philosophical problems of science, technology and production is formed, and in which the interpretation of the interrelation between man, society and nature is worked out. The STR has integrated various principles, aspects and factors of the historical process and become a sort of crossroads and point of active interaction of philosophical, natural-scientific, social and humanistic knowledge.

Modern science and technology have given mankind powerful instruments. Small wonder that people today are more and more frequently saying that the scale of man's influence on nature may with all justification be compared to geological forces. There has, therefore, never been such immense interest in science, among people of all walks of life as today, in everything linked with its development when the power of scientific knowledge is unfolding with unparalleled clarity.

However, the gigantic forces that man now has at his disposal may be used and are being used both for the welfare of society and to its detriment.

For that reason, among the cardinal social and, at the same time, humanistic tasks mention must be made, first, of the task of foreseeing the social, economic and ecological consequences of human activity; and second, the task of mastering, controlling and directing the process of scientific and technological development.

The protection and improvement of the life of man and the

environment has become the most general and important problem of the varied world of today. Scientists, more than anyone else, are aware of the dangers threatening man and nature in the conditions of the STR. And it is only natural that today they are focusing their attention on a problem vital to mankind—that of averting a thermonuclear catastrophe and ensuring a lasting peace. The deepening of international détente is linked with the solution of such problems as protection of the environment, the rational and fair use of natural resources, co-operation in science, technology and culture.

Conscientious research workers cannot ignore the striking contrasts of the twentieth century when the fruits of modern civilization are actually enjoyed by a comparatively small part of mankind, when elementary living conditions are still not ensured on the greater part of the globe.

This is a time when the atmosphere of business-like co-operation, positive discussion of major problems of the day, is becoming the prevailing trend in international scientific intercourse. There is no doubt that this trend is in many respects connected with the general process of détente and with the consolidation of what is now called the 'spirit of Helsinki'.

Scientists in all countries are concerned with how to make the grand achievements of the STR serve all mankind.

Our experience shows that the development of science and technology can only be directed, on the one hand, on the basis of knowledge of the inner laws of scientific and technological development and, on the other, of a profound understanding of the link between scientific and technological progress and social progress.

It is only on the basis of a comprehensive, genuinely synthetic approach to an assessment of the STR in its inseparable connection with fundamental social processes that it is possible adequately to determine its essence and its historical importance.

The principal characteristic of our interpretation of the STR lies in recognition of the regularity according to which the revolution in spiritual and material production is linked with the social revolution.

In their analysis of the industrial revolution of the eighteenth and nineteenth centuries, Marx and Engels showed its connection with social revolutions, the series of which culminated in the emergence of the new, capitalist mode of production. The dialectics of the modern epoch manifested itself in the fact that in a number of countries the socialist revolution had preceded the STR, thereby ensuring the necessary social conditions for its implementation. In its turn, the STR serves in these countries as the means for developing and multiplying the achievements of the social revolution. In other countries, however, the STR anticipates the social revolution, prepares the material conditions for it and aggravates the main contradiction of capitalism. This distinction in the sequence of the STR and the social revolution is one of the concrete manifestations of our epoch's main contradiction. In it is manifested the diversity of the historical process of our day. However, in both these variants of social development the STR is part and parcel of the creation of the material prerequisites of community society. The revolution in the productive forces is giving rise to a new material and technical basis adequate to the communist mode of production. Therein lies the unity of the historical process today.

The monistic interpretation of this unity has found its theoretical and practical expression in the formula of the development of socialist society that has determined the task of organically combining the STR with the advantages of socialism.

The problem of directing the development of science and technology and giving society control over the utilization of their achievements requires an all-sided analysis and generalization of the laws governing scientific and technological development and an elucidation of the role and significance of the different factors determining present-day scientific and technological progress.

It is important to have a profound and comprehensive understanding of these issues not only for purely intellectual consideration but also for working out the strategy of scientific development, for correctly selecting the main directions of scientific quests, for resolving the problem of planning and managing scientific work.

The STR is characterized by the fusion of two currents—the scientific revolution and the technological revolution—and represents a qualitative change in the development of society's productive forces.

The scientific revolution is the focal element and most vivid expression of progress in science. The problem of the character, essence and structure of the scientific revolution attracts considerable attention among scientists and the public at large. However, the essence of this revolution can only be laid bare on the basis of a definite philosophical concept of science.

Naturally this has led to the emergence and spread of various theories each claiming to be the 'philosophy of science', to establishing and unravelling the laws and factors determining its progress. It is noteworthy that positivist theories, only recently widespread and in vogue, no longer have the trust of scientists. Likewise, scientists reject positivist attempts to repudiate the philosophical understanding of science on the pretext of rejecting the metaphysics of 'eternal' problems and the 'traditional' methods of philosophical thought and turn the logico-methodological problems of philosophy over to individual sciences. The attempt to convert philosophy into an analogue of a concrete science, to strip it of its philosophical content proper and use it as an ancillary discipline serving science, has proved to be of no avail. Despite all its claims, positivism has been unable to give an adequate explanation of the emergence of new knowledge and of the essence of science as a social phenomenon, and to bring to light the actual laws of its progress. The 'halo of intellectual respectability', which Bertrand Russell spoke about relative to positivism, no longer exists in the eyes of the overwhelming majority of scientists and philosophers seriously studying epistemological problems. The crisis of the positivist theory of science has today become a reality.

The attempt of linguistic philosophy to accomplish a revolution in philosophy, to convert it into an analysis of the language of science, has not been successful either. On the one hand, it is part of the above-mentioned tendency to deprive philosophy of its ability to assess science and influence its development. In this tendency, it must be noted, there has been an attempt, deliberate or not, to evade such assessments and, at the same time, ignore social regulative influences on science, to remain under the protection of romantic prejudices about scientific knowledge being possibly independent of society.

In the light of the complex and contradictory progress of scientific knowledge, the theories that regard scientific development as a simple linear cumulative process have likewise proved to be untenable.

Our analysis of the development of science, particularly of the essence, character and structure of scientific revolutions, is founded on the philosophy of dialectical materialism. The content of this philosophy consists of the teaching about the most general laws of the development of nature, society and human knowledge. With this are connected three basic functions of dialectical materialism relative to science.

First, the function of generalizing knowledge and bringing to light the underlying general principles. The purpose of each science is to uncover laws, and it, therefore, engages in generalization because laws are deduced from a comparison and collation of date. A specific feature of philosophical generalizations is that they are of a universal character. They embrace quantitative and qualitative changes, evolutionary and revolutionary, the identity, distinctions and contradictions of phenomena and processes.

Second, philosophy is the means of synthesizing knowledge, the methodological centre of communication and reciprocal influence of different scientific disciplines. Some fundamental categories of the cognition of the world develop as categories of philosophy and natural science, such as matter, space, time, motion, causality, quantity, quality and so on. While neo-Kantian philosophy draws a dividing line between the natural and social sciences, dialectical materialism links all spheres of knowledge, bringing to light their common foundations. The objective process of the integration of knowledge generated by the requirements of the modern epoch and, in particular, by those of the STR is given adequate expression in the philosophy of dialectical materialism. On the basis of the interaction of the natural, technical and social sciences and philosophy, scientific complexes are today vigorously taking shape for the study of the biosphere, socio-economic development and the problems of man.

Third, the scientific-heuristic role of philosophy. We regard the philosophical interpretation of scientific data to be knowledge of the ever deeper and more general links and forms of interdependence of phenomena. This philosophy does not claim to resolve specific problems of concrete sciences, but it develops in close contact with the latter, reveals in its concepts and categories the logic of scientific cognition, a process that is profoundly dialectical in essence and content and serves as the methodological foundation of the Marxist concept of scientific development.

The main features of this concept may be summed up as follows:

It analyses the entire range of social factors and relations determining the motive forces and laws of scientific development. But our comprehensive approach rules out the eclectic theory of the equal operation of many factors, recognizing the decisive role of social productive activity in the development of all aspects of man's social life, including scientific development. The necessary stimuli and conditions were created in the process of such activity as in the harnessing of nature, and science emerged as a social phenomenon. But, in the final count, the world outlook and the way of thinking typical of a given epoch, and the corresponding structure of science, its principal range of problems and its methods of resolving scientific problems, were worked out on the same basis.

From these positions the Marxist concept of science considers the role and significance of the various factors influencing the process of scientific cognition and the work of individual scholars and teams of scientists, being guided by the principle of the concreteness of truth, which is the main principle of dialectical logic.

The Marxist concept of science attaches great significance to the problems of the inner logic of the development of knowledge, to questions of the relative independence of the development of the forms of man's spiritual activity. In the process of scientific cognition innumerable internal contradictions arise in science itself, for instance, the contradiction between established theories and the data of an experiment, between new and old concepts and ideas.

The study of inner contradictions of this kind in science is of great importance for understanding the ways and laws of the development of scientific knowledge, including the mainsprings of scientific revolutions.

In bringing to light the objective logic of the movement of scientific knowledge, the Marxist concept of science shows the dialectical link of continuity and discontinuity, continuity and revolutionary transformations in scientific progress.

The problem of scientific revolutions, which today attracts so much attention, is neither sudden nor new to Marxist philosophy. As far back as the turn of the century Lenin studied the main features of the modern revolution in natural science.

One of the cardinal results of trying to understand the revolution in natural science was the conclusion that the progress of scientific knowledge is subordinated to general dialectical laws, which manifest themselves in a specific way in scientific development. Scientific revolutions take place naturally and are prepared by the preceding evolutionary, gradual development of knowledge. Whatever the duration of the evolutionary transformations of scientific knowledge, they inevitably give way to a revolutionary break-up of established views and traditions, and are again followed by evolutionary development that prepares the ground for another revolution.

An analysis of the history of science shows that its modern state is the result of a whole series of scientific revolutions that have been breaking out ever since the emergence of natural science.

Scientific revolutions are landmarks on the road of scientific progress. During these revolutions the creative aspects of human thought manifest themselves most strikingly, the most fundamental theoretical generalizations are made and the very method of scientifically explaining and seeing the world changes. In one way or another, the post-revolutionary development of science is nourished by the juices of the preceding scientific revolution.

But this interpretation of the correlation between evolutionary and revolutionary processes in scientific development is by no means accepted by everybody. The revolutionary and evolutionary periods in science are at times much too sharply drawn away from one another and are counterposed to one another. For instance, in Thomas Kuhn's *The Structure of Scientific Revolutions*,[1] which is widely read, the question of the continuity of scientific knowledge is linked solely with the evolutionary periods of scientific development. The link between the evolutionary period in the development of science and the scientific revolution is essentially lost.

When the inter-revolutionary periods in the development of science are reduced to an algorithmic solution of theoretical and practical problems, it is difficult to understand the scientific revolutions themselves, for then the origin of the new ideas that ultimately lead to a revolution remains unclear. It turns out that these ideas arise not as a result of the natural development of knowledge but, so to speak, out of nothing.

Scientific revolutions are inconceivable without preparatory work during the evolutionary stage of the development of knowledge. However tranquilly the evolutionary development of knowledge proceeds, it is accompanied by the discovery of phenomena which are hard to explain from the standpoint of prevailing theoretical concepts. The discovery of these phenomena leads to the penetration into science of the ideas and methods of 'neighbouring' spheres of research, to a quest for new experimental methods. Existing theoretical ideas are enriched with concepts that may subsequently clash with them.

Revolutionary ideas are born in diverse spheres of natural science. But in one way or another they spring from a cognitive situation taking shape in the whole of natural science. A scientific revolution is a means of settling the contradictions in natural science; it is linked with a struggle between scientific schools, with the change of interdisciplinary boundaries, with the restructuring of the organization of science. An analysis of the nature of scientific revolutions clearly brings to light the fact that science is an integral formation, whose different components are linked and interdependent.

In considering the development of science, attention is sometimes drawn solely to the inner 'mechanisms' of scientific creativity, while the social and cultural-historical context of the genesis of scientific ideas is ignored. This approach is obviously inadequate for explaining the nature of scientific revolution.

The essence of the Marxist philosophical interpretation of the STR lies in showing not only its impact on society and man but also the decisive influence of society and man on the STR. In Marxist literature society and man are shown not only as an object of the STR but also as its subject, as its creative and driving force. Here the processes of the STR are characterized not only from the standpoint of what society undergoes as a result of the STR but also in the context of what society and man do for the purposeful development of the STR. Moreover, the Marxist approach to the STR accentuates not the tasks that the STR sets society[2] (such accentuation separates the STR from society, places it over society as a force that does not belong to society) but what tasks society and its need for further economic, social and spiritual development put before science and technology and determine the revolutionary character of their development.

In their efforts to work out theoretical concepts of the historical development of science many Western scientists stress that the content of the problems being resolved by society is determined by science itself, that available scientific knowledge shows the direction in which science must advance and what cognitive problems it must work on. This determination of the prospects for cognition by knowledge already obtained, of scientific problems by problems already resolved, exists of course. But it is by no means the only factor. There is also the dependence of new problems of cognition on the character of the tasks maturing in social practice.

However, the existence of this dependence is entirely ignored by many Western scientists. They absolutize the inner logic of scientific progress and put forward and uphold the concept that science and scientific progress are autonomous. According to this concept, science does not depend on society and its problems. In this concept the immanent link of science with social factors and the social context of its development is doggedly denied in spite of, one might say, the fact that the existence of this link is absolutely obvious. In the autonomous concepts of the functioning and development of science propounded by Polanyi, Hagstone, Storer and other historians of science, scientists are regarded as people guided solely by the immanent impact of scientific ideas and the inner logic of scientific study.[3]

The viability of this viewpoint was shown, for instance, by the proceedings of the international conference on the social impact of modern biology (London, 1970).[4] At that conference scholars of the older generation, the eminent scientists M. Wilkins, D. Bohm, G. Monod, M. Kaplan and J. Bronowski strongly upheld the concept which had, it would seem, been disproved by life, that science is an autonomous institution both nationally and internationally, that it is a sort of 'State within a State' capable of dictating the terms for using its achievements. Bronowski went so far (in a quite utopian fashion) as to urge a total demarcation between science and governments in all countries, the disestablishment of science in the sense in which the churches have been disestablished and have become independent of the State.[5]

Of course, under the influence of reality this viewpoint is growing increasingly archaic. Many Western scholars, particularly of the younger generation, today consider that without analysing the processes of scientific development in the general context of socio-historical development, it is no longer possible to understand and describe phenomena such as the formation of many state institutions designed to direct science, and the enhancement of the dependence of the institutions of science on factors linked with politics, financing, the economic situation, manpower resources, and so forth.

This is evidence that even in Western theoretical thought the prevalence of science's autonomous status in society no longer goes unchallenged. However, the concept that recognizes *solely* science's institutional dependence on society is gaining currency. It recognizes science's dependence on society only as regards the organizational form of the functioning and development of science. They believe that society influences scientific progress in this and in no other way.

However, a comprehensive analysis of the STR proves that science is a product of social development not only as regards the organizational forms of its existence but also in its content and, ultimately, even as concerns the theoretical and cognitive forms of its development.

By and large, the dependence, the objective law under which mankind passes from resolved problems of scientific cognition to new problems (both in content and, ultimately, in form) is dual. On the one hand, it is internal, maturing in the very 'body' of science; on the other hand, it is determined and stimulated by the processes of social practice. The development of science signifies not only an evolution from what we already know but, at the same time and to an even larger extent, an evolution to what we want to know. It signifies the dialectical unity, the link and the mutual conditionality of both these 'alternatives'. For that reason Kuhn's striving to release science from external stimulation, from practical aims, comes into conflict with the objectively existing determination of the content of scientific knowledge by the needs and problems of human progress in the sphere of practical knowledge of reality.

When the structure of the STR is analysed, the question naturally arises of the correlation between its two components—science and technology. The change in the nature of these components is universally known. On the one hand, there is the 'scientification' of production, a process that fills it with knowledge and turns it into the 'technological application of science', from a simple process of labour into a scientific process.[6] On the other hand, science itself—in this case the natural sciences—becomes technological, i.e. it rests on the technical basis of experiment, on the experimental-production basis of the laboratory and is filled with concrete executive work. The mutual penetration of science and technology is particularly apparent at the stage of the embodiment of scientific discoveries in production, this being the key phase of scientific and technological progress. Needless to say, these changes do not erase the functional distinctions between scientific and direct productive labour, or the social distinctions between its subjects. However, this refers to the mainsprings of the revolution in science and technology, to the succession of the links of scientific and technological progress.

In the case of the industrial revolution of the eighteenth and nineteenth centuries, this starting point was the development of the new technique, namely, working machinery. But in the epoch of the industrial revolution the development of machinery which revolutionized production was evidently the result mainly of an empirical quest. In the modern technological revolution the development of new machinery is the result initially of scientific research and then of development. Consequently, it is not machinery but science that is the initial element with a revolutionizing influence on production. This enhances the significance of fundamental science, which gives rise to fundamentally new, abstract-theoretical discoveries and ideas.

From this is hastily drawn the basically liberal-technocratic social theory that elevates the élitist concept of the primacy of scientists in the future 'post-industrial society'. The reactionary Utopia of the right wing of the technocratic theories about the future, alongside the élitist theories about the domination of a selected managerial-scientific meritocracy, underscores the primacy of machinery in the social process: such are the theories about the totally automated machinery of social development propounded by Z. Brzezinski and others.

But in the eighteenth century, too, science played a large part in the development of machinery and industry. It is considered, not without grounds, that in the eighteenth century knowledge acquired scientific form, that it was systematized in the shape of laws and principles and became science.

The most important thing is that in resolving the question of the motive forces of the scientific and technological revolution Marxism sees the way out of the dilemma of the science-technology contradiction in laying bare its social roots. As seen by Marxism, the revolutionizing of production is called forth by the need for new knowledge and a new technical solution of the problems of development posed by the social process. Social practice is the driving force of the STR. Engels wrote:

> If society feels a technical need, this moves science forward more than a dozen universities. The whole of hydrostatics (Torricelli and so on) was brought to light by the need to regulate the mountain streams in Italy in the sixteenth and seventeenth centuries.[7]

The history of the development of science produces a mass of examples where problems arising in practice lead not only to the elaboration of new theories in established branches of science but also to the creation of fundamentally new areas of scientific knowledge. Suffice it to recall, for instance, that thermodynamics emerged in response to the practical need for improved steam engines, and that the scientific chemistry of the eighteenth century was the answer to the practical problems linked with combustion, dyeing and other chemical processes that in the course of the industrial revolution of the eighteenth century became the objects of practice and production.

The need for space machinery gave an unparalleled impetus to the development of mathematics, physics, chemistry and biology. In their turn the successes in space exploration opened up new possibilities for astronomy, geology, geography and other fields of knowledge.

We could also mention the complex order and life-giving impulse received by mathematics and various departments of physics and chemistry from computer technology.

The Marxist answer to the question of the role played by social practice makes it possible to remove the division between science and technology as a pseudo-problem and raise the question of a dialectical unity between science and practice, the overlapping and conventionality of the boundaries between them (in the form of the interaction and mutual penetration of science and technology, scientific research and production, theoretical, managerial and executive work, the productive and reproductive elements in all forms of labour, and so on). This approach permits, without making absolute any aspect of this unity, a successive study of the stimulating impact of technology on science in resolving problems of production, and also of science on technology when scientific knowledge requires the material embodiment of its discoveries and a technical solution of its problems.

The STR has confronted mankind with many fundamentally new problems, which it had never encountered before or which it had encountered but not in such an acute, voluminous, and sometimes dramatic form as today. The bone of contention in the clash between various theories is the understanding of the man–machinery–nature relationship.

In the struggle between different theories a special place is occupied by the problem of man. The significance of this problem grows steadily as the humanistic tasks, whose fulfilment mankind links with the STR, become ever clearer. Outside Marxist philosophy, this problem is becoming the focal point of both scientistic and anti-scientistic or anti-technocratic tendencies in assessing the progress of science and technology. The consonance of the development of science and technology with man's development and prospects is determined by the harmony of the STR with mankind's humanistic ideals and hopes. On behalf of man, technology and science are irresponsibly accused of a demoniacal capacity for destruction, of combining creative with destructive elements.

Unlike the majority of anthropological interpretations of scientific and technological progress, Marxism-Leninism has always concentrated on the human problems of the development of knowledge and technology, and has accentuated them in its teaching on the historical process. In this, essentially speaking, lies the Marxist sense of the concept of the contradiction between the productive forces and social relations, of the teaching concerning social revolution and the part played in it by the productive forces, concerning the technical basis of production and the material and technical base of society, concerning the practice, activity and decisive role of the masses in the historical process.

Marx showed that the social conditionality of science and technology has a specific human measure—the function of the implements of labour and the spiritual and material means of man's purposeful activity.

In this sense the intercourse between man and machines and man's activity in the sphere of the production of knowledge necessarily implies man's development, which can, of course, be profoundly contradictory. Marx's critique of capitalist organization of production laid bare its dehumanizing character.

Beginning with the simplest forms of the transfer of labour operations to machinery and ending with the automation of production and the creation of non-machine-like (bionic, chemical, biological) methods of production, the successive replacement of the 'technical character of labour'[8] comprises the history of man's steady ascent to more developed activity. As in the problems of science and technology, Marxism showed the integrity of production and social conditionality in this process and also its humanistic content. The historical inevitability of the settlement of the question of the orientation of the STR in or against the interests of man is thereby determined. Socialism, which brings this integrity to materialization, thereby comes forward as the decisive factor of the historical process, a factor giving it a humanistic orientation.

In the theories of optimistic technological determinism as expounded in the theories of industrialism and post-industrialism (W. Rostow, R. Aron, D. Bell, A. Etzioni, Z. Brzezinksi and many others) the development of science and techno-economic growth determine the historical process in one way or another up to the full and absolute predetermination of all its present and future spheres. This erases the very problem of man, his role in history, his social and intellectual development. P. Drucker's formula that civilization is unquestionably always formed by technology[9] clearly mirrors this attitude.

The reverse side of technological optimism—technophobia, the negation of the positive part played by scientific and technological development in history—takes analysis to the other extreme, to the sphere of the antinomies of culture, to abstract anthropological reasoning, to a clash between man and machines, between man and social institutions (I. Illich), to a conflict between the conscious and the unconscious (S. Freud), between violence and non-violence (H. Marcuse), between alienation and freedom (E. Fromm), and so forth. In other words, both the Utopia of technological determinism and the anti-Utopia of anti-scientism are evidence of the untenability of a philosophical reflection of the social significance of the STR from anti-dialectical and anti-materialistic positions, evidence of their inability to understand the moral and humanistic foundations of social development.

In the context of the STR the most acute problems arise in the relations between man and his natural environment. It cannot be said that there have been no crises in this sphere previously, that in remaking nature man had never caused damage to the environment. But the present ecological situation is indeed unique for its significance to our planet's entire population and for the character of the problems that it puts before society. It is quite natural that scientists and public figures in many countries have tried to formulate these problems, and that they have proposed ways and means of their possible solution. Indeed, philosophy cannot hold aloof from the problems worrying mankind. For that reason one of its pressing tasks is to understand the specifics of the present state of the relations between man and nature and to assess the changes that have taken place in these relations under the impact of the STR and also the tendencies that come to light in that sphere and make it possible to judge the prospects for the development of the present ecological situation.

What are the primary principles of Marxism's approach to its analysis of the interaction of society and nature?

Both the individual and society as a whole exist and develop solely in their interaction with nature. On the level interesting us here we can distinguish three basic interrelated forms of the relationship between man and nature.

Primary is the biological relation. Man is part of nature, a living organism in a state of metabolism with the environment, going through individual evolution, a biological cycle of life. As a living organism and as a biological species, man belongs entirely to nature. Of course, with society's development, this biological relation with nature is increasingly mediated by social relations and does not exist outside them. But just as the stability of chemical compounds and processes is the condition for the emergence of ever more complex biological formations, so the availability of biological structures and vital processes in the human organism is the condition for the existence of suprabiological socio-cultural forms of human existence. For the first time in its history mankind is today faced with the problem of preventing actions and processes that may seriously upset the normal state of the biosphere and thereby undermine the natural conditions for social being.

Another mode of the relationship with nature characteristic only of man is the practical relationship mediated by the manufacture and use of implements of labour. Whereas for an animal the obtaining of food and other activities are moments of its biological vital process, for man material labour is the determining foundation of his social life. Changes in the relationship between man and nature depend, above all, on the development of the implements of labour, on changes in society's technological basis.

In the history of man we can distinguish several qualitatively unique phases of the relationship between nature and society, depending on the level of development of material production and, chiefly, of the implements of labour.

The first phase is the appropriating economy. It existed at the very commencement of human history, when people lived by gathering, hunting and fishing. Man did not as yet exercise any essential influence on nature. But this economy was what gave rise to the species *Homo sapiens*, to the first forms of its social organization and the creation of key implements of manual labour.

The second phase began with the emergence of agriculture and livestock-breeding, i.e. with the transition from an appropriating to a producing economy. The revolution in production linked with this transition is now called the Neolithic revolution. This was a colossal upheaval that laid entirely new foundations for the existence of human society, fundamentally changed the entire way of people's life and created the conditions for the transition to a class society. During the many millenniums of existence of slave-owning and feudal society, agriculture (and in some countries, live-stock breeding) remained the basic form of production activity.

At the early stages of history man influenced nature chiefly through various forms of agricultural production. The influence of such forms of activity as handicrafts, mining and construction (with the appearance of towns) on the environment was insignificant compared with the cultivation of land and the utilization of domesticated animals. This period witnessed the cutting down of forests, the digging of canals, the building of various irrigation systems, and so on. Man exercised a fairly strong and sometimes detrimental influence on the environment. It is surmised that some deserts appeared because man destroyed natural complexes. But in remote history such instances were local and did not assume a global character.

The third phase began with the industrial revolution. This revolution, linked with the development of capitalist production, enabled the bourgeoisie to create colossal productive forces. Under capitalism industry subordinated agriculture technically and economically for the first time; people began to influence nature mainly through industrial production, which greatly increased both man's possibilities for remaking nature and the productivity of human labour.

Lastly, the attitude of cognitive human thinking towards objective reality is of great significance to the life of society. In the broad philosophical context, the cognitive attitude is only relatively independent, being subordinated to practical relationships. 'It is in the measure that man has learned to change nature that his intelligence has increased'.[10] But without accumulating experience, without developing knowledge of which science is the highest form, man would have been unable to perfect his practical activities, improve machinery and promote production.

Industry gives society a dynamic technical basis that is capable of constant change and progress. With the appearance of machine production science began to exercise an increasingly revolutionary influence on production. Knowledge is vital for any human activity because it is useful and presupposes modification of an object in accordance with its properties and links. But until the appearance of machines, mainly empirical knowledge and day-to-day experience was used in production. Although science had existed in embryonic form since ancient times, its direct link with production was extremely limited.

With the development of machines, the conscious application of the natural sciences to production became technically possible, and science gradually evolved into a productive force. This gave production quite new and unprecedented possibilities for progress. Capitalism utilized these technical possibilities.

However, capitalism linked the creation and utilization of machines solely with their economic efficiency, disregarding the need for protecting and improving the natural environment. It was capitalism that produced the theory that nature is an inexhaustible fountain of resources and a bottomless reservoir into which all the waste of society's vital activity could be thrown. But with the growth of the technical armament of human labour this theory grew increasingly obsolete, while with the development of the STR it became plainly dangerous. It is not the STR as such but its capitalist use that has now placed mankind before the threat of an ecological crisis. Capitalism is extraordinarily wasteful with regard to the natural resources of man himself and of the environment.

From this standpoint we can assess the trends in recent literature that have taken shape in the discussion of ecological problems.

One of these trends moves in the channel of 'technological pessimism', which considers progress in science and technology as the source of all misfortunes. To some extent this viewpoint was supported in papers by D. Meadows and others, who propose halting the development of technology and the growth of production in order to create a 'balanced economy' that would allegedly avert an ecological catastrophe. But this is not a way out of the situation, and this proposal did not win the support of most scientists.

Many realistically thinking scientists suggested a resolution of the problem facing mankind through co-operation between different socio-political systems, through a joint quest for solutions meeting the interests of society. It is stated quite plainly that the spontaneous mechanism of the capitalist market economy cannot cope with the problems that arise, that elements of planning must be introduced, that use must be made of science, of the prevision of the natural effects of man's productive activity. This is suggested by scientists who are far removed from Marxism. For instance, A. Peccei and A. King declare that the modern structure of the world has grown obsolete, that private and state capitalism are stale, that we are compelled to develop something else.[11]

M. Mesarovic and E. Pestel wrote in their paper for the Club of Rome that the economy must be managed by directives and not with the aid of market mechanisms.[12]

The global character of the problems facing mankind has made it necessary to engage in 'global modelling' of the future and working out the 'scenarios' of different variants of socio-economic development. While by no means rejecting the importance of forecasts of this kind, we should like to emphasize that a harmonious relationship between society and nature can be established and the 'metabolism' (i.e. the process of interaction) between nature and social man can be placed under conscious control only with the establishment of public ownership of the means of production and the abolition of exploitation and social antagonisms. This is a fundamentally new guideline of society in its attitude to nature, a guideline that is implicit in socialism.

Socialism was first established in Russia, which was economically far behind the level achieved by the developed capitalist countries. For that reason it had to make up for lost time quickly, to accelerate the development of the economy with priority for the development of heavy industry-a high price for technological progress. Under these conditions the problem of protecting the natural environment did not, of course, receive prominence, and development of industry proceeded mainly without taking this effect into account. For that reason ecological problems exist in the Soviet Union as well. As soon as socialist society grew strong and stood firmly on its feet, the regulation of the relations between society and nature became one of the chief orientations of State policy and public activity. This, too, showed the true nature of socialism, a system that does not live solely by the requirements of the day but pursues far-reaching aims, concerning itself with the future of the people. This concern manifests itself, in particular, in the policy of preserving the natural environment and rationally utilizing natural resources, a policy that is consistently pursued by the Soviet Union. Suffice it to say that 11,000 million roubles have been budgeted for the preservation of the environment under the tenth Five-Year Plan.

The STR unquestionably marks the beginning of a new phase in the relationship between society and nature, when, in order to preserve the natural conditions of its existence, mankind must establish harmony with nature and place the 'metabolism' between them under conscious control. Socialism is the system that regards harmony between society and nature, remade in the interests of man, as one of its cardinal social values. This guideline is entirely consonant with both the requirements of the present phase and the new relationship between society and nature.

The solution of ecological and social problems is today fusing as never before. The machines created by the STR may be used for the global destruction of the natural environment. But it is precisely these machines that can also help to solve ecological problems. The question is where and how to develop and direct them. The requirements of the present phase of the interaction between society and nature and the interests of mankind require that science and technology should develop in a way that would reduce and soften the negative effects of the functioning of existing machinery, promote the broader utilization of secondary raw materials, improve the technology of wasteless production, and so forth. In other words, new machinery must meet not only economic but also ecological requirements.

Today science's social functions are also changing. Whereas formerly the practical application of natural scientific data in production was corrected solely by the economic sphere—technical and technological innovations had to be economically profitable in order to become widespread—today, in addition to economic efficacy, two new requirements, which are social requirements made of science, are growing increasingly clear-cut. First, science must open the way to the creation of machinery conforming to ecological requirements and standards. Second, science must work out ways of forecasting the long-term effects of human activity.

A social system can build rationally and it can systematically control its relationship with the natural environment only with the aid of science. Of course, this concerns not some definite science but the entire range of sciences, for to resolve ecological problems it is necessary to combine and integrate the data of the natural, technical and social sciences.

Let us sum up. At the dawn of his existence man was in unity with nature. But this was initial and primitive unity, when oppressed by the difficulties of the struggle for existence man was entirely dependent on the elements, and in his consciousness he still did not separate himself from them; this is mirrored in mythology, for example. Such was the first phase.

With the rise of civilization man was increasingly separated from nature. But from the very outset this was a contradictory, antagonistic process. On the one hand, man separated himself more and more from nature, creating his own cultural environment. True, agriculture was still the main form of man's production activity, while his technical base consisted of primitive implements of labour powered by himself or animals, and this process was slow. But it was sharply accelerated with the emergence of industrial production and the accompanying development of urbanization, the spread of the urban way of life and so on. Capitalism led man's confrontation with nature to extremes, to a glaring opposite, to a crisis state. Capitalism spurred not only social antagonisms, but also the antagonism between society and nature, alienating not only man from man but also man from nature. These contradictions are concentrated in the economic, demographic and ecological problems that today confront society.

But an antagonistic contradiction matured not between nature and society as such, but between nature and capitalist society, whose socio-economic mechanism has been stimulating a rapacious attitude to nature in the course of centuries.

The contradiction between man and nature, a contradiction that has reached the bursting point under present-day capitalism, thus matured and developed throughout the development of class society. Such was the second phase.

But there undoubtedly was another side to this contradictory process of development: gradually, step by step, man harnessed the forces of nature, learning to subordinate and use them in his own interests. As social production developed, social man gave shape to increasingly closer and varied links with nature, because in his activity he used more and more properties of matter and its laws, increasingly diversified forms of matter and energy. Machinery not only separated man from nature but also linked him to it by the most diverse channels. Man differs from nature precisely through his deep penetration into and not through his separation from nature. The dialectics of this process also manifests itself in this.

The combination of scientific and technological achievements with socialist social relations creates the conditions for achieving unity between man and nature on a new, higher foundation—technical, scientific and social. Consequently, the communist social system removes not only social antagonisms but also the antagonism between man and nature, and asserts harmony between them under society's conscious control. This is a sort of negation of negation that is typical of the present phase of civilization. Consequently, the interaction between nature and society develops also in accordance with the laws of dialectics.

Our concept of STR includes the recognition of the growing role played by the social sciences.

In this connection the social sciences today have to state in the name of what, and how, the achievements of science and technology can be used for the best purposes, and which social conditions correspond to these purposes.

Today therefore the effectiveness of scientific research to a considerable extent depends on the close co-operation of the natural, technical and social sciences. The unity of these three main branches of present-day scientific knowledge is determined by the comprehensive character of the problems, the solution of which is included on the agenda by the requirements of social development. Without such co-operation it is impossible to solve comprehensively the problems of scientific and technological progress, the development and rational location of the productive forces, the construction of large industrial and agro-industrial complexes, the improvement of towns and villages, the creation of optimal living and working conditions, and environmental protection. On the whole it can be said that the solution of the socio-economic tasks is impossible without the participation and close co-operation of specialists of the most diverse fields of knowledge.

Such co-operation can be exemplified by scientific substantiation of the principles and methods of the planning and organization of socio-economic construction.

Leonid Brezhnev noted at the twenty-fifth Congress of the Communist Party of the Soviet Union (CPSU) that academic institutes, jointly with the ministries and departments, had drawn up the draft Comprehensive Programme for Scientific and Technical Progress and Its Social and Economic Effects for 1976–90. We continue the work on this programme. It is to be an organic part of current and long-term planning. It is designed to give the orientations without a knowledge of which the economy cannot be managed successfully. It should give a general concept of socio-economic development for a long time to come.

In our work on the comprehensive programme we base ourselves on the long-term orientation of economic policy which was formulated by the twenty-fourth and developed at the twenty-fifth Congress of the CPSU. This is an orientation towards the steady rise of the people's living standards and cultural level on the basis of the dynamic and balanced development of social production and the enhancement of its efficiency, the speeding up of scientific and technological progress, the growth of labour productivity and the utmost improvement of the quality of work at all levels of the national economy.

Guided by these tasks while working on the comprehensive programme we shall have to take into consideration the following interconnected tasks.

The comprehensive programme includes a number of directions of scientific and technological progress or separate large interindustrial and industrial programmes, such as the fuel and power complex, mechanical engineering, materials, transport, communication, agriculture, consumer goods production, housing construction, education, public health, etc. Many variants are being worked out for each direction, for each programme. It will be necessary to determine the qualitative change in science and technology and give a socio-economic appreciation of each of them. The technical and socio-economic assessment should be based on scientific criteria of efficiency and quality. Large-scale and complicated calculations are carried out in order to compare various variants and to substantiate the most effective scientific and technical solutions.

Secondly, the optimal correlation is to be determined between the long-term programmes that will bring results only in a relatively long period and the measures that will provide practical results soon and contribute to the national income. Marx warned that a correct distribution of investments between long-term and short-term objects of the economic turnover under socialism should become a major principle of planning.

Thirdly, while assessing new equipment and technology we take the maximum saving of manpower resources and improved working conditions because we remember Marx's statement that saving labour is the main economic law of communism. In our understanding, saving labour does not mean saving working time alone, it means making work as easy as possible, doing away with arduous physical labour and monotonous and exhausting operations. To make labour the prime need for all and not only a necessity, it should be made more attractive and ever-more creative.

Fourthly, the comprehensive programme should contain a scientific substantiation of the optimal correlation in the growth rates of the production of means of production and of that of consumer goods so that the growth of the first department creates the material basis for continuous technological progress and the development of the second department secures a balance between the growth of the population's cash income and the amount of consumer goods and services.

Fifthly, the requirements of environmental protection are taken into consideration for each direction of scientific and technological progress. The improvement of the use of natural resources is not only a scientific and technical but also a social problem of paramount importance. The question is what natural surroundings man will have to live in, what natural resources society is and will be able to use. Our approach to the envisaging of future development differs from the Western global models of the world in that it is comprehensive, concrete and well founded.

Soviet scientists have considerably advanced social science and particularly the theory of planning and improving social processes on the basis of the generalization of the experience of creating and developing a socialist society. The Soviet Union was the first country with a planned organization of the entire national economy which ensured an organic combination of the centralized plan with the effective economic self-dependence of enterprises and amalgamations. New methods of prognosis began to develop recently on the basis of the systems analysis of socio-economic processes, comprehensive use of the achievements of the social and natural sciences, mathematics and statistics, wide use of computers.

Notes

1. Thomas Kuhn, *The Structure of Scientific Revolutions,* Chicago, Ill., The University of Chicago Press, 1970.
2. Precisely these tasks are dealt with by Western theorists. See, for example, W. McElroy, 'The Rise of Fundamental Research in an Advanced Society', *American Scientist*, Vol. 59, No. 3, 1971, p. 294–7.
3. Y. Ezrahi, 'The Political Resources of American Science', *Science Studies* (London), Vol. 1, No. 2, 1971, p. 117–33.
4. *The Social Impact of Modern Biology*, London, 1971.
5. ibid., p. 241.
6. Karl Marx and Friedrich Engels, *Works*, Vol. 46, Part II, 206 and 208 (Russian translation).
7. ibid., Vol. 39, p. 174.
8. ibid., Vol. 23, p. 384–5.
9. P. Drucker, *The Age of Discontinuity, Guidelines to Our Changing Society*, p. xiii, New York and Evanston, 1968.
10. Friedrich Engels, *Dialectics of Nature*, p. 234, Moscow, 1964.
11. *Science* (Washington, D. C.), Vol. 185, No. 4145, 1974, p. 19.
12. M. Mesarovic and E. Pestel, *Mankind at the Turning Point*, p. 97, New York, 1974.

The unity of the social and the natural sciences

Sava Ganovski

Academician, Bulgarian Academy of Sciences, and President, World Association of Philosophy

The article by Radovan Richta examines in detail the problems of the scientific and technological revolution and the social sciences; the essence and role of the scientific and technological revolution; its consequences for the two social systems, the socialist system and the capitalist system; its place in the relations between society and nature, etc.

The profound changes and social conflicts accompanying the scientific and technological revolution do indeed pose a series of extremely important problems and questions. I shall touch here on the problem of that revolution and certain other problems connected with it.

The scientific and technological revolution is, by its very essence, a complex and many-sided phenomenon. Its integral character is reinforced by the fact that it acts on more or less all the areas of society while being at the same time subject to its vast ascendancy. At first sight it seems (and the ideological arsenal has at its disposal a number of theories which try to emphasize its merits) that the scientific and technological revolution is a purely technical phenomenon and that it develops only on the basis of discoveries in the natural, technical, and mathematical sciences. This conception is extended by the point of view which asserts that the scientific and technological revolution 'subsists', so to speak, only in the area of society's productive forces and produces no revolution in particular save in the area of material production. And from here we end up with conclusions which make the role of scientific and technical personnel absolute in the complex of social development and in the technical and technological factors of this development. Such a technical, technological and technocratic approach portrays the scientific and technological revolution erroneously. Its partisans examine this revolution unilaterally and, despite the fact that its role seems exaggerated, they present it as a limited phenomenon.

Naturally, the essence of the scientific and technological revolution cannot be reduced solely to this or that scientific invention, even a very important one, or to its orientation towards scientific and technological progress. Rather, its essence rests in the reorganization of the entire technical foundation, the whole technology of production, beginning with the exploitation of material and energy processes and ending with the system of machines, the forms of organization and management, and the place and role of man in the production process.

In its most generalized form, we can say that the scientific and technological revolution itself conditions the specific level of this development. It has been made possible by the higher degree of social production.

The scientific and technological revolution is a universal process in terms of its form and content, vast at any given moment, but perceptible in an increasingly profound way from the cognitive point of view. It is not only connected to material production but is also manifest in all the other areas of society: it is not just a question of science and technology—it is at the same time a struggle, a politics, a diplomacy, and it has ideological, moral, pedagogical, psychological, biological, and other aspects. This complex also includes the problem of the relations between the scientific and technological revolution and the social sciences, and the correlation between them.

This influence is bilateral; on the one hand, the social sciences influence and at the same time are themselves a component of the essence of the scientific and technological revolution; and, on the other hand, this revolution, in its turn, has a reciprocal influence on the social sciences.

It is these processes of the reciprocal action of the different branches of the sciences which play a decisive role; in particular, this amplifies the role of the social sciences as much in regard to the solution of practical problems as to the ideological struggle which is emphasized in terms of the development of the scientific and technological revolution in the competitive conditions obtaining between the two opposed social systems. Examined under the social and economic aspect, the main point is that science becomes the guiding element of the productive forces. This means that the unity of scientific knowledge and productive activity becomes a powerful motive force behind production.

The transformation of science into one of the guiding factors of social evolution appreciably extends its social functions. Scientific activity acquires the quality of independent, socially necessary work, assuming more and more a mass character. In its most generalized form this process can be characterized as a phenomenon of the growing intellectualization of social work. Science acts as much on production as on the producer, on his physical and spiritual forces and needs, on his product. The production, and the turning to account of constantly renewed knowledge becomes a factor in the accelerated rejuvenation of all areas of social life.

Despite some common points of support, the scientific and technological revolution, as it really functions, can be subdivided conceptually into two types: the scientific and technological revolution under capitalist conditions and the scientific and technological revolution under socialist conditions. And this is so if we recognize that the social conditions, within whose framework the scientific and technological revolution is produced, make a deep imprint on it.

We must not forget that the scientific and technological revolution develops differently according to socio-economic and socio-historic conditions and leads to new and different social consequences depending on their characteristics. But in this case we must agree that the social conditions, i.e. the social reality, is not a foreign body: it is not simply an environment or a condition for the development of such a revolution but rather an inseparable part of its essence, its law. Seeing that capitalism and socialism are, as social systems, radically opposed to each other, it follows that the essence of the scientific and technological revolution, its scientific, technological and social aspects, will not be the same under capitalist and socialist conditions. In reality, the scientific and technological revolution, as Radovan Richta emphasized in his article, leads to an unparalleled recrudescence of the contradictions between the social character of production and the private mode of appropriation—contradictions which can be resolved only by the establishment of new and improved production relations, such as socialist production relations.

According to historical experience, the development of the scientific and technological revolution in capitalist society leads to the exacerbation of social antagonisms given the fact that the radical defects of the bourgeois social regime receive new forms of expression paralleling the traditional forms.

The connection between the scientific and technological revolution and social progress is neither synonymous nor linear by far. It is a profoundly dialectical relationship of unity and mutual influence. In order for science to be able to develop in a varied manner and in order to put its attainments into practice in the interests of the entire society, it must be assured of well-defined social conditions which capitalism cannot guarantee. These conditions can be created only by reorganizing society on a socialist basis. It is precisely this fact which defines the nature of the social consequences of the scientific and technological revolution in different socio-economic systems.

Within the framework of socialism, the scientific and technological revolution has at its disposal, first, a new economic basis—socialist property and the socialist mode of production; second, it has a new political basis, which is so because the direction of technical and scientific progress is assumed by the new, popular state; third, it has a new ideological basis—the scientific basis of Marxism-Leninism; fourth, it has a different aim and radically different functions—the increasing rise in the level of the workers' material well-being, the construction of an advanced socialist society and the accumulation of foundations by which to pass to a higher level of socialist building. On the other hand, it is only on the basis of the accelerated development of science and technology, scientific planning, and the management of social processes that the radical tasks of building the new society can be completed.

One of the most important tasks of socialist building at present is that of the organic union of the advantages of socialism with the attainments of the scientific and technological revolution. In this regard the problem of studying the mechanism of the link between the social system and the character, orientations, and rhythm of the scientific and technological revolution occupies the central place. In other words, it proves necessary to discover the essential aspects of the social system itself and to see to what extent its particularities and specific advantages define the development of the scientific and technological revolution. The socialist mode of production, by its very essence, by its nature, makes possible the maximum utilization of the attainments of science and technology in the interest of the progressive development of society. Nevertheless, the conversion of this possibility into reality does not come about automatically; it assumes enormous organization work and the oriented management of the processes of socio-economic development. It is the Communist Party which, armed with the scientific theory of social development, fulfils this historic mission as the organization and orientation force.

It is well known that the admission of the necessity of the social management, scientifically established, of social development is a basic conception of Marxism-Leninism, and in practice, it is produced in the course of building the advanced socialist society, the communist society. Short-term as well as long-term planning and forecasting are inseparable elements of this management process. Elaboration of such a scientific programme naturally assumes enormous investigative work in order to establish the forecast for economic, scientific and technical development, and imposes new demands on the entire complex of the social sciences.

Unfortunately, there are certain ideologues in the West who misunderstand the essence of the scientific and technological revolution, above all, the difference in its consequences for capitalist and socialist societies. Some see in the image of science, technology, and the scientific and technological revolution what must in their opinion resolve all the contradictions of capitalism (though they are more and more disappointed in their hopes); on the other hand, others think that these phenomena represent in themselves a calamity for society and social development. For this reason, they succumb to pessimism and deception, and find themselves in an impasse when it comes to the problems of society. We in socialist society see in the scientific and technological revolution a powerful ally in the struggle to build the new—socialist—society. This is why the proper focusing of the problem concerning the essence of the scientific and technological revolution is an indispensable component of thc study of the Marxist-Leninist conception of it.

The Marxist-Leninist position that scientific and technical progress not only fails to cure capitalist society by itself, but on the contrary deepens and exacerbates all its contradictions, is confirmed by the historical experience of the last decades. The wave of crises flooding capitalist countries can serve as a striking example of the deepening of the general crisis of the capitalist system accompanied by a rise in discontent at the heart of all the layers of society. The active socio-political struggle recruits new forces coming from many factions among the intelligentsia, scholars, and students. All this confirms capitalism's incapacity to profit from the possibilities of the scientific and technological revolution in the interests of society and the impossibility of acceding to stable social progress of the management of the economy within the framework of the capitalist system.

In studying present-day theories of the scientific and technical revolution and in revealing their apologetic essence, it must not be forgotten that capitalism, despite its contradictions, makes the best of evolutionary conditions—for example, the measures taken by the bourgeois state to avert ecological catastrophe. To this end, in capitalist countries, the responsibility of those who wish to make their fortune and realize the profits of production, is brought into question as much as the essence of private enterprise capitalist activity, etc. In studying the phenomena of social life, the Marxist-Leninist tradition cannot ignore these kinds of problems. Under the conditions of the contemporary ideological struggle, research and profound critical analysis, starting from Marxist positions, are increasingly indispensable as are new tendencies and orientations of bourgeois economic and sociological theories whose appearance is conditioned by the contradictory development of the scientific and technological development in capitalist countries.

The complex of problems raised by the scientific and technological revolution is extremely broad and varied. The conscious, correct orientation of these processes demands a profound and multilateral analysis of scientific and technological progress, the elaboration of a well-founded policy, and the implementation of organizational measures guaranteeing the conditions most favourable to scientific and technological attainments in the interest of socio-economic progress.

The study of these problems is closely linked to the solution of very important practical questions—such as the definition of social and economic levers to stimulate the scientific and technological revolution under the conditions of socialism. Another series of problems demanding careful study is connected with the discovery of the social consequences of the scientific and technological revolution in social relations and of the reciprocal action between the scientific and technological revolution and structure; it is also connected to the relations of production, those of a juridical and administrative order, and to social planning as much on the level of isolated enterprises as on that of entire branches of production on a regional scale, and of co-societies joined in co-operation between socialist countries. No less important is the problem of the technical and economic guarantee of the organic union of the attainments of the scientific and technological revolution with the advantages of socialism. The analysis of the effectiveness of scientific and technical projects in new technology has a special role here; it is above all a question of knowing whether the effectiveness of social production corresponds to the grafting of progressive changes on to the 'technology-production' system.

No less desirable would be the study of the important current problem of the influence of the scientific and technological revolution on the world economic process and on the character and orientation of the international division of labour. Undoubtedly the scientific and technological revolution speeds up the rhythm of the internationalization of production and contributes to the deepening of the processes of integration.

I cannot stress enough that the scientific and technological revolution, with its dynamism and results of primary importance, shakes up the mystical, ideological—or whatever—picture of social reality. It both demands and confirms a correct, scientific, and, in fact, dialectico-materialistic understanding of history, social life and evolution. In rendering the social processes dynamic and by infiltrating into all the spheres of society, the scientific and technological revolution also renders their essence and correlation, the very social reality (object of social sciences) complex. In this way the social sciences are faced with an object subject to profound changes and having complex, dynamic components. This is a profound, dynamic consideration which forces us to deduce that the social sciences are faced with new tasks which must be raised to a higher level of their development.

The dynamic of the social processes, brought about by the scientific and technological revolution, gives the social sciences the task of soaking up, through all its components and elements from 'top to bottom' and 'bottom to top', the dialectico-materialist methodology. The role of methodology in the functioning and movement of the social sciences grows not only by virtue of the above-mentioned circumstances but also because research in the social sciences must have a heterogeneous character in order to succeed in increasingly complicated tasks.

The scientific and technological revolution occurs in the social sciences in the form of a specific integration process conditioned by the union of the efforts of different humane and social sciences and aimed at the resolution of the complex problems of social development. I can state with certainty that recently we have particularly felt a tendency in the social sciences towards the collective study of this or that problem by different branches of socio-scientific thought and towards the obvious co-ordination of social research with the analysis of socio-political philosophical problems and social processes and phenomena. This tendency corresponds to the fundamental character of the study of problems which are the object of socio-scientific thought, such as the problems of: criteria; the typical aspects and stages of the evolution of advanced socialism; the planning and forecasting of the development of the national economy; the increase in efficiency and the latest paths to the intensification of material production; the problems of the development of socialist democracy under the conditions of socialist society; the evolution of the processes of economic integration in co-operation with countries having the socialist system; the socialist mode of life and the paths to its improvement; the particularities of world evolution at the present time; the specificity of the processes which are now produced in the developed countries, etc. The heterogeneous method can give optimum results only in the presence of a proper scientific-methodological base.

A very important problem of the social sciences, related to the scientific and technological revolution, is that of the study of the ideological aspects of this revolution, in particular that of the disclosure of the methodological vicissitude from the scientific point of view of different social theories based on the erroneous and faulty examination of the social functions of the scientific and technological revolution. The incapacity of various ideologues to disguise the sombre perspective facing the human race—as in the case of the capitalist world—is reflected in different ways in the political thought and concepts of the bourgeoisie. We are dealing with: the romantic, 'anti-cultural' theory (from the pessimistic point of view on behalf of technology and even science) of the necessity for 'zero development' or the anti-human appeals to stop the rise of 'developing countries', appeals which spring up where there have been different anti-scientific theories such as the 'theories' on 'de-ideologization', convergences, and the unique industrial society; on the components of economic development; on technocracy; on the post-industrial society; on the distortion of the working class as a revolutionary force, etc. Richta correctly stresses the fact that the perspectives of danger, linked to the scientific and technological revolution, are conditioned on the social plane; and that for all humanity, the dazzling perspective of this revolution open with the victory of the socialist relations of production. In actual fact Marxist-Leninist social theory allows us not only to explain correctly social processes but also to control them from the scientific point of view; it allows us to assert that the scientific and technological revolution acquires its truly humanitarian meaning only under the conditions of an advanced society.

The scientific and technological revolution creates the premises that allow the reduction of important forms of human activity—science, technology, production—to a unified system. The formation of this union leads to long-range consequences for each of its components and reinforces their correlations.

In so far as the social sciences are concerned, the transformation of the sciences into a single, direct force means, for example, that they have the important task of not only developing a conception of the world, cultural thought, and the moral traits of the human personality, but also of participating directly in the increase of material production, the improvement of social relations, and the management of the social processes.

For example, bringing to light the essence, role, and importance of the scientific and technological revolution is a task which goes beyond the possibilities of a single science, be it social, technical, or natural: it demands the union and mutual co-operation of all these sciences.

In addition, it must be noted that the solution of methodological problems plays an important role in the study of the scientific and technological revolution. Only the firm support of the methodological principles of Marxism-Leninism allows one to lay the foundations for answers to questions as radical as those concerning the theory of the scientific and technological revolution, as those concerning its essence, essential orientations, socio-economic consequences, etc., and to define its basic notions and categories. The analysis of the scientific and technological revolution cannot be limited by the frameworks of social access and that of the natural sciences. This analysis assumes profound philosophical generalizations which have naturally allowed the notion of the 'scientific and technological revolution' steadily to penetrate the structure of present-day philosophical knowledge of Marxism-Leninism. Philosophers, economists, sociologists, representatives of the other social and natural sciences must do still more work in order to create a complete scientific Marxist-Leninist theory of the scientific and technological revolution, of methodological importance for the solution of concrete political, economic, and ideological tasks in the process of building the advanced socialist society, the communist society.

At its core, the scientific and technological revolution is the result of the union and action of technology and science. However, it must be immediately stressed that it is as much a question of the natural, technical and mathematical sciences as of the social sciences. Long-term plans for the national economy, currently established in our socialist countries, give the social, technical, and natural sciences the task of speeding up research in this area. It is a question of looking for the roots of the denominative fact and of the demand currently addressed to the development of the sciences, i.e. the realization of the union, the unity of action between the natural and the social sciences which is, in itself, one important condition for the further development of contemporary science; the resolution of new problems; the correct clarification of the role and perspectives of the scientific and technological revolution; and the resolution of man's problems: the reciprocal relations between society and nature, the problems of peace, etc.

The importance of socio-scientific thought is much greater today than in the past. It is losing its national narrowness and is acquiring a global character. We are witnessing the birth of the process of the unification of social and natural history.

The scientific and technological revolution challenges the social sciences not only in posing new tasks for them of enormous importance in theory and practice, but also in strictly reinforcing the integration processes in the humanities and their behaviour towards the natural and technical sciences. By uniting the processes and phenomena of the different spheres of human activity into one system, the scientific and technological revolution reinforces the importance of the heterogeneous approach to the problems of social evolution. In fact, no large problem concerning the national economy can be resolved today without the reciprocal action of the natural and the social sciences. For example, the contemporary solution to the problem of the efficiency of social production assumes an approach which takes into account technical, technological, economic and social terms (components). It is an invisible social process like that which, objectively today, tends to increase the efficiency of production. With the creation of the large industrial complexes of our times, we must bear in mind not only questions of a technical, technological and economic nature but also all of the social, psychological, ecological and other consequences of the functioning of these complexes, i.e. we must examine these questions within the framework of the social, technical and natural sciences combined.

Under the conditions of the scientific and technological revolution, the role of the correlation between different processes and phenomena grows larger, which reinforces the importance of the heterogeneous approach to all the other problems and consequently the importance of the reciprocal action of the social, natural, and technical sciences and their organic unity. This will allow them to exert greater influence to increase the efficiency of social production, improve the conditions of life and cultural evolution, and guarantee a true contemporary analysis of the phenomena of the scientific and technological revolution. Social practice conforms increasingly to long-term forecasting, hence the theories that social life is not subject to rational forecasting are rapidly losing ground.

The whole course of the evolution of contemporary science and practice tenaciously insists on the necessity of studying these problems.

Above all, it must be pointed out that the problems concerning the importance of studying matters related to the *rapprochement* and co-operation of the natural and the social sciences, are posed with much acuteness at scientific conferences and symposia, where new possibilities for the application of fruitful research—general as well as fundamental and practical theory—are emphasized, opening up new horizons before the various sciences, in particular the natural and the social sciences.

The integration of the sciences assumes the presence of a complex of both social and theoretical premises. These premises include: (a) increasing the role of social practice in order to complete the tasks whose resolution necessitates a systematic co-ordination of effort on the part of the representatives of the different sciences; (b) the society's fitness to guarantee such co-ordination on the organizational level; (c) the mutual, potential, theoretical preparation of the sciences included in the integration process; (d) the presence of respective theoretical integrators (philosophical and others) capable of co-ordinating and synthesizing the respective sciences.

The paths of the development of the natural and the social sciences were relatively independent during the last decades. It is only now that the necessity radically to change that path has arisen. We have recently seen qualitatively new tendencies in the reciprocal relations between the social and the natural sciences take form, which allow us to confirm with conviction the abovementioned idea about future co-ordination. These tendencies have been brought to life by the qualitative changes in the general strategy of contemporary scientific evolution, conditioned, first of all, by the scientific and technological revolution.

The integration and differentiation of the sciences—here are two processes of scientific evolution which, to a great extent, are mutually conditioned. Although the very character of the integration and differentiation processes of science and the proportion between them is defined primarily by the character and conditions of this or that epoch, these same processes appear in the unity, expressing two different, and, at the same time, necessary sides of scientific progress.

In the past, the isolation of the research efforts of the two principal science groups—the natural and the social sciences—was in well-defined agreement with the tasks of social practice and material and technical progress. Now the situation has changed. Today, it is necessary to effect the maximum *rapprochement* as well as co-ordination of research efforts on the part of the representatives of the most developed areas of scientific knowledge.

In breaking with these isolated traditions, many obstacles have been encountered, not only of a scientific and organizational but also of a psychological nature. Particularly enormous are the difficulties in attaining the integration of the reciprocal unity between the natural and the social sciences. The prolonged isolation of the development of natural history from that of social history, the qualitative differences of affected and often exaggerated research objectives, corresponding to whatever emphasis was placed, have created tenacious, rigorously and even absolutely opposing stereotypes between different scientific groups. It is for this reason that it is particularly important today to bring about, in general, a dialectical materialist statement of the problem and, in particular, a statement concerning the processes of the reciprocal action of the natural and the social sciences as well as the elaboration of the mechanisms for realizing the optimization of these processes.

Still, what are the scientific systems at the basis of the integration of the reciprocal action of the natural and social sciences? In order to answer that question, we must first of all consider the fact that past and present contacts and reciprocal action between the natural and the social sciences were and are realized by means of philosophy. It is precisely philosophy which, as a basic theoretical mechanism, fulfilled to a certain extent the function of liaison with regard to the two scientific groups.

At present, this liaison function is performed more or less by a series of specialized scientific theories, paralleling philosophy, which attempt to establish directly or indirectly increasingly broad theoretical bridges between the given sciences. Foremost is cybernetics, the common systems theory, information processing, and organization theory, which naturally do not and cannot take the place of scientific philosophy.

These modifications in the general strategy of scientific evolution are caused by the enormous complexity of the objectives included in social practice and, above all, by the necessity of a controlled influence on given objectives in order to manage them effectively.

It is the cybernetic conception of self-management which serves as the basis for the complex of scientific notions related to the study of complicated systems of organization. The role of this conception in the formation of reference points in contemporary scientific evolution is enormous. At present, the new stages of the scientific and technological revolution are rightly linked to the development and actualization of the ideas of self-management.

Until now, the principle of self-management has been elaborated, first of all, from the point of view of the perceptual tasks of the practical and natural sciences. Now, the tasks posed by the whole of contemporary scientific knowledge insistently demand systematic study and the application of the common principles of self-management to the study of a system as complicated as that of human society.

This is, in effect, one of the most important current scientific problems, for on its solution depends, in the final analysis, the effective realization of the cardinal tasks of managing society and regulating its reciprocal action with the environment. We have every reason to assert that the success of the integration process of the natural and the social sciences depends to a great extent on the general solution of a given problem, considering that the principles of self-management include the important ideas of contemporary scientific knowledge. Furthermore, the systematic application of the principle of self-management to human society will allow not only the reinforcement of 'the powerful current of natural history towards social history but also the orientation in the opposite direction'.

On the other hand, it is important to consider the fact that the principles of self-management, and systems and cybernetic ideas in general, cannot by themselves explain social reality. These principles can only pose new problems for the social sciences, trace new aspects, approaches and perspectives for investigation but no more. The primary function of explaining social reality can be filled only by the theory which reproduces the fundamental specificity of society, the fundamental laws of its origin and further evolution. It is common knowledge that this theory is the theory of historical materialism.

Thanks to the discovery of the fundamental laws governing the functioning of society, historical materialism has played a decisive role in the theoretical self-determination of the natural sciences within the science system. At the same time there arose the possibility of the 'dialogue' between social history and natural history as close partners in equality capable of defining their methodological limits, which represents one of the essential existential conditions for the reciprocal action of fruitful integration between these two groups of sciences. The application of the organizational principles of society is not summoned up to replace sociological generalizations but to shed new light on them and 'inscribe' them in a broad intertheoretical perspective resulting from the establishment of general, well-defined properties of a goal organized in a heterogeneous manner. On the other hand, these general principles of self-management can be applied consecutively to the analysis of society only in the presence of a sociological theory capable of expressing the nature of society as a particular system with its own specific laws governing the functioning of evolution; such is scientific philosophy, the philosophy of dialectical and historical materialism.

The scientific and technological revolution poses for the social sciences a problem of primary and vital importance for the existence of humanity in general—the problem of peace, the struggle against the outbreak of a new, thermonuclear, war. Social history, which as a whole is concerned with the past, present and future of humanity, must by its action and practical realization indicate clearly and categorically to humanity the danger hidden in the scientific and technological revolution. Naturally, this danger in no way stems from the very essence of the scientific and technological revolution but from the specificity of imperialism which can use scientific and technological attainments for bellicose ends, destructive war. With much zeal, social history must support the efforts of socialist countries, all the progressive forces in the world, the action of reasonable political men in capitalist countries until peace triumphs in the whole world. The problem of peace—the only alternative for the present and future of humanity—is included in the 'prognostic' functions of natural and social history taken together. Here, as with the solution of fundamental problems, scientific social history is faced with non-scientific social forecasts, and occupies an important place in the struggle on the ideological front of our times.

In a relatively short time the scientific and technological revolution brought about sudden changes in the details of the reciprocal action between society and nature. As in the past, the exchange processes between society and nature are at the basis of reciprocal action but the rhythm and character of these processes have been impetuously modified; they have become dynamic and large-scale, corresponding to the level of man's equipment in his work activity, with increasingly improved and powerful technical means. It has proved possible to transform not only isolated objects in nature but also all of nature on the earth's surface, at a faster rate than during the prehistoric period.

Recently, special attention has been paid to the need to improve the technology of social production with the aim of guaranteeing a more complete and varied treatment of mineral resources in order quickly to reduce the harmful reciprocal action of wastes on the environment. To this end it has been proposed that the study and installation of technological procedures be actively undertaken, which would allow the reduction of wastes and their maximum re-use; it would also allow the introduction of a system using water in a closed cycle. Socialism puts at humanity's disposal a real solution not only to economic but also to ecological problems.

The opinion expressed by certain scholars regarding the mutual possibilities and twofold consequences of men's work as a function of the society in which the activity is produced, acquires particular importance. In a society built on the principle of private property, work is not only a constructive but also a destructive factor. This is revealed as much at times of intense ecological crises and the growing production of men's destructive means—different kinds of armaments—as in the destruction of open spaces.

In technology, thanks to its reciprocal action which has expanded with the sciences, the materials created and energy sources, new forms of production, radical changes in technological procedures begin to play a primary role. This is particularly apparent in the new approach to the problem of exploiting natural resources. For centuries, men thought that natural resources were inexhaustible. However, at the moment man exerts a very strong influence on nature. Technology, the instrument in man's hands in his reciprocal relations with nature, was able to develop for years by exploiting certain resources while, in most cases, ignoring or squandering others. Now such an approach to natural resources is not only unjustified but also dangerous because, leaving aside the problem of wasting resources, it is also linked to the problem of environmental pollution. This forces us to approach the problem of technology and the management of the exploitation of natural resources in a new way. We are also approaching the problem of environmental pollution in a new way by examining it from the point of view of the undesirable losses suffered by man in terms of indispensable and precious materials. Technology has the essential role of solving the problem of the exploitation of valuable wastes. Unfortunately, the opinion apparently still persists that such a technology would be indispensable only as a safeguard for the environment. It is rarely examined from the point of view of the rational exploitation of natural resources, as the means for such exploitation.

It is precisely this aspect of the new technology which is attracting the increased interest of engineers, scientists and economists today. The installation of non-wasteful technology is becoming one of the most important practical undertakings in the struggle to exploit natural resources rationally. It is a question of production technology which repeatedly uses raw materials through the constant recycling of reverse production. What we habitually call pollution is nothing other than the non-utilized substance, the transport and treatment of which requires the expenditure of energy and work. The scientific and technological revolution provides us with the means to solve this problem.

Depending on the degree of evolution of technological means in industry and the rural economy, the destructive consequences of work activity at the heart of society—in chaotic evolution—acquire an increasingly serious and massive character, hence the need to effect a radical change in social relations, wipe out private property, and pass to a conscious regulation of all social processes. It is only under these conditions that the constructive part of men's work activity, directed towards their own good, can have a varied evolution. Work must become a constructive factor as much in the relations of human society as in its relations with the entire environment, which, in principle, is impossible where production is subject to the race for big profits.

Having said that, the problem of the reciprocal action between society and nature acquires a sharp political character under the conditions of the scientific and technological revolution. It is not by chance that workers in capitalist countries are making more and more frequent protests in order to protect nature from the abuse of monopolies. This proves that people are seeing with increasing clarity the reactionary role of capitalism with regard not only to social phenomena but also to environmental protection.

The methodological analysis of the problem of the reciprocal action between society and nature shows the need for its diffusion, primarily from the point of view of the history of the evolution of the relations between them.

In terms of its origin society is linked to the surrounding nature of the earth. Society is part of the earth as a detail is part of a totality. This is why the proportion of society and the natural environment in which it exists can be examined from the methodological point of view on the level of general laws governing parts to the whole. However, it must be added immediately that man is such a specific detail of the earth's natural environment, which, depending on the degree of man's development, defines to an increasingly large extent the properties and tendencies towards a change of the whole—nature.

Every complete system is always heterogeneous in terms of its composition—hence the roles of the parts in integrating the whole are not equivalent. In any totality of details one can always distinguish one detail or group of details on which the properties of the totality primarily depend. For this reason it is not altogether necessary that this detail (or details) be dominant on the level of mass and volume in terms of the other details. The role of the parts in defining the specific system as a whole is filled, according to the rule, by the subsystem which is most dynamic and mobile in the changes of the whole.

Science and practice will show that the study of natural conditions and their exploitation must be approached in a heterogeneous way because these conditions represent in themselves a systematic totality whose laws of development must be known and taken into account with increasing awareness depending on the degree of the rise in practical importance of society for nature.

The most adequate complete characteristic of natural conditions is reflected in the notion 'biosphere'. It is related to that part of the earth's surface where life, which it transforms qualitatively, exists. The biosphere was established as a system long before man's appearance on earth and its mode of existence is a process of permanent synthesis and destruction of organic substance, important for man. All of man's productive activity is based on the utilization of this global process of nature and here I can say without exaggeration that the future of civilization depends entirely on man's ability to know, exploit, and subsequently optimize the processes of synthesis and destruction of the organic substance of the biosphere. In studying any system, the essential thing is to understand its self-development on the basis of its own immanent conditions. This is why it would be altogether logical to see what was characteristic for the development of the biosphere before the appearance of human society. This will allow us better to clarify the situation created later by man's appearance and will enable us to apply certain methodological principles to the solution of this problem.

As a complicated heterogeneous system which includes many spheres in a varied, aggregate state of substances, the biosphere represents in itself a system which develops by itself and possesses the property of self-regulation. The role of the integration factor is filled by a vortex (circulation) of substances and energy (geological and biological), at the side of which each of the biosphere's components, appearing slowly and gradually, is included in this system's self-regulation mechanism by fulfilling a defined function in correlation with the other parts. In the course of the differentiation and organization of substances on the earth's surface, not only were different spheres formed but also different paths for their reciprocal action. Hence with the appearance of thousands of organisms, there arose not only different species but also the reciprocal mode of action between them. Gradually, different co-societies of plant and animal species were created, closely linked to the exchange processes of the surrounding environment and between them—called biogeneses.

However, the intensity and breadth of exchange processes in nature are limited, first of all, by the natural dimensions of the elements which participate in it and, secondly, by the regulation of the number of organisms of each species and other species, either co-operating or living on their own.

If one of the species, by virtue of temporary favourable conditions, separates itself from the fixed limits of the existing system of ties with other species, sooner or later, the equilibrium will be re-established, thanks to growing pressure on the part of the other competitors who are struggling to survive. Thus, we see how a permanent state of dynamic equilibrium is maintained in the biosphere, and at the same time, we observe a gradual, progressive equilibrium: the power of the fertile layer of the soil increases the amount of free oxygen in water, the air rises and the organisms themselves improve in the process of natural selection.

Self-regulation in the biosphere creates conditions which maintain not only the existence of the organic world but also the equilibrium of its more vast reproduction.

With the appearance of human society, the guiding role in the development of the biosphere became more and more an anthropogenic factor. What then are the modifications which the biosphere has undergone since man's appearance?

One of the important indicators of social progress lies in man's mode of life. The extraordinary rhythm of social evolution; man's evergrowing influence on the natural conditions of his existence; the positive and negative consequences of scientific and technical progress—these things clearly show that we must understand by the term production not just the production of things, merchandise, and services, but also the production of man's very needs, and furthermore, the production of the conditions of human existence, i.e. the conditions which define the face of contemporary society. In other words, this means that the term production must also be understood as the production of man's mode of life and the mode of his social communication.

Man, like every living thing, needs a permanent exchange of substances and energy with the surrounding environment. However, this process is assured man as much directly as collaterally by the instruments of work. It is well known that man, through work, expresses indirectly, controls and regularizes the exchange processes between himself and nature. These are the very general details of work, proper to each of its social forms. By placing the instruments of work between himself and the objectives of nature, man eliminates the rise in the intensity and breadth of the exchange processes which were once barriers; the exchange processes are now produced more through the intermediary of the instruments of work than by the organism of man. At this point the level of progress of the productive forces becomes the only process limiting exchange between society and nature. The possibilities for improving productive forces are unlimited while the time allowed for the creation of technical installations, whose level rises without interruption, constantly diminishes thanks to the evolution of the sciences. The rhythm of technical progress is appreciably speeded up because of the vast application of electronics and computers. To this we can add the role of automatic processes for information retrieval, storage, manipulation and exploitation, which previously passed primarily through man's brain and nervous system. In fact, the elimination of the last biological limits of information processing in the form of the natural possibilities of the human brain, is one of the essential traits of the scientific and technological revolution. During the last twenty years the rhythm of the evolution of certain countries has been six times more rapid. Under these circumstances we are seeing the increasingly tenacious appearance of the limit to existence to which we previously paid no attention, or the meagreness of the biosphere's compensatory possibilities.

Because of the breadth of its activity, humanity has become a global factor. This is clearly indicated by comparing the quantities of substances and energy put into motion by man's activity with the analogous processes of nature on the earth's surface. To this end it should be remarked that the processes which frequently develop on the earth's surface, which we presently call natural, are in fact to a great extent conditioned by the influence of man's activity.

For example, the modifications now produced at a faster rate in the composition of the atmosphere, seas and oceans, considering the constantly growing amount of industrial waste and other refuse poured into them, undoubtedly exercise an influence on the dynamic of the aeration processes of forests, the formation of condensation and other phenomena which have long been considered useful. At the moment, the great danger is the pollution of the oceans by fuel oil because of the discharge of oil into the sea, tanker accidents, etc. Consequently one-fifth of the oceans' surface is polluted by oil. One of the most alarming dangers lies in the fact that industry, given the present stage of technology and the goals of progress, can exhaust the two atmospheric layers of oxygen reserves in 180 years.

Innumerable factors of environmental pollution can be cited which unfortunately do not reflect the essential aspect of man's influence on the environment—its qualitative aspect. It lies in the fact that man, through his productive activity, draws off chemical elements from their natural bonds, transforms them into tight compounds, previously unknown to nature. Many of these compounds are generously applied by industry to the rural economy and daily life. Most of them are in natural bodies of water and the atmosphere and accumulate in the organisms of plants, birds and animals which man uses to feed himself.

Human activity is not only one of the powerful factors but is also a new factor in the natural environment from a qualitative point of view. We can say without being mistaken that it is at the same time the most dynamic factor. It is obvious that the earth's surface, under man's influence, has never before undergone such sharp, brutal changes in so short a time than at the present.

A disproportion has been established between the rhythm of the evolution of the natural environment as a whole and the rhythm of the evolution of human society—part of that whole. As a result, the channels of the inverse relation between the part and the whole are not formed except as the mechanisms of chaotic (spontaneous) self-regulation. Thus, our further evolution will lead to even greater discord in the ties to the whole, which will inevitably have a negative influence on the correlation of the other parts of the whole and will contribute to its progressive destruction.

Present analysis of the reciprocal action between nature and society inevitably leads us to the conclusion that there is a serious contradiction between the unlimited possibilities of the increase of man's productive forces and the limited compensatory possibilities of the biosphere. More recently the natural environment succeeded on the whole in regulating the problem of neutralizing the wastes of civilization; but the further increase in industrial wastes will lead to catastrophic changes in this natural environment. We are in the presence of a dramatic transition from quantitative to qualitative changes. Ecologists forewarn that further pollution of nature not only threatens the planet's living population but also involves as a consequence a brutal reorganization of all the parts of the biosphere.

How will all of this end, as far as the further evolutionary tendencies of society are concerned? The changes in the natural environment are already so serious that we cannot ignore them. The biosphere is man's only source of food and raw materials. At the same time, the further evolution of humanity along present lines will lead to a dangerous increase of elements seriously threatening the biosphere. It is, in effect, a very complicated problem and attempts to solve it are naturally not equivalent.

In most of the capitalist countries various specialists are proposing to lower the rhythm of progress, seeing in this measure the only possible way out. This point of view as well as others like it are wrong and unacceptable.

The only correct position is that of the scientists who consider the reorganization of the quality of production technology starting from the contemporary evolutionary tendencies of the sciences and technology. At the same time they emphasize the need to eliminate the social relations of private property which are in principle incompatible with the tasks of the rational exploitation of nature.

Naturally, it would be utopian to think that man will renounce the advantages provided by scientific and technological progress. Under the present circumstances progress must be examined as an objective law of the evolution of society. From the moment this is so and the needs for natural resources continue to grow, the only correct solution to the problem concerns the nature of progress itself. It is not the diminution of progress which man must decide upon but the elimination of its unilateral character, i.e. scientific and technological progress must be completed by progress in the relations between man and the natural environment. Similarly, the rise of science and technology demands improvement in production management and all of social life which will allow the solution of the imposing tasks of reconstituting and protecting the natural environment in an optimum state for production needs and for the survival of a healthy and sound generation.

In the struggle to increase production efficiency and raise qualitative work indicators, we also come across the problem of the ecology of technology, a more economic expenditure of natural resources. It is only in this way that we can overcome the steep deficit in natural resources, materials, and work and guarantee a faster rhythm of the evolution of the national economies of our countries. Increased production efficiency will also allow us to receive supplementary means to protect the environment.

It is obvious that the problem of protecting nature and rationally exploiting natural resources must be solved as part of the problems of the national economy as a whole; it depends directly on the qualitative indicators of production and all of social life. As a result, we have a system of interdependent links which qualitatively define one another: quality of production—quality of the environment—quality of life. The correlation of the links is complicated and clearly not rectilinear. We must find the heterogeneous solution to the problem of the rational exploitation of natural resources made possible by the mutual efforts of the natural and the social sciences.

In order to complete this task successfully, we must study the whole complex of knowledge of the natural environment and create a general theory of the reciprocal action of nature and society, i.e. we must find the fundamental methodological principles of the construction of that theory. The first methodological principle lies in the basic position of Marxist philosophy which considers society as a complex material system which develops by itself according to objective historical laws. The law which conditions the self-regulation of this system is that of the obligatory correlation of production relations with the character and evolutionary level of production forces.

According to the rule production relations are examined as relations between men in the process of producing material goods; they are underestimated only in their form of expression of the production relations towards the natural conditions or, more precisely, towards natural resources. For this reason the relationship between society and nature cannot be examined without considering all social relations in their totality, primarily those which appear in the production of material goods, as far as its mode is concerned. In the eyes of the owner, the means of the production of goods are important only as the source of the direct guarantee for increasing profits. Such a point of view is related to all the objectives of nature without exception and its destiny depends on what profits they do or do not bring. Naturally, the maintenance of the natural equilibrium, the compensation for products taken from nature, the preoccupation for purely natural compounds—all this does not serve as a source for realizing profits nor, it goes without saying, does it furnish marketable products. Considering the establishment in certain capitalist countries of State control over the natural environment and penalties for pollution, there is the recently felt need to have a 'pure' technology, well thought out from the ecological point of view. Thanks to the intervention of the State, construction of installations to purify and exploit wastes has begun; but here again we see the problem of profits. However, under capitalism, the drop in profits as a consequence of 'ecological' expenses can only be compensated for by the promoters by raising production prices. This measure engenders a new wave of inflation with all the crisis phenomena which stem from it.

It is obvious that capitalism is incapable of guaranteeing interior (economic) as well as exterior (ecological) conditions for the evolution of society in the era of the scientific and technological revolution. New contradictions stemming from the social importance of the state of the natural environment and the private form of the appropriation of natural riches must be added to capitalism's contemporary contradictions. This contradiction existed in the past as well, but it has become more acute in recent times. The present level of the rise of social production very much necessitates a change in the nature of the relations with the natural environment, a change which is indispensable not only for the consumption but also for the protection and reproduction of natural riches. Further increase in production is impossible without the guarantee of such a state of the natural environment by society itself.

In other words, the problem of guaranteeing the normal functioning of the natural environment is posed for all the spheres of contemporary production. This problem can only be solved in a heterogeneous way given the fact that all sections of the natural environment are subject to man's influence. Man's yearning towards the natural environment, apparent from the beginning of his existence, gradually increased with the help of the instruments of work, has finally reached such a degree that the problem of the modification of the entire biosphere surrounding man is on the agenda. Will these modifications take place drastically to man's own detriment or will men succeed in subjecting nature's reconstruction for their behalf?

This task can be completed only on the basis of the collective ownership of the means of production. It can be said that just as in primitive society the necessity for the collective ownership of the means of production was dictated by the low level of the evolution of the productive forces, the same holds true for the contemporary situation: at a time when powerful productive forces and their chaotic utilization in the interest of different private owners threaten nature and man because of their destructive consequences, the necessity for the re-establishment of the collective ownership of the means of production has become a basic demand of the logic of the rise of social progress. Recently ideas of this kind have begun to issue forth from bourgeois scientists who consider themselves far from being Marxists. For example, the eminent American economist and ecologist Barry Commoner has quite clearly formulated the contradiction tearing apart contemporary capitalist society:

We already know that present-day technology, being private property, cannot survive long if it destroys the social richness on which it depends—the ecosphere. Hence, the economic system which is based above all on the primacy of private interests and personal accounts becomes less effective and capable of disposing of this patrimony which is important from the social and vital point of view. Hence this system must change.

By way of conclusion I cannot fail to stress the fact that the decisive problem at this moment is not that of cutting down the process of transforming nature and restraining the process of the scientific and technological revolution. The sciences must resolve a much more difficult practical problem—the realization of conscious control over the social processes so that the human community can permanently utilize the enormous forces unleashed by nature and regularize its evolution in reciprocal action with nature. As Radovan Richta says in his article, society's task is not to 'clip the wings' of science and technology but just the opposite. It is in society's vital interest to develop in every way its attainments in order to embrace the entire complex of problems touching on the reorganization of nature and society. At the same time, the most difficult and complex scientific tasks are linked, in the final analysis, to the necessity of guaranteeing priority for problems concerning all of man's evolution within a single science system—social history and natural history.

We scientists can have the most diverse conceptions on the philosophical, religious and political plane; but when it comes to problems as important as safeguarding peace, the vital environment, the struggle against pollution of nature, the harmful effects of the scientific and technological revolution, we must apply ourselves to reaching a correct solution by studying these problems on a broad front. Our different conceptions of these problems must not prevent us from marching together in the struggle to protect humanity and civilization, and the proper scientific regulation of the relations between nature and society. I am convinced that in this way we will do our social duty to man, history and science.

Social sciences and human development

Aurelio Peccei

Club of Rome, Italy

The literature on the scientific and technological advances and the social sciences points to the serious imbalance which exists between two very large and different but ever-more closely interrelated fields of human knowledge and enterprise. One concerns the natural systems and what our species has been and is able to do in using them in order to emerge from and transcend other forms of life on earth. The other field treats the human system and its organization and functioning, which not only must be satisfactory in themselves but must also be kept in constant harmony with the natural systems.

In the present stage of human evolution, there is, however, little doubt that, essentially thanks to what is generally termed the scientific and technological revolution, our proficiency and accomplishments in the first field outrank by far those in the other. The fact that our knowledge and understanding of the natural systems' phenomena and laws have progressed much more rapidly than our insights into the sphere of societal and human matters in general—which have indeed lagged behind—indicates that a grave cultural distortion affects our proud civilization. Herein lie the roots of the almost intractable world problematic besetting mankind. It would be unwise to hide from ourselves that the disparity between techno-scientific and psycho-social developments is growing, not diminishing, and hence that, if remedial action is not taken, the human predicament is bound to deepen.

The situation is aggravated by the uneven unfolding of the scientific and technological revolution itself. The monstrous folly is well known, namely that its main thrust is in military-related sectors—which according to authoritative estimates occupy 40 per cent of the world's scientists. As a matter of fact, it is largely due to the obsessive efforts deployed in this direction that far more advances have been made in the branches of what constitute 'science' in the restrictive Anglo-Saxon conception—physics, chemistry and engineering, and their industrial applications—than in other disciplines or directions. As a consequence, the human environment and existence have been transformed radically, but in a deformed and dangerously lopsided manner.

In taking stock of the amazing developments wrought by the scientific and technological revolution so far, it is readily realized that they have opened up to our understanding most of the secrets of the physical world, without having given us a reasonably equal grasp of and mastery over the processes of life. We are still devoting much more time, talent and treasure to discovering what happens inside the atom and in the galaxies than in the earth's biosphere. Yet, this tiny film of soil, water and air covering our planet provides the only habitat fit for life as we know it, and, therefore, is our most precious patrimony and the base for the continuation of our species in the future. Though previously we in our millions and billions exploited all available natural resources, and more particularly this unique asset, as if they were limitless, we now realize how reckless this has been, and understand that our interference in the webs of life on which we depend has become so intensive and massive that it may exceed nature's regenerative capacity. It is to be hoped that the chilling awareness that this can become a fatal boomerang threatening our own existence may at last prompt us to develop the life sciences on a par with the sciences of matter.

Although of grave concern, these are by no means the only imbalance acquired or generated by the scientific and technological revolution in its torrent-like rush forward, which tends to de-stabilize the entire human system. Others are due to the presumption and arrogance generally born of success. We are tempted to focus research and development activities more on what can be done than on what ought to be done, or to foster prestige projects or those serving specific sectoral interests, namely projects of the already powerful, rather than those geared to meeting the real, pressing needs of mankind.

The re-equilibration of the scientific and technological revolution and its purification from such serious deviations are quite evidently a precondition for the sane conduct of human affairs. Therefore, even before approaching the relationships of the scientific and technological revolution with the human sciences, a thorough and critical analysis of the scientific and technological revolution itself is indispensable.

This is all the more necessary because of the unprecedented growth of the world population, which may be expected to double from 4,000 million to 8,000 million by the beginning of the next century. The growth in numbers of human society and its ever-greater technologization are each other's multipliers. The two phenomena have already reached orders of magnitude which are totally new in human experience. If they are not kept under control and if their further development is not planned, and if the consequences are not thoroughly assessed in advance, it is only to be expected that the stupendous but anarchical onrush of the scientific and technological revolution will throw mankind totally out of balance, submerging it under an enormous tangle of problems far bigger and more complex than those it is finding difficulty in solving today.

One of the conclusions that should be drawn even from a cursory overview of the scientific and technological revolution is, therefore, the imperative necessity that thc present phase of blind reliance on it, and on any kind of 'progress' it is capable of engendering, should now be followed by a new phase, based on careful assessment and planning on a global scale. This will mean concerting the science policies of the various human groups (presently none of them has a clearly defined long-term science policy), in such a way that this extraordinary and unparalleled human enterprise can be pursued purposefully and responsibly.

Let me add, though, a final caveat about the scientific and technological revolution. However satisfactory may be the conception of its policy, planning, strategies, research, forecasting and assessment, this will not suffice if good management of these activities is not assured. Today, for example, the amount of technology easily and cheaply, even freeely, available in the world is much greater than what is actually put to use, and this is often made in an erratic, sloppy manner. This is true not only in agriculture, but also in many branches of industry, public administration, and other sectors which are in great need of modernization. Appropriate technologies, pertinent know-how, reliable methods of organization, simple and efficient planning procedures, as well as training courses and facilities open to lower, middle and top management personnel are obtainable virtually for the asking.

None the less, the entire spectrum of the scientific and technological revolution's activities is, by and large, grossly mismanaged not only in the developing but also in the developed countries. There is no lack of instances of nations, nay entire regions, where performance or productivity are well below their scientific and technological potential. Most of us have first-hand experience of this. Therefore, to trust essentially, all the same, as many seem to do, in the scientific and technological revolution itself as the source of remedies to each and every problem is naïve and misleading in many ways. The least that can be said is that, if a quality jump is indispensable to our capacity to conceive rationally our techno-scientific endeavour, this is no less true of our capacity to implement efficiently our designs in practice.

What is even more troubling at this turn of history, however, is not so much the runaway, unpredictable and even dangerous character of the scientific and technological revolution and our capacity to put some order in it, but our capacity or lack of capacity to make the human system evolve in parallel to it, let alone, as will ultimately be necessary, prepare in advance for it. The present inability of the human system to adapt swiftly enough to, and still more to anticipate and regulate the impact of, this metamorphic revolution, which it itself fosters, is proof of the cultural mismatch and disorder of our time that I have mentioned. Our generation is indeed the witness and a party to a gigantic case of schizophrenia, which is crippling everything human. This epochal derangement can be cured only by re-establishing a sound, healthy cultural equilibrium within modern man himself.

Let me start however by saying that to catch up with the scientific and technological revolution and to keep pace with the profound changes that it is continuously bringing about all over the place, and finally to put these grandiose processes under control, no doubt the human system must undergo a fundamental transformation. A few adjustments here and there would be utterly insufficient. The tempo and nature of the changes occurring in everything else and the extreme alternatives for good or evil they open up for mankind as a whole suggest that much stronger medicines are required. In essence, they tell us that some sort of basic mutation in our modes of being and governing ourselves and our planet is becoming a life-and-death question.

A concentration of efforts in the social sciences is certainly indispensable to this effect. This would be a means of clarifying issues, of envisaging what alternative plans for societal change seem feasible and of charting the way to enact them. It would, however, be illusory to rely essentially on the results obtainable by developing the social sciences and applying their recommendations—even if all-out development were possible—if this is not accompanied by a change in the hearts and minds of people.

Although there is a growing body of literature on the relationship between the scientific and technological revolution and the social sciences, this is a matter so vast and complex that deeper and broader investigations are needed. The overall objective of such investigations is to lead not only scholars but also decision-makers and those in a position to influence public opinion to grasp the fact that it is a question of vital importance to redress the presently unbalanced situation.

I have particularly appreciated Richta's essay included in the present publication, although some of the points made are in my view based on an excessively stereotyped vision of capitalist and socialist societies. These are themselves in flux and far from uniform, and the world also comprises other types of society, which escape any formal categorization. The diverse cultural patterns which represent and substantiate the identity and individuality of the very many and different human groups populating the Earth constitute one of mankind's most precious assets, not least for survival, and in their very diversity call for a variety of socio-political organization forms. These cannot be reduced to one or two models. To be sure, none of the types of socio-political organization in existence can claim to be perfect, and none is suitable anyway for general export to the world as a whole. On the other hand, in practically all respects, future requirements will be quite different from those at present. Hence a plurality of experimentations in this sphere and a cross-fertilization of different solutions seem to be the most fruitful steps that can be made in this period of transition.

At the same time, we should never forget that the scientific and technological revolution has altogether changed the size and power of mankind, and its capacity to communicate, travel and influence the condition of every corner of the world, making of human society an integrated, global system vitally depending on an organic web of interconnections and interdependencies. 'Unity in diversity', as referred to mankind, is not a mere slogan nowadays, even if many who use it just pay lip service to it or envision the two terms of the formula not in association, but separately, sometimes recognizing the system's unity and sometimes its diversity, as may be expedient.

One of the common errors made in the application of the social sciences is to try to adapt such a new, global reality to our preconceptions, if not prejudices, coming from a quite different past, and not vice versa. Thus it is widely admitted that a host of modern problems—predominantly of a security, development or environmental character, for instance—have grown and expanded so much to become planetary in themselves, or at least their consequences have, and that they intermingle so inextricably as to form a global problematic. Yet, having made a basically correct analysis of situations which are without precedent, most of us try to combat them by methods that are patently old-fashioned and inadequate. By all means, global problems should be met with an adequate 'global' approach, but we try instead all sorts of 'international' approaches which are a far cry from the right answer. In fact, they attempt to reduce something which is inherently 'world wide' into segments cut to suit our 'national boundaries'.

With the present political fragmentation embodied by some 150 egocentric national States, among which devergence of aims and tensions are greater than solidarity and co-operation, the result is that this new category of overarching problems, not being seriously attacked, grows further in intensity and complexity, leading not to a stalemate but to sure defeat involving in one way or the other all nations and peoples, whatever their political organization or degree of development. Now, since the world community of States and each of its members are profoundly affected by the scientific and technological revolution—an enterprise which by nature is a propellent to globalism—it is time for all of them to accept its logic. And the social sciences should show the way. In other words, while taking into account the world's multiple cultural strains and value systems, it is for them to show that these are facets of an organic whole, and that the parts and the ensemble must be rendered mutually compatible.

An example that this higher level of vision and understanding has unfortunately yet to be attained is provided by one of the hot issues of the moment—the new economic order which should be adopted by the world. Not only is this essentially viewed as an 'international' order—rather than a 'world' order—but the nations negotiating it do so in the optics of their sovereign interest, turning a blind eye to the state of the world economy which may ensue, although all of them will heavily depend on it in the final analysis. Moreover, other, important nations are totally absent from such 'world' negotiations. The outcome cannot but be a bad, precarious compromise. To satisfy immediate national aspirations and at the same time safeguard the long-term interests of mankind is like squaring the circle. The time to make a choice, as mentioned, is now.

The debate on the world economic order is good, anyway, because it may serve to introduce new dimensions, ideas and vistas into our thinking. It will make abundantly clear that the wealth of the planet, in both resources and space, has to be shared more equably among all peoples now that the number of humans has doubled in the last fifty or sixty years and is all set to double again. It will also stress that the knowledge and wealth produced by human ingenuity and work should be distributed more evenly, with more consideration to human needs and less to other factors which now are given the leading role. It is for the social sciences to absorb and disseminate these orientations and concepts and use them in devising how such a teeming and powerful mass of mankind should organize itself not only to ensure that its different components can co-exist, but to permit each and all of them to develop without prejudice to the others now alive and those yet to be born.

In another sense, I feel that the present ways of analysing the new human condition and how to strengthen it and make it more rewarding are at fault. They miss the fact that the key element is the individual, the billions of individuals who are the true protagonists of the human venture. Both the physical and natural sciences, and the social sciences, as well as any combination thereof, tend instead to focus on environment, technologies, structures and generally non-human things, and to find solutions starting not from within but outside the human being.

A similar error of vision is made even when attention is addressed to questions related to human development. Development is currently equated with the reasonable satisfaction of human needs, not with the improvement of people's quality and capacities, which alone can offer to one and all a real possibility of self-fulfilment. Hence, the current optics should be reversed. From a conception of needs-oriented development, we have to move to one of the development of the human potential, if for no other reason than because man's needs can be satisfied (and restrained, if necessary) only by man himself.

These erroneous attitudes are reflected in the approach to one of the most difficult and delicate parts of the problematic. Population problems are almost always considered in their quantitative aspects, which relegate people to the rank of economic units, biological entities, demanders, consumers, and by and large problem-raisers, while the qualitative aspects are largely neglected. Yet the quality of the members of a community, the people of a town, the citizens of a nation and the inhabitants of the world is much more important than their sheer number.

The business of human life has indeed become so complicated that the downward trends of human fortunes—what the Club of Rome calls 'the predicament of mankind'—can be counteracted and reversed only by a rapid, intensive cultural evolution capable of making our generation all over the planet, and those who will follow, fully adapted psychologically and functionally to the new world in which we and they will have to live. If this evolutionary jump is not made, the human problems will remain fundamentally unchanged, fundamentally unresolved.

Let me now conclude with a few questions. Why the scientific and technological revolution? What objectives should the social sciences set themselves? What is the place of the human being, of our species, on the planet? What are their 'finalities'? I do not think that appropriate answers can be found in the direction along which we are now moving. We seem to take pride in considering man somehow separated from nature, and his goal that of achieving a complete mastery over it and transforming it in order to install a dominant human system all over the earth, and possibly beyond. However, is not man just a part of nature and should he not strive to become its best part and be its best defender? I submit that such questions have now to be asked, because our species has acquired so decisive a power, for better or worse, that its destiny depends essentially on how it will use it in the context of its terrestrial habitat.

The impact on man and social structure

Gordon Rattray Taylor

United Kingdom

We are invited to consider the impact of science and technology on society and to suggest ways of reorienting both of these so as to enable man to attain his goals and preserve his values.

I shall argue that technology, while conferring inestimable benefits in the short run, has created severe social problems in the long run—and in particular the loss of social cohesion which is occurring in many parts of the world. The impact of science, which I take to mean a body of theoretical knowledge and understanding of process, is rather different. Of course, if we did not have a corpus of scientific knowledge, we should not have an advanced technology; thus, indirectly, science is responsible for the problems evoked by technology. But before coming to these, let us consider for a moment what direct effect science, taken as a whole, may have had on man.

On the one hand, by revealing the universe to be ordered—to be nomological—it has banished the evil spirits, taboos, unpredictable and vengeful deities, and cosmic influences which made man fearful and insecure. It has saved much time hitherto wasted in spells, incantations, prayers and rituals. But, on the other hand, it has undermined or even destroyed, for many people, the sense of personal identity and worth, and the hope of immortality. Man finds himself, apparently, merely a superior kind of machine. Before the end of the century the brain sciences may reveal, or claim to reveal, man's behaviour as wholly determined, thus depriving him of free will, and this may prove, for many people, a grievous burden to bear. Moreover, if men are only superior machines, it is alarmingly easy to regard them as expendable and even more tempting to manipulate them. The brain sciences are making such manipulations easier and more far reaching, and before the end of the century the control of the mind may be most serious of the problems confronting us.[1] A recently published book reveals how, already, in the United States and elsewhere, subliminal stimuli are incorporated in television programmes and advertisements and that the United States Government has refrained from intervening.[2]

Technology: bane or boon?

The benefits bestowed by modern technology are numerous and widely advertised. Conveniences like electric light, more numerous possessions, speed and ease of communications—these are all so well known that I need not emphasize them. At the same time I do not want to seem to underrate them. Greater production efficiency provides society with a surplus which makes it possible to have increased leisure, to educate, and indeed to pursue science itself. In a country with a low living standard, the advantages bestowed by technology are substantial. We all need food, shelter, clothing and so on. It is when these basic needs have been met and technology begins to provide more and more trivial benefits that the problems chiefly arise. If we look around industrially advanced countries today, people do not invariably seem happier than in the past, and often they are no healthier. Environmental wastes produced by technology have created new forms of illness and fostered others: between 80 and 90 per cent of cancer is now attributed to environmental influences.[3]

So let us look in a more systematic way at the debit side of technology's balance.

The work process itself has become more monotonous and frustrating. Men need a task which challenges their skill and their wits—it is not a pleasure to play tennis against a very inferior player. The production line must be designed for the slowest and least intelligent man. Man needs variety: subdivision of labour, however, has minimized variety. (The strike, of course, breaks this monotony, which may account for some 'wild-cat' strikes.) Centralization of planning and decisions reduce the challenge to the ordinary man's wits. And man needs to be valued, to have a personal identity. In technological society he tends to become a cipher. In contrast, in many pre-industrial societies, work is relatively varied, the pace is flexible, and so on.

Urbanization can be regarded as a product of technology. Industrial needs drew people together in the first place, and without modern technology, a city of a million or more people could not be kept going. But human beings need contact with nature and occasional solitude. Young people require to work off steam in physical activities, which is much easier in the country than the town. So they turn to violence and vandalism—or withdraw by using psychotrophic drugs.

Technologically based industry has disrupted the family. For instance, long commuting times often separate a father from his children at the critical age of 1 to 2 years. The disruptive effect of paternal absence has been explored in detail by H. B. Biller.[4] Technological society also tends to separate the mother from her infant—in the United Kingdom a woman's job must be held open for six months if she leaves to have a baby. And the same forces have weakened the extended family, and attachment to place. Easy communications induce children to move far afield, and the extended family becomes steadily more scattered. As a further result, the cultural transmission of techniques (e.g. cooking, child upbringing) is weakened, and educationists step in and attempt to fill the gap.

Group structure is the least understood, yet one of the most important points. Societies are normally composed of coherent groups—communities of which the village used to be the prime example—in which personal identity is recognized and pressures are exerted to deter people from anti-social acts. If too isolated, such groups become oppressive. But normally they are embedded in larger structures, linked by super-ordinate groups in a complex network. Modern mobility, plus the centralization of administration, which new means of communication have made possible, has worked to create a 'mass society' in which people feel themselves to be isolated individuals controlled by an élite. As Kornhauser has clearly shown, this opens the way to dictatorship.[5] At first the effects of easy communication seem beneficial because they free people from local pressures, but as the trend proceeds, they find themselves isolated. The loneliness of the big city is almost proverbial.

Technology has certainly made war more disastrous: it now affects all, not only civilians but neutrals and primitive peoples.

The population explosion, our greatest problem, is with us purely because of technology—I need not spell out why. The run-down of the world's non-renewable resources is also due to it. And so is environmental pollution. This is now well recognized, so I will not dwell on it.

When we are starving, we do not brood but struggle for food. But when all our pressing material needs have been met, we begin to ask, 'What is it all for?' Existential despair is exposed. Gustav Jung recalled how often his patients were men of 40 or so, who had established themselves in life and were deeply disturbed because they had begun to ask this question and to look for an overriding purpose. The moral would seem to be that industrialized societies should taper off the application of technology—but no one knows how to. They are trapped in a consumption machine.

On all these trends one comment must be made: they are occurring too fast. Human development contains a natural limiting factor, the human life-span, which has more than doubled in the last century. Children can adapt to new ways, old people find it harder. Rapid social change makes it difficult to plan one's life and probably leads to social costs which we have not yet begun systematically to assess. (Here is a valuable research area.)

In sum then, man has psychological needs as well as physical needs. We have found in technology a marvellous system for meeting his physical needs, which at the same time frustrates many of his psychological needs. In proportion as we escape from physical slum conditions, we find ourselves in a psychological slum. Our lives are varied; we acquire status through what we do rather than what we own or how we were born; we are faced with challenges which give scope to our mastery drive and have a degree of autonomy in tackling them. Some of us may find it hard to perceive the extreme frustration and the sense of hopelessness which afflicts many of those in industrialized societies and which is becoming manifest in the developing world as well. To reorient technology so as to reduce this frustration and sense of impotence is not merely desirable. It is imperative.

Herman Kahn of the Hudson Institute, despite his generally optimistic standpoint, has admitted: 'A respectable case can be made that the threat of catastrophe is real enough so that we must control and limit technology even at the cost of progress and growth.[6] And as long ago as 1955, a much greater man, the mathematician John von Neumann, asked in an article in *Fortune* magazine, 'Can we survive technology?'

Substitutes and distractions

But to break out of the groove we are in is particularly difficult for at least three reasons which deserve our close attention. First, we have developed the trick of providing manufactured objects which cater for frustrated needs, so that the more frustrated we are, the more we depend on the system. For instance, we buy an indoor rowing machine or bicycle to give us the exercise our daily life fails to provide, as we sit at our desk. To a visitor from Mars, the sight of a man rowing away at nothing would look like pure lunacy, not far removed from digging holes and then filling them in. We take pills to give us the tranquillity that in a pre-industrial society most people took for granted. Perhaps the most obvious and widespread example of substitute satisfaction is the purchase of objects to bestow status. In unspoiled societies, status—or to be more precise, prestige—is derived from the contribution one makes to society. Even in industrial society, doctors, sportsmen and a few others are valued because people know and understand what they contribute. Many others have to buy costly objects which they may not really want, or live in a lavish style, just to prove to the world that they have achieved something. Conspicuous consumption has reached a ridiculous pitch. In the United States one can now buy a machine which will choose for you which tie to wear. You tell it what suit you are wearing and what you are going to do during the day and it tells you what tie to wear!

The marginal utility of such status-bestowing objects decreases rapidly. A black-and-white television set may make a real difference to one's life. To substitute a colour set makes a smaller contribution. To be able to change channels without rising from one's chair is so trivial that I for one would not waste money on it. I can also survive without an electric carving knife, to pick one more example. Taken as a whole, labour-saving devices are in danger of making life too easy. There is a satisfaction in climbing to the top of a mountain which cannot be gained by driving to the top in a car.

Therefore, I believe we should consider whether the social scientists can tell us how to reorder our status and prestige sources.

The second factor is rather similar: the demand for distractions. If the working week is frustrating, we demand rather intensive experiences in our week-end leisure to compensate. Some of us like to travel at 100 miles an hour on a motor cycle—a pure waste of petrol and machinery in the practical sense. The whole apparatus of sex-exploitation is, of course, catering to the demand for distraction. So are many books, films, plays and television programmes. A person whose job absorbs him has to be more selective in such areas.

The third factor I call the technical trap. If you have worked all week and acquired money, you may as well spend it on something, even if the pleasure to be gained is less than the unpleasure caused by your boring job. Your preference might have been for more leisure or a more interesting job, even if the pay was less, but your life circumstances may not offer you this option. In the professional field, people often do accept a less well-paid job because of compensating advantages. For low-grade workers it is usually harder. Could we not change this situation—provide more options?

The need for autonomy

I would now like to draw attention to a profound human need which I believe has been much underrated and sometimes overlooked. Human beings have a deep desire to manage their own affairs, to achieve independence. They do not simply want to be given food, shelter and so on: they wish to achieve the satisfaction of their own needs by their own efforts. A minority may lose heart and settle for a parasitical existence, but most people find dependence degrading. Those who do so settle find themselves permanently dissatisfied: not knowing the reason for their frustration, they may go on asking for more and more. In the jargon of psychiatry, this is known as the need for autonomy.

It is often argued that decisions will be better made if they are centralized and made by high-calibre individuals who have studied the situation closely. But even if this were true—and it often is not true—it is still the case that efficiency is not, or should not, be the only criterion of social organization. Dostoievsky understood this:

> And how do these wiseacres know that men want a normal or a virtuous choice? What has made them conceive that man wants a rationally advantageous choice? What man wants is simply independent choice, whatever that independence may cost and wherever it may lead.

And of course this is what current demands for devolution are about. The separation of the Basques, the Welsh and others, like the fragmentation of Africa into national States also arise from it.

A word now much in vogue in the West is 'élitism'. This refers, I believe, to the feeling that too many decisions are taken by small groups of 'experts' (who may indeed be better qualified to judge) without much reference to those their decisions will affect. As one woman said in a letter to a British newspaper recently: 'I don't expect my opinions necessarily to prevail—but I do want to feel that they have at least been taken into consideration.' It is notorious that state servants tend to feel that they know better—they probably do—but that is not the point.

A cardinal mistake of reformers over the past half century, I suggest, has been to underrate the demand for autonomy: to assume that the satisfaction of material needs took precedence over all else. Yet we have heard people, looking back on the past, say: 'We were much poorer then but we were happier.'[7] Now that, in many countries, material needs are being adequately met for the great majority, the demand for autonomy begins to be heard increasingly loudly. Eventually this trend will sweep centralism away or, in some cases, will provoke a dictatorship.

However, the term 'autonomy' is sometimes misunderstood. It does not mean that people wish to battle away in isolated individualism. One does not lose autonomy if one voluntarily follows a leader or if one voluntarily works co-operatively with others in a group. Nor does it mean that one must take every decision oneself. People wish to make the decisions they feel competent to make, and are usually delighted to delegate responsibility in matters where they feel at a loss to competent persons, provided they feel confident that such persons understand their objectives and have their interests at heart. (We see this in fields as diverse as going to the doctor or making an attack in war.)

This being so, the function of the State is, I suggest, to maximize autonomy by creating the conditions in which a maximum number of decisions can be peripheralized.

I believe it is also useful to look at autonomy in developmental terms: I mean individual development. Initially the child is completely dependent: it has little or no autonomy. Its development is, or should be, a steady progress towards independence. The child needs to discover, 'What I do makes a difference'. This discovery is the basis of self-respect. If the child fails to make such progress, it remains in a passive state, demanding satisfaction but making little effort to achieve it. Moreover, if it fails to change the environment constructively, it may fall back on changing it destructively. Little boys who stamp on the sand castles of other children develop into bigger boys who pull up young trees, spray paint or try to prove that they must be taken account of by even more destructive actions.

At puberty (traditionally) the child passes to adulthood, from dependence to independence, from the influence of the supportive mother to that of the more demanding father. In 'uncivilized' societies, this transition is clearly marked by solemn ceremonies and the individual is left in no doubt about the kind of behaviour expected of him, or her. In 'civilized' societies, however, we delay the assumption of adulthood and obscure the transition, leaving the growing individual in doubt of his or her manhood or womanhood. Hence the adolescent need for danger and defiance, from 'doing the ton' on motor cycles to crime or political protest. Others may seek autonomy in drug experience, eccentric behaviour or clothing, and so on.

One's control of oneself—one's body and one's emotions—can be demonstrated by acts of physical bravery (hence the toughness of the initiation ceremonies of many pre-literate societies) and risk-taking generally. Modern society, by constantly seeking to protect people from risk, often forces them into creating risk, which they may do in an anti-social or self-destructive fashion.

I can clearly remember how anxious I was, at 19, to finish my formal education and 'get out into the real world'. I can therefore sympathize, to some extent, with those young people who, having no academic bent, raise hell during their last year at school. To a much larger extent, education needs to become education for autonomy rather than education for a role in the productive system or for academic ends.

Many parents, one must add, fail to manage the transition from the mother's to the father's influence for their sons, and sometimes exert no parental influence at all.

The assumption that the State knows best, the assumption that people should be protected from risk (I except invisible risks), the assumption that people want only material comforts, form a complex which I have christened 'nannyism'. It was the United Kingdom that invented the nanny, which perhaps explains why the United Kingdom today seems so deeply committed to nannyism.

Nannyism permeates not only Parliament, which attempts to govern by promulgating more and more rules of behaviour with more and more penalties for being naughty, but also is rife in local government, the courts, and so on. But I do not believe it is purely a British problem: the United Kingdom is merely an awful warning.

A major obstacle to the reintroduction of local autonomy is the obsessive desire for consistency which many people display. Like the commanding officer who cannot bear to see some boots treated with polish and others with dubbin, and who wants even the flowers to stand to attention, many bureaucrats are deeply shocked by what they term 'anomalies', and work with an enthusiasm worthy of a better cause to iron them out. Trade unions have often exploited the administrators' horror of local variation to raise wages or improve conditions. Many people tend to see apparent inconsistencies as injustices, without stopping to ask if there are local conditions or preferences which explain and justify them. In reality, since local conditions vary so widely, variation of treatment is only sensible. It is unification and centralization which most often generate injustices. To which the official response often is: 'We can't cater for minority cases.' Exactly so. Our educators should make these matters better understood.

Unfortunately, it is possible to condition many people to dependence by minimizing their opportunities to 'stand on their own feet'. We could call such a population heteronymous. It would be a useful piece of social research to establish measures of autonomy/heteronomy and to scale contemporary societies accordingly.

Violence and crime

I turn now to the connection between technological society as we know it and anti-social behaviour. I predict an increase in such behaviour; many people think that increase has already begun. It is all very well to say, as some people hasten to do, that there have been periods of extremely violent and anti-social behaviour in the past. The question is rather why today, in societies with elaborate social services, comprehensive systems of education, and far higher living standards, violence should still exist. It is not in any simple sense the most oppressed who perform such actions: often it is young people from comfortable backgrounds. I would venture to suggest that it is psychological slums, which need not also be physical slums, which produce anti-social individuals.

I have already drawn attention to the numerous frustrations that are associated with technological society—and it would be easy to lengthen the list—but I would only point out that frustration leads to aggression, as Dollard and Miller showed many years ago.[8] However, there are more specific links between society and behaviour.

It is generally agreed that super-ego formation depends upon the introjection of a father figure in the earliest years of life, approximately between 18 months and 2 years. At this age children go to sleep, in many homes, before the father returns from work. We have been told a great deal about the importance of the mother to the child. Now data is coming in about the role of the father.[9] Of course, it is not only absent fathers, but weak, inconsistent, alcoholic, or woman-dominated fathers who provide inadequate images for super-ego patterning.

Of course, even an adequate father can produce an anti-social child if he exhibits anti-social behaviour himself. Thus, the development of a cynical social ethic—possibly as a by-product of the scientific approach—can also contribute to anti-social behaviour.

But it is not only the family group but also the community which is important. Many years ago I pointed out the significance of what I called the 'assessment group': that is, a group of one or two thousand people, each of whom has some personal contact with or knowledge of at least half the rest of the group.[10] Such groups are bound by powerful mutual ties. Urbanization as well as increased mobility tend to break up such groups. As Charles Wright Mills has said:

> The growth of the metropolis, segregating men and women into narrowed routines and environments, causes them to lose any firm sense of their integrity as a public. The members of publics in smaller communities know each other more or less fully, because they meet in the several aspects of the total life routine. The members of masses in a metropolitan society know one another only as fractions in specialised milieux. . . . In every major area of life, the loss of a sense of structure and the submergence into powerless milieux is the cardinal fact.[11]

In an organic society, intermediate structures intervene between the individual and the State, and Kornhauser has stressed their importance, as did Durkheim before him. 'Our political malaise thus has the same origin as the social malaise we are suffering from. It too is due to the lack of secondary organs intercalated between the State and the rest of society.'[12] It is these organs which check the abuse of power. It is a cardinal error to identify the State with society.

Finally, the largest groups have to be tied together into a world community. But I suspect there is an upper limit to the size of organizations, and I fear the world has passed that limit. In the last analysis, the limiting factor is the number of personal contacts which an individual can sustain. Technology can bring him in contact with more people, but his personal capacities set a limit to what he can handle, in human as well as in intellectual terms. Many leading political figures seem to be near that limit already.

Much has been written on anomie—the sense that one does not belong to society and hence that one is not subject to its laws. I shall not go over that ground again, but shall content myself with making one point which has perhaps been neglected. This is the sense of attachment to place. We all know that proverbially, 'there is no place like home'. It is a characteristic of the mind to attach emotions to concepts. Yet we treat this as trivial, moving people from one place to another to suit the needs of industry or defence or of planning or dispersion of urban population—not to mention migration. We disrupt local loyalties and a sense of commitment as if they were of no special or personal importance. Attachment also develops to organizations and the world is full of clubs and groups which meet to recall common experiences. We also treat these attachements as trivial in our planning. (It is true that in moving urban groups, planners are now beginning to attempt to keep communities together by moving whole streets of families to similar streets. But this is still fairly unusual. In the United Kingdom recently, the government has altered local boundaries for reasons of administrative convenience with total disregard for the clearly expressed wishes of those concerned.)

The social and psychological sciences have amassed a great deal of knowledge on all these matters. I am discouraged by the fact that so little is done to apply it. Now that delinquency can be predicted with 95 per cent accuracy at the age of 8, Professor Cortes in Washington has proposed a controlled experiment in assisting families where parenting is poor.[13] It has not been done.

Using technology selectively

The broad conclusion to be drawn from this analysis—however much its details may need modifying—is surely that we should use and apply technology in a more selective manner. There is already some popular appreciation of this fact. There are movements oriented to the growing of food less intensively and to its preparation by traditional methods instead of using prepacked, preprocessed 'convenience food'. There is considerable interest in jobs demanding craftsmanship and individual taste, such as pottery, silversmith work, photography and so on. Repairs and improvements in the home are often undertaken by the owners or tenants, not simply to save expense but as a satisfying outlet for the human constructive and manipulative urge. Communes have been established with the aim of a simpler but more satisfying life style. Numerous small magazines flourish in the service of this idea.

While such initiatives reveal the emergence of a demand for a post-industrial life style, they do not tell us very much about how to achieve it on a larger scale. Experimental communes rely on power tools and pumps made by industry and, in the event of illness, draw on the medical services, specialized drugs and supplies provided by large organizations. They rely on the police, the law and Parliament to protect them from various kinds of exploitation and also to maintain relations with other nations. They are also dependent for all but the simplest education, for books, specialized information and much else.

Nevertheless, they see the goal. How to give realistic expression to these aspirations is a main task for us. I would not presume to produce a blueprint—I could not—but I would like to put forward a few ideas for further discussion.

THE ACCOUNTING SYSTEM

It is now widely recognized that the usual economic indicators fail to measure, or even indicate roughly, welfare. They do not even trace economic activity at all accurately. If a farmer grows most of his own food, what he consumes does not figure in the gross national product (GNP). If he becomes a specialist, growing a single crop, selling it, and buying in his other needs, the GNP is increased by these transactions.

Furthermore, the true economic cost of different sections of the economy is not revealed. In comparing motor-vehicle transport with railways, the numbers of people killed and injured, and the medical and police effort consequently made necessary, are an important part of the equation. There are other unlisted costs, such as pollution, damage to buildings, etc. The cost of road building and maintenance, together with traffic lights and signs, etc., should also be taken into account. (According to a recent British statement, large commercial vehicles in the United Kingdom receive a concealed subsidy of £1,700 per annum.)

This is absurd enough. But the situation becomes even more ridiculous when we consider that the cost of traffic accidents and injuries appears as an addition to the GNP, when it is really a diversion of resources in support of something no one wants. If it is true, as I have argued, that a large part of crime and delinquency is due to our social pattern, then the cost of coping with it is likewise a cost, a debit, and not an addition to the GNP. The manufacture of armaments is a still more clear-cut example of something which appears as an addition to GNP but which does little to increase welfare. The cost of anti-pollution measures is another.

Thus, we need not only better indicators of welfare but a national (and eventually international) system of accounting which depicts welfare more realistically. The Organization for Economic Co-operation and Development's *List of Social Concerns* and the American *Towards a Social Report* represent attempts to formalize indicators of welfare on a large scale, but it has been left to the Japanese, to the best of my belief, to sketch out the basis of a new national accounting system. In May 1971, a Net National Welfare Measurement Committee was established as a sub-committee of the Economic Council of Japan. The sub-committee reported in March 1973 but its path-breaking report has not received much attention or discussion.

Entitled *Measuring Net National Welfare of Japan*, it proposes to modify GNP so as better to indicate economic welfare, while measuring social advance in non-monetary forms. Another important concept it advances is the 'stock approach' in which social capital is made to include such intangibles as public parks and good orchestras. But the main advantage of the 'stock approach' is that it eliminates the illusory benefits of rapid obsolescence. If the average life of a motor vehicle is halved, the number of car sales must double, over a period of years, to maintain the stock and the increased sales appear as an increase in GNP. The stock, however, has not changed. I suggest that this Japanese initiative should be followed up widely.

Of course, such measurements may show us what we are doing and where we are failing, but they do not tell us how to do it. I believe the core of the problem is this.

THE LOCUS OF DECISION

Individuals have to operate within the system as they find it. The man who lacks status and buys a status object (which as a functional object he does not want) is doing the best he can. He cannot (except in the longest run) change the system so that his status is functional. The man who buys a rowing machine to provide him with exercise might prefer to row a real boat but circumstances rule this out. And individuals cannot purchase clean air, freedom from noise, etc. in small packages. The nearest they can get is to buy a country cottage, but many cannot afford this solution and if they could the advantages would be dissipated.

Increasingly, therefore, we leave it to regional authorities or to the State to give effect to presumed 'ideal demand'. On the whole, this solution is only weakly effective. Little is known about what people would want in ideal circumstances. State bureaucracies are cautious and slow to act. They find it difficult to empathize with the needs of the ordinary man. They are greatly given to thinking that they know best. They cover up mistakes. Since we live in a world of increasing bureaucracy, it would be wise to study how it functions and to seek to devise means of offsetting its faults. A search through the British register of research in the social sciences failed to reveal a single project concerned with the functioning of large contemporary bureaucracies. There must be a good deal of know-how, however, in the top echelons of industrial concerns and in management studies which could be brought out and applied. (Of course, bureaucratic institutions often call in management consultants, but this is in the name of 'efficiency' as seen from within. It is efficiency as seen by the customer, or victim, which I am concerned about.)

In any case, centralization has become a habit. Much of its strength is derived from the human appetite for consistency—inconsistent decisions being seen as injustice, at worst, and as untidy at best. We need to devolve decisions—or rather, we need to create the conditions in which it is possible to devolve decisions. The more we centralize and concentrate the physical structure of society—the houses, factories, amenities—the more we necessarily centralize the administrative structure. A vicious circle is established.

In sum then, we are faced with the superhuman task of changing the structure of society and the structure of socio-political ideas.

Conclusion

In drawing to a close, let me summarize some of the more practical and immediate suggestions I have made.

We should conduct research into the social and human costs of social change to establish what, in any particular set of circumstances, is an optimum rate.

We should study how bureaucracies actually work and how to make them more responsive to human demands.

We should develop the technologies of small scale and decentralization.

We should establish measures of autonomy/heteronomy and apply them to contemporary societies.

We should investigate the 'ideal demand' or scale of subjectively felt, personal needs.

We should study how to modify economic indicators so as to give a less misleading picture of economic progress.

We should develop indices of net national welfare which could give expression to the 'quality of life' and non-economic factors in general.

Above all, we should apply what we know about personality development and social structure.

It might be worth investigating why it has proved so difficult to do this. Perhaps governments, being concerned with power, are ill-suited to the task.

The penalty of ignoring the dominating nature of technology may be to court disaster. The late Professor Arnold Toynbee, after his study of some thirty great civilizations, reported that a culture which becomes obsessed by a particular technique is likely to perish.[14]

Notes

1. S. M. Farber and R. H. L. Wilson, *Control of the Mind*, McGraw-Hill, 1961.
2. W. B. Key, *Subliminal Seduction*, New American Library, 1973.
3. Dr J. Higginson, director of the International Cancer Research Agency at a conference in Manchester, 19 September 1975.
4. H. B. Biller, *Paternal Deprivation*, D. C. Heath, 1974.
5. W. Kornhauser, *The Politics of Mass Society*, Routledge & Kegan Paul, 1960.
6. H. Kahn and B. Bruce-Briggs, *Things to Come*, Macmillan, 1972.
7. J. Seabrook, *City Close-up*, Penguin Books, 1971.
8. J. Dollard, *et al.*, *Frustration and Aggression*, Yale University Press, 1939.
9. Biller, op. cit. See also J. McCord *et al.*, *Journal of Abnormal and Social Psychology*, Vol. 64, 1962, p. 361–9; E. M. Hetherington, *Journal of Personality and Social Psychology*, Vol. 4, 1966, p. 87–91.
10. C. W. Mills, *The Power Elite*, Oxford University Press, 1956.
11. G. R. Taylor, *Sociological Review*, Vol. 40, 1948, p. 57–65.
12. E. Durkheim, *Professional Ethics and Civic Morals*, Glencoe Free Press, 1958.
13. J. B. Cortes and F. M. Gatti, *Delinquency and Crime: A Biopsychosocial Approach*, Seminar Press, 1972.
14. A. Toynbee, *A Study of History*, Oxford University Press, 1947.

Part Three

Social functions of science and science policy

Science, technology and social progress

D. M. Gvishiani

Corresponding Member of the Academy of Sciences, USSR

To broad circles of scientists throughout the world, and to numerous segments of the world public, it is becoming increasingly evident that the present epoch deserves the reputation of being the epoch of the most radical break-up of social relations in the history of human civilization, the epoch of the radical reconstruction of all social forms of vital human activity, the epoch of the most varied and exceptionally complicated social changes which are closely interwoven and which are proceeding at a pace hitherto unprecedented. At the same time there is a growing conviction among many scientists of the world that one of the most specific features of the present epoch is the indissoluble connection, the very close interaction and mutual conditioning of contemporary social transformations with the revolutionary transformations in science and technology, which, too, are of a most radical nature, unprecedented complexity and so strongly intertwined that they have, in essence, merged into a single stream which has been given the specific name of the 'scientific and technological revolution' (STR).

This revolution has acquired world-wide dimensions, and has become a global process of modernity, although it should be noted right away that it is unfolding in different social forms with different and, at times, even opposite social consequences. Nevertheless it is on the scientific and technological revolution that the whole world today pins its hope as it carries out this revolution ever more broadly and deeply, and ever more strongly feels its impact. It is with this revolution that various countries (and undoubtedly with good reason) link the general prospects of their further social and economic transformations. It is this revolution that is considered, and rightly so in general, to be the most radical means of accomplishing the major social tasks that have been put on the agenda by the contemporary process of social development. For all the difference in understanding of the criteria and indicators of social progress, for all the difference and even opposition of aims and consequences of the STR in societies with differing social systems, this revolution is everywhere regarded as a major factor of contemporary social progress, as a global phenomenon of historical development of fundamental social significance in its depth and sweeping breadth.

Elucidation of this significance, analysis of the functions the STR performs and which it can, in principle, perform in the social progress of the present epoch presupposes, naturally, the identification of the characteristic features of the STR, its essence and main directions.

The features of the STR

I think I am right in saying that there is still no more or less broadly recognized scientific idea of the characteristic features of the STR phenomenon, of its essence. To date, a considerable number of different variants of theoretical interpretations of the said phenomenon have been put into scientific circulation. But more often than not these are confined to disclosing only certain aspects of the STR, to reducing its essence, consciously or unwittingly, to one or another individual revolutionary process in the sphere of modern science, technology or to their interaction with society. The attempt to characterize the specifies of the STR, and to disclose its essence, inevitably confronts the following difficulty. The STR process is unusally varied. It includes a multitude of separate revolutionary transformations, qualitatively new phenomena both in the systems of science and technology themselves and in the various systems of their interaction with one or another sphere of social practice. Moreover, the interaction, the intertwining of various revolutionary processes and phenomena in various spheres of human activity embraced by the STR process, is so close that it is not difficult to find one or another dependence by means of which the greater or lesser totality of scientific and technological transformations can be reduced to an individual revolutionary transformation.

Of course, the essence of the STR does not boil down to one or another factor, not even major scientifc discovery or direction of scientific and technological progress. It consists of the reconstruction of the entire technical basis, of the entire technology of production, beginning with the utilization of materials and energy processes and ending with the system of machines and forms of organization and management, the place and role of man in the production process. The STR is the fusion of the major forms of human activity into a single system. These forms are: science—theoretical cognition of the laws of nature and society; technology—the complex of material means and experience in the remaking of nature; production—the process of creating material wealth; management—methods of rational correlation of advisable practical actions in the process of accomplishing production and other tasks.

Fashioning of this unity leads to far-reaching consequences for each of its components and strengthens their interconnection.

In science, it is the processes of the interaction of its branches that begin to play the determining role, with the role of the social sciences particularly increasing—both in the solution of urgent practical problems and in the ideological struggle which is intensifying in connection with the development of the STR under conditions of the competition of the two opposing social systems. In the socio-economic aspect, the main thing is that science is becoming the leading element of the productive forces. This means that the unity of scientific cognition and production activity acts as a powerful driving force of production.

In technology, thanks to the closer ties with science, it is the creation of new materials and sources of energy, the emergence of new kinds of production activity and the radical change in existing technological processes that begin to play the paramount role. This is particularly strikingly apparent in the new approaches to the utilization of natural resources.

In truth, revolutionary changes are taking place in production—both in its social structure and in its material and technical basis. As a result of the all-round automation of production and management and the creation of technical means performing not only mechanical but also logical operations, man's labour functions are changing radically. A major social effect of such transformations is that the material and technical conditions are created for overcoming the basic distinctions between mental and manual labour, between labour in agriculture and in industry, between town and country, between the non-productive and productive spheres.

Finally, management becomes one of the main forms of activity closely linked with all the key spheres of social life.

In the most general form, it can evidently be said that the STR is a radical qualitative transformation of the production forces; it is the transformation of science into a direct productive force, and correspondingly—it brings revolutionary changes of the material and technical basis of social production, of its content and forms, of the character of labour and the social division of labour. These processes are influencing all spheres of social life, including education, living conditions, culture and the psychology of people. While it exerts increasing influence on the socio-economic advance of society, the STR is itself predicated on a definite level of this advance. It became possible only thanks to the high degree of the socialization of production, that is, to the process creating the objective preconditions for the transition from the capitalist to the socialist mode of production. The STR has intensified the objective necessity of this transition and is consequently becoming an important factor of the contemporary world revolutionary process.

The Marxist-Leninist doctrine of the social development is fundamental to understanding the essence, the socio-economic conditions and the social role of the STR. The concept 'mode of production' introduced by Marx as the key concept in the materialistic understanding of history, disclosed the connection between the development of production and social progress and made it possible to substantiate his conclusion about the inevitability of the transition from capitalism to a classless society.

The STR is a law-governed process in social development, the essence and main features of which were anticipated by Marxism. A thorough analysis of the developmental tendencies of the capitalist economy enabled Marx to formulate the main conditions for production's further development: the connection between the progress of production and social progress; the transition to a balanced development of social production of the basis of social property in the means of production; the conversion of science into a direct productive force designed to transform the whole process of production. 'If the process of production', wrote Marx, 'becomes the application of science, then science, contrariwise, becomes a factor, a function, so to speak, of the process of production.'[1]

The transformation of science into one of the leading factors of social development considerably expands its social functions. Scientific activity acts as an independent socially necessary type of labour which increasingly acquires a mass character. In the most general form, this process can be characterized as the intellectualization of social labour. It affects the aggregate producer (his physical and intellectual capabilities and requirements), as well as his product. Production and the application of continually renewed knowledge are becoming a major factor of the accelerated introduction of innovations in all spheres of social life.

Main conceptions of the social significance of the STR

As we have already noted the STR unfolds differently in different socio-economic and socio-historical conditions, leading to qualitatively different social results. In capitalist society, as historical experience shows, the development of the STR leads to the aggravation of social antagonisms, to the fact that the fundamental vices of the bourgeois social system acquire new forms of expression alongside the traditional forms.

While accelerating the process of the socialization of the means of production, the scientific and technological revolution at the same time deepens and aggravates the contradictions inherent in State monopoly capitalism and engenders new contradictions characteristic particularly of the present stage of its development. It is, first of all, the contradiction between the unprecedented possibilities of vastly improving the well-being of mankind and radically bettering the conditions of the working people. It is the contradiction between the social character of modern production and the State-monopoly system resulting in a colossal waste of resources and wealth. It is not only the deepening of the antagonism between labour and capital; it is also the antagonistic contradiction between the interests of the overwhelming majority of people and the ruling financial oligarchy.

In socialist society, the STR is characterized by qualitatively different features, consonant with the nature of the socialist mode of production. The STR has made possible the accelerated utilization of the reserves in socialist society. L. I. Brezhnev, General Secretary of the Central Committee of the Communist Party of the Soviet Union (CPSU), in his report to the twenty-fifth Congress of the CPSU, noted:

> We Communists proceed from the belief that the scientific and technological revolution acquires a true orientation consistent with the interest of man and society only under socialism. In turn, the end objectives of the social revolution, the building of a communist society, can only be attained on the basis of accelerated scientific and technological progress.[2]

Socialist property, the planned organization of social production (the active participation of broad sections of the working people in the organization and management of the economy)—all this makes possible maximum utilization of the achievements of science and technology in the interests of man.

Interpretations of the essence and social consequences of the STR is the arena of a sharp struggle between Marxist-Leninist and bourgeois ideology.

In recent years many bourgeois ideologists and theorists have been making various kinds of attempts to answer the question: What is the future of human society, whither is humanity going as a result of the headlong advance of science and technology?

Up to the end of the 1960s, more or less optimistic ideas about the future predominated in bourgeois literature on the subject. These expressed the hopes of the ruling class to utilize the scientific and technological revolution for resolving the economic and social contradictions of capitalist society. These hopes are crumbling.

For the past five years, the capitalist system has been pounded by a world economic crisis.

The crisis spread simultaneously to all the main centres of the world capitalist economy. Besides an absolute drop in the volume of the gross national product and in industrial production, it is characterized by structural crises of a world scale—the energy and raw materials crises; falling-off of production in various branches of the manufacturing industry; general curtailment of housing construction. The current inflation is intensifying the crisis processes. Stoked by constantly mounting military expenditure it has reached dimensions unprecedented in peacetime.

Beginning with the 1970s, bourgeois theorists began to revise their views on the STR and its consequences. Today, even the works of the optimistically inclined bourgeois theorists show a strain of doom and melancholy. Only fifteen to twenty years ago, it was said that it was impossible to foresee the consequences of scientific and technological progress. Not so today; in most writings it is admitted that many of its consequences, disastrous in the opinion of bourgeois authors for mankind as a whole, are already evident.

Today, Marx's words, enunciated by him long before the present STR, are very much to the point:

> Machinery, gifted with the wonderful power of shortening and fructifying human labour, we behold starving and overworking it. The new-fangled sources of wealth, by some strange weird spell, are turned into sources of want. . . . Some . . . may wail over it; other may wish to get rid of modern arts, in order to get rid of modern conflicts.'[3]

Thus, the truth of the proposition of Marxist-Leninist theory that scientific and technological progress by itself is not only incapable of curing capitalist society, but that, on the contrary, it deepens and aggravates all its contradictions, is borne out by the historical experience of the past few decades.

It is not surprising that bourgeois theorists in these conditions try to elaborate their ideological interpretations of the STR. They uphold two basic conceptions which on the face of it appear mutually exclusive but which actually perform similar functions. The first of them can be called a melioristic, or relatively optimistic, one; the second—a pessimistic one.

The essence of the first conception boils down to the assertions that capitalism is an adequate form of the development of the productive forces, that profit-making is the most effective form of ensuring rapid scientific and technological progress, that the STR solves all social contradictions of capitalist society and makes radical socio-economic transformations unnecessary. Varieties of this melioristic conception are the 'convergence theory', and the theories of 'industrial society', 'post-industrial society', and 'the technotronic age'.

The conception of social pessimism indirectly records the real contradictions of modern capitalism but presents them as the contradictions of 'technological civilization' in general, that is, in fact accepts the premises of the theory of a 'single industrial society'. A characteristic feature of this ideology is the reduction of all social evil, engendered by capitalism, to the spontaneous development of technology. The latter is given as the root cause, the culprit for the ecological disaster threatening mankind, for increasing alienation, etc.

Thus, on the face of it, these two trends of assessment of the social consequences of the STR by Western theorists belong to opposite poles in their interpretation of the social significance of this revolution. But it is not difficult to see that these seeming oppositions rest on a single invisible foundation. They absolutize the social significance and social functions of the STR, endow the STR with ability independently, without any interaction whatsoever with the general social conditions of its unfolding, directly, 'in pure form', so to say, to influence people and the social forms of their being. In one case, this purely independent influence of the STR on society is a positive one, in the other, negative. But in both cases it is a matter of the purely independent influence of the STR on social life, an influence that is divorced from a social base and social conditions.

This dissociation of the STR process from the social and economic nature of society is a characteristic feature of the interpretations of the STR current in the capitalist world.

At the eighth World Sociological Congress the French sociologist Alain Touraine stated: 'The STR engenders contradictions which go beyond the framework of the contradictions between the exploiters and the exploited.' Another sociologist, Dahrendorf (Federal Republic of Germany), proceeding from his conception that the life of mankind is a struggle for survival and justice, contended that in view of the crises that have gripped the contemporary (read, capitalist) world the struggle for survival is put in the forefront. According to Dahrendorf, the STR is, on the one hand, the cause of the crises, but on the other, it is able to provide the means for overcoming them. But the class struggle must make way for the struggle for survival, while class interests make way for the interests common to all mankind.

With such a position, the theorists inevitably working under the aegis of capital, either elevate the STR to the status of a panacea for all the social misfortunes of today, or accuse it of being wholly responsible for all these misfortunes. And with some of these theorists their absolutizing of the social significance of the STR goes so far as to equate the contemporary world revolutionary process with the present scientific and technological upheaval. R. Tortrot, for instance, calls the STR 'the real revolution of the twentieth century'. It is precisely with this revolution that he links the entire process of present-day social development and declares that 'only such a less harsh, but permanent and consistent revolution as the STR furthers steady social development and the improvement of living conditions'.[4]

Marxists reject both the melioristic and the pessimistic position as being one-sided and theoretically untenable. The STR is unable to solve the economic and social contradictions of an antagonistic society and to bring mankind to a state of material abundance without radical social transformations of society on socialist lines. Naïve and utopian too are the leftist notions according to which a just society can allegedly be built by political means alone, without the STR.

The connection between the STR and social progress is by no means simple or straightforward. It is a profoundly dialectical relationship of unity, of interconnection. The all-round development of science and the practical application of its achievements in the interests of the whole of society calls for certain social conditions which capitalism cannot ensure. Such conditions are created by the transformation of society on a socialist basis. It is this basic fact that determines the varied character of the social consequences of the STR in different socio-economic systems.

The STR and the social revolution are not, as it were, two parallel revolutionary processes of our time. In our country, the STR historically coincided with the entry of our society into the stage of developed, mature socialism. The importance of further unfoldment of the STR, of utilization of its results in creating the material and technical basis of communism is determined by that.

The theoretical and political significance of the conclusion drawn by the twenty-fourth Congress of the CPSU about the need to combine organically the achievements of the STR with the advantages of the socialist economic system lies in the fact that it was based on knowledge of the laws of social development in the present age. This fundamental line of the CPSU has been given further concrete definition by the twenty-fifth Congress.

The efforts of Soviet scientists are centred on comprehensive research into the fundamental questions of the STR, its socio-economic conditions and consequences, on mapping the concrete ways of realizing the social programme of development of Soviet society laid down in our country.

Great attention is paid to problems of the STR in other socialist countries. The close ties between the scientists of the Council for Mutual Economic Aid (CMEA) member countries enable them to disclose the content of the pressing problems of the STR more successfully and deeply. Of great importance, in particular, was the international symposium of scientists and specialists of the CMEA member countries and Yugoslavia, 'The Scientific Revolution and Social Progress', held in Moscow in 1974.

The STR and control of the processes of social development

The problems posed by the STR are numerous and varied. Conscious and purposeful control of the process calls for a profound and thorough analysis of the tendencies of scientific and technological progress, for the elaboration of a scientifically based policy and the carrying out of organizational measures to ensure the most favourable conditions for applying the achievements of science and technology in the interests of social and economic progress.

The necessity of conscious purposeful control of the social development processes is the corner-stone of Marxism. Subjecting the anarchy of production in the capitalist society of his day to criticism, Marx showed that the capitalist system creates the material prerequisites for the planned organization of production and of the entire social process.

Opponents of Marx maintained that private enterprise, competition, the pursuit of maximum profit are the main driving forces of any society. Social development, in their opinion, cannot, generally speaking, be planned. It is controlled, in the words of Adam Smith, by an 'invisible hand'. This reference to an 'invisible hand' directing social processes expressed belief in the spontaneous regulation of the capitalist economy. Such regulation was possible, indeed, 100 years ago, although it engendered economic crises, mass unemployment and other cataclysms. But today the question of the purposeful, planned development of social life, not only on the scale of individual countries but to a certain extent on the scale of the entire planet, is posed by the STR so insistently and sharply that even some bourgeois theorists are compelled to admit it. The present rate and methods of capitalist production accelerate the processes of the exhaustion of natural resources, and ecological imbalance and pollution of the biosphere, with results that create dangers for mankind. Those dangers can be averted only through the judicious organization of all economic activity and the purposeful control of social development.

In our examination of the problem 'the STR and social progress', we cannot but note that in our times it is no longer sufficient to discuss this problem only in the context of its 'social consequences'. The STR is, in a certain measure, increasingly acquiring a *direct* social character; its movement spells a definite and direct social change, although naturally not an unbounded one, in all spheres of social life. This means that the advance of science and technology in our times should be considered not simply as a leap forward in the development of the productive forces of society, but as itself a specific social process evolving by means of techno-scientific transformations, as a process which does not simply exert influence on other elements of the social organism but to a certain extent directly shapes them. Thus, the STR appears as the practical realization of the social and cultural potentialities of science and technology, as the embodiment of their socio-cultural transforming power.

Scientific and technological progress shapes the special culture in which we live, transforms the character of social relationships in society and strengthens their rational, objective elements. Not only direct human communication, but also the character of the management of production and society change in this direction. The intensive rationalization and technical character of the process of decision-making results in these decisions also becoming more adequate.

The STR also substantially changes man's relationship with nature; and what is most important here, along with realization of the impermissibility of upsetting the ecological balance, is a radical change in the structure of this environment itself which today includes such fundamentally new elements as outer space and the sea bed.

By transforming culture, including the relationship between people and with nature, the STR thereby changes man himself, although not so directly and radically as the over-emotional critics of 'technological civilization' claim.

Examining the STR as a specific socio-cultural process, we arrive at the conclusion that assessment of it only in the polar context of 'optimism-pessimism' is not complete. For the question now is not what will the STR give man, but what sort of a person will it make him, what kind of culture, life style, communication and interaction with nature will it shape? And here it is impossible to remain with positions of prophecy or Utopia. Not only the forecasting of the tendencies and consequences of the STR, but also their regulation, i.e. the implementation of a scientific and technological policy, is becoming an urgent task of mankind.

The regulation of socio-cultural processes presupposes the solution of a number of complex and heterogeneous problems, including sociological ones. In our view, the main and most difficult thing today is to determine the range and the limits of the possibilities of controlling science and technology.

As we resolve this problem we must answer the question whether, and to what extent it is possible to overcome those limits which objective (cultural, economic, political) social processess put on the purposeful regulation of the STR in the interests of man.

Maybe the objective deterministic character of scientific and technological progress shaped by its own laws, by economic rationality and political interests, is so effective that actual regulation is simply a matter of forecasting, and remedying to the maximum degree possible the undesirable consequences of the STR? In other words is the notion about alternatives of scientific and technological process a myth? But any discussion about a scientific-technological policy without such alternatives becomes idle talk and wishful thinking.

The most rigid limits of controllability run within science and technology themselves and are determined by the immanent logic of their development, which directs them to one side or another. If this represents an objective, deterministic sequence of scientific discoveries and technical inventions, the inevitable succession of one stage after another of scientific and technological progress which we witness in the history of science—then one can speak only very conditionally of a scientific-technological policy as a choice. In this case it is a policy of pre-apprehension and escapism, and not a policy of decision in the true sense of the word.

When examining the history of science and technology, scientists sometimes underestimate the danger of the intellectual-psychological phenomenon that the historical process is always more straightforward than when it was at its inception and that, following the logic of simple retrospection, it is impossible to discover in it 'points of a historical choice', alternative ways of development. The past history of science bewitches logicians and historians of science by its consistency and immutability, prevents them from perceiving the alternative ways of developing scientific thought that existed but were not realized, the technical conjectures that remain conjectures (because, let's say, of the absence of social necessity). However, there is every reason to assume that just as the history of human culture as a whole has given us an alternative set of different types of culture, so too science, its organic part, potentially had and has alternative scientific 'subcultures' which are not realized in parallel only because of the special cumulative nature of scientific development and because of the absence of certain social conditions.

Now, it can be said that a number of the basic conditions, essential for identifying and utilizing the alternative ways of developing science and technology, are taking shape in our times.

It is, first of all, the achievement of social potentialities, and the possibilities of science itself as a social institution. The sharp increase of technical and economic resources, a trained scientific personnel, the purposeful organization of scientific research, etc., allow one to say that modern science can do what 'classic' science could not permit itself to do—to elaborate simultaneously alternative scientific and technical paradigms, to use different scientific approaches, testing the social consequences of each and then choosing the one that is closest to the humanistic values of our times.

Secondly, an alternative way of developing science and technology becomes not only possible but to a certain extent necessary because contemporary society begins to make different social demands upon them, in essence, alternative demands. Today, this manifests itself only in a vague form of censure of disillusionment in the results of current scientific and technological development (the ecological crisis, the energy crisis, etc.), in the form of statements that modern science is 'not moving in the right direction'. But already tomorrow these demands may (and should) be formulated more precisely, as new social criteria and cultural standards within the framework of which science and technology must advance.

Thirdly, the possibilities of a rational and humanistic policy in science and technology are augmented by the fact that the most economically developed countries can allow themselves to reckon less now than before with the purely economic criterion of the assessment of investments in science and technology. This gradual release from rigid economic rationality, which is an essential condition for securing other, non-economic human values, is already taking place in the socialist countries, where any technical innovation is assessed first from the standpoint of the immediate interests of the individual and only then from the standpoint of economic profitability. A change in the mass mind in this direction is to be observed also in the Western world where the purely utilitarian approach to science is being subjected to increasingly sharper criticism. It is obvious that mankind, having received the possibility of choosing not necessarily the 'most profitable' science and technology, is radically extending the boundaries of possible scientific and technological paradigms and investments.

Fourthly, the purely sociological aspect. It is absolutely clear that the implementation of a policy in science and technology calls not only for scientific, economic and social prerequisites but also for major cultural prerequisites. They also appear, in our opinion, in contemporary society, at the most diverse cultural levels. Modern scientists ever more deeply realize social limits, the cultural role of science as a means for attaining human values, their own moral responsibility, expressed in the strong sense of civic duty of scientists and their readiness to take an active part in decisions connected with the social application of their sciences.

The STR in the conditions of socialism

When analysing the aforementioned social conditions of scientific development, we must not confine ourselves to just their abstract postulation, to examining them in a 'social vacuum'. This must be done in concrete socio-economic conditions. The objective tendency of scientific progress, both its inherent demand and its social effect, still more sharply poses the problem of the need for radical changes in the social organization of society itself, which must rest on scientifically based principles.

Today, it is becoming increasingly obvious that the rational, purposeful regulation of social processes can be ensured only given the proper social organization of social life.

Under socialism, forms are created which correspond to the STR for realizing its achievements. This is attained through the organic combination of the achievements of science and technology with the advantages of the socialist system of management.

Of fundamental significance in this connection is the proposition in Leonid Brezhnev's report to the twenty-fifth Congress of the CPSU that

> the success of the scientific and technological revolution and its beneficial effect on the economy, on all aspects of society's life, cannot be ensured through the efforts solely of scientists. It is increasingly important to draw all the participants in social production and all the links of the economic mechanism into this historic process.[5]

The party's policy of accelerating scientific and technological progress greatly contributed to the fact that our economy in the Ninth Five-Year Plan period developed at high and stable rates. This is very indicative. Not individual achievements and effects, but a high level of social production, the fullest satisfaction of the material and cultural needs of society, steady improvement of the people's welfare—it is this that determines, in the conditions of socialism, the nature and content of the socio-economic processes connected with the STR.

A major indicator of social progress is the individual's way of life. The tremendously increased pace of social development, mankind's continuously intensifying impact on the natural conditions of its existence, the positive and negative consequences of scientific and technological progress—all these facts make it abundantly clear that production is not only the manufacture of things, goods, services but also of human needs, and what is more of the conditions of human existence, that is, the conditions which determine the physiognomy of contemporary society. In other words, production is also the production of the way of life of peoples, and the modes of their social intercourse.

In the conditions of socialism, economic and scientific technological progress is not an end in itself. It is subordinated to the objectives of the building of communist society, the all-round development of the individual, fuller satisfaction of material and spiritual needs. The socio-psychological processes under way make for a change in the very structure of the requirements of society's members, for the increasing predominance in it of spiritual and moral values. This does not mean, of course, that socialist society considers the satisfaction of the material needs of its members as something of secondary importance. On the contrary, from the viewpoint of Marxism-Leninism satisfaction of the material needs of the working people is a mark of the level of development of both society as a whole and of its individual members. Nevertheless, as material needs are more fully satisfied, spiritual needs make themselves more insistently felt. The growth of the productive forces is accompanied by the perfecting of socialist democracy, by political and cultural progress.

Seeing that, under socialism, economic and scientific and technological progress is not an end in itself, the realization of the STR calls for a special, in its essence integrated, scientific and technological policy, which was laid down by V. I. Lenin. Its essence consists of principles of regulation of scientific and technological progress such that not one or another scientific result as such, taken by itself, is regarded as of paramount importance, but rather an integrated approach is taken, which relates scientific and technological progress to questions of social development when this progress is subordinated to the tasks of social progress and proceeds from the State plan of development of material production and spiritual culture. A striking embodiment of this policy, aimed at ensuring the unity of scientific, technological and social progress is the tenth Five-Year Plan and Comprehenive Programme for the Socio-economic and Scientific and Technological Development of the USSR up to 1990, drawn up in our country. This long-range plan provides for the extensive utilization of the achievements of science and technology in all spheres of the national economy and of social life.

A most characteristic feature of the comprehensive scientific and technological policy pursued under socialism is that the development of the social sciences and the practical application of their achievements in the socialist countries is given the same attention as the progress of the natural and technical sciences. The development of the social sciences under socialism constitutes one of the major conditions of the complex and contradictory process of the creation of new forms of social life. The social sciences serve as a theoretical basis of guidance of the development of all aspects of human activity in socialist and communist construction. In the conditions of the STR the role of the social sciences increases immeasurably, both as a result of science becoming a direct productive force and because of the new processes that socio-scientific thought undergoes as it is drawn into the process of the STR and as it becomes an active participant in that process.

The transformation of science into a direct productive force which is conformable to the social sciences means that not only the moulding of the world-outlook, intellect, moral make-up and spiritual world of the individual becomes the major task of the social sciences, but also direct participation in the development of material production, in perfecting social relations and regulating social processes.

As to the diversity of the processes of that essential and at times, one might say, fundamental renewal which the social sciences undergo as they become part of the STR process, the most important of them appear to be the following.

To begin with, not only does the STR place fundamentally new tasks, which are also of great practical significance, before the social sciences; it also sharply intensifies the integration processes in the humanities themselves and in their relation to scientific and technical knowledge. By tying together the processes and phenomena in various spheres of human activity into a single system, the STR inevitably augments the importance of a comprehensive approach to the problems of social development. In fact, not a single important national economic problem can be solved at the present time without the close interaction of the natural, technical and social sciences. For instance, solution of the problem of the efficiency of social production presupposes today a comprehensive approach which takes into account the technical and economic and social 'components' of the unprecedentedly intricate process that the efficiency of production has objectively now become. Also in the case of the establishment of large modern industrial complexes, account must be taken not only of technical and economic factors but of the entire totality of the social, psychological, ecological and other consequences of the functioning of these complexes. The creation of new towns and the expansion of existing ones likewise can be properly tackled only given the close co-operation of representatives of the natural, technical and social sciences. Without such co-operation it is quite impossible to develop urbanization in such a way as would ensure people optimal living and working conditions. It would be no exaggeration to say that it is in the interaction, integration and mutual enrichment of these three major spheres of science that its role as a revolutionary factor of contemporary social progress is most fully realized.

Secondly, in the social sciences the STR proceeds in the form of a specific integration process of the humanities such that the unification of various social-scientific disciplines for solving problems of social development form a single complex. It can be said that of late there is a growing tendency in the social sciences towards joint elaboration of related problems by representatives of various spheres of the social sciences, towards strengthening of the ties of special relation with the elaboration of general theoretical problems of social development, with the analysis of philosophical and general methodological problems of the scientific cognition of social processes and phenomena. This tendency and these ties are particularly characteristic of the elaboration of such problems of social-scientific thought as the establishment of criteria, the features and stages of development of mature socialism, economic planning and forecasting, enhancement of efficiency and further intensification of material production, transformation of the social structure of socialist society, further development of the economic integration and co-operation of the countries of the world socialist community, the social way of life and ways of perfecting it, specific features of contemporary world development, of the current processes underway in the developing countries.

Thirdly, the STR process in the social sciences is ever more intensively proceeding in the form of their methodological and conceptual 'reconstruction'. Economic, psychological, sociological and other research undertakings are characterized of late by the increasingly broader application of mathematical methods which in turn greatly increases the validity of their propositions for the solution of one or another problem of social development. The social sciences are more and more actively adopting systems, cybernetic, theoretico-information, mathematico-statistical and mathematico-logical approaches and methods of scientific cognition, interpreting them in accord with the specific characteristics of social processes and phenomena. Computers are being used on an increasing scale for research into and processing of social information. The spheres of application are expanding, and the methodological efficiency of the simulation of social processes and phenomena, as well as of social experiment, is increasing.

Perfecting the methodological arsenal of the social sciences has become an imperative under the conditions of the STR. This is dictated by the unprecedentedly increased complexity of the problems of social development being put on the agenda both by the gigantic practical transformations taking place within framework of the STR and by the stupendous achievements of science and technology generated by this revolution. This problem, we might note, is a common one for the whole of modern scientific thought. The growing complexity of the problems of theoretical and practical mastery of reality, to the solution of which mankind is now quite able to apply itself, is a characteristic feature of mankind's historical development in indissoluble connection with the revolution in science and technology it is carrying out.

The growing role of science is accompanied by the increasing complication of its structure. This process is reflected in the rapid development of applied research, research and development, and experimental work as links connecting fundamental researches with industrial production, in the growing role of complex interrelated researches.

In the conditions of the STR the interconnection between various processes and phenomena increases which itself makes a comprehensive approach to any major problem still more essential. In connection with this, the close interaction of the social, natural and technical sciences acquires particular timeliness as well as their organic unity, which can to an ever-increasing extent have an influence on heightening the efficiency of social production, raising living and cultural standards, and ensuring a comprehensive analysis of the STR.

The solution of such problems—qualitatively new, in scope and character—as the raw materials and energy problems, outer-space exploration and harnessing the resources of the world ocean, eradication of the most dangerous and widespread diseases and protection of the environment, integrated development of regions and long-term planning require considerable expenditures of time and material resources. Besides, the number of factors that must be taken into account when tackling problems of such magnitude, the degree of uncertainty and the risk in the event of an unsuccessful solution, are so great that it becomes difficult to depend on intuition and experience alone.

The development of science and technology, which has produced these problems, in its turn gives mankind new means for solving them. These include utilization of the totality of methodological principles known as systems analysis.

The application of systems analysis as the methodology of investigations and determining the principles of the realization of decisions taken, and a large number of alternative decisions at that, makes it possible to reduce considerably the time needed and to exploit more effectively material resources when working out complicated projects. Not only quantitative but also qualitative methods of research are characteristic of systems analysis, which makes it possible to most fully elicit, regulate and structure problems whose solution requires the use of modern technical means and available resources.

For more than a decade now in a number of countries, research into systems of different levels and scales, including attempts to create and analyse development models and their dynamic behaviour, is being conducted on the basis of systems analysis, as is also the elaboration of models of the economy of countries, models of economic planning, dynamic models of energetics, models of the influence of energetics development on the environment. Research investigations centred on devising methods of regulating social and economic development in the conditions of the STR, taken in the international global context, are growing in number. Their appearance has contributed to the increased role played by methods of research and control, especially by the systems analysis method, which now has at its service the achievements of mathematics, and the theory of management and computers. This has made it possible to tackle major interrelated programmes of geographical dimensions.

The importance of practical problems to be solved from positions of systems analysis, their gigantic scale and, as a consequence, the enormous cost of each experimental verification of an analytical solution, make it necessary to find rational ways of integrating the efforts of specialists and concentrating the labour and material resources of many countries on the main directions of systems analysis.

In this connection, the establishment in 1972 of the International

Institute of Applied Systems Analysis is of considerable interest. Its members include the academies of sciences and national committees of fourteen countries, among them both developed capitalist and socialist countries. The institute is open to membership of other countries.

The research conducted by the institute to work out and develop methods of a systems approach, universal for a broad range of the most sophisticated means for solving complex problems, strengthens the existing system of international scientific contacts. The application of systems methods makes it possible to improve substantially existing methods of research and analysis and to ensure more reliable forecasting and assessment, and more effective control of the social and other consequences of scientific and technological development.

The STR and world development processes

Being a global process of our times, the STR exerts an increasing influence on world economic processes, on the character and direction of the international division of labour. Undoubtedly, the STR considerably accelerates the rates of the internationalization of production, deepens integration processes, making possible the formation of powerful international economic complexes within the framework of which particularly favourable conditions for the application of scientific and technological achievements take shape.

The Comprehensive Programme of Economic Integration by the CMEA countries graphically demonstrates the immense advantages of the integration processes developing between the socialist countries which make available to each of them the most favourable conditions for utilizing the achievements of the STR and for perfecting the structure of their national economy.

The STR introduces considerable prospects in the position of the developing countries in the world economic system. In principle it opens up before the young States real possibilities for overcoming the gap in the development levels between them and the developed countries. But many big obstacles still stand in the way. What particularly prevents the developing countries from actively participating in the scientific and technological transformation is the absence of their own scientific and technical potential (only 4 per cent of world expenditure in research and development, and 6 per cent of scientific personnel fall to the share of these countries).

Of fundamental importance at the present time is the struggle of the developing countries for national sovereignty, the fight to be arbiters of their natural resources, for equitable conditions of international trade and scientific technical exchange, for the establishment of a 'new international economic order'. This struggle has the full support of the socialist countries.

The STR deepens the process of internationalization in science, technology and in the economy still further. The tendency towards international détente that has become apparent of late, the consistent course of the Soviet Union and of all the countries of the socialist community, based on the principles of peaceful coexistence of countries with different social systems, makes it possible to expand mutually advantageous ties in the interests of all mankind.

The differences in the ideologies and social systems of the socialist and capitalist countries are not an obstacle to the development of business relations between them provided, of course, that they are built on principles of sovereignty, equality, non-interference in internal affairs and mutual advantage.

It is generally known that as science progresses co-operation in research as a requisite condition of its development acquires increasing importance. Scientific and technological progress sharply intensifies the social character of the production of new knowledge.

Progressive scientists of the world are well aware that science can make the most effective headway only given broad international co-operation, the promotion of rational co-operation in research, a constant exchange of scientific and technical information and experience, the expansion of direct contacts between scientists and specialists of different countries.

In the past half-century, when research investigations have reached an unprecedented scale and an extraordinary variety of forms, the problem of co-operation has acquired particular importance. Today, international co-operation makes it possible to use the international division of labour to further the development of each and every country. But there is another broader aspect of international contacts that must be emphasized. Many of the scientific problems now on the agenda are so global in character that their solution affects the interest of all countries and peoples. International scientific co-operation is, therefore, not only the most advisable form of carrying out investigations that require large material and manpower resources, not only an effective means of developing the scientific potential of each of the participating countries; it is also a necessary condition to finding the solutions to these problems.

Thus, international scientific and technological co-operation serves two goals: on the one hand, it makes it possible, through the international division of labour, for each country to use world achievements which are of universal importance, and, on the other, it serves as an instrument enabling scientists of different countries to pool their efforts towards the solution of common, global problems.

The process of détente and the affirmation of the principles of peaceful coexistence of States with differing social systems, is a necessary condition of the organization of international co-operation for solving current problems of social development. Peaceful coexistence is becoming a constantly operating factor shaping relations between different socio-economic systems on a long-term basis.

It is quite obvious that peaceful coexistence, being a requisite of international co-operation, develops and gains in strength through active economic, technological, scientific and cultural co-operation of States with differing social systems. This is an instance of dialectical interconditionality. The expansion and deepening of international scientific and technological co-operation is a necessary condition to imparting a material content to the process of détente.

These concepts were affirmed in the Final Act of the Conference on Security and Co-operation in Europe when the participating States expressed the conviction that 'scientific and technological co-operation constitutes an important contribution to the strengthening of security and co-operation among them, in that it assists the effective solution of problems of common interest and the improvement of the conditions of human life'.[6]

When we speak of the main interest of society, of the vital requirements of people, we naturally have in mind, first and foremost such cardinal issues as ensuring peace and disarmament. Nobody can escape these issues. We are in solidarity with those who see in the scientist's attitude to these problems a manifestation of his social responsibility, his personal responsibility for the well-being and socio-economic progress of peoples and his further responsibility that the achievements of science and technology are not directed against the interests of humanity.

The struggle for social progress is indissolubly connected with the struggle or peace, for deepening the process of détente, and for embodying it in concrete forms of mutually advantageous co-operation between States.

Notes

1. K. Marx and F. Engels, *Works*, Vol. 47, p. 553, Moscow (in Russian).
2. L. I. Brezhnev, *Report of the CPSU Central Committee and the Immediate Tasks of the Party in Home and Foreign Policy. XXVth Congress of the CPSU*, Moscow, 1976, p. 56.
3. K. Marx and F. Engels, *Selected Works*, Vol. I, p. 500–1, Moscow, 1969.
4. R. Tortrot, *La véritable révolution de 20e siècle (la révolution scientifique, ses conséquences pratiques, ses conséquences philosophiques et le grande problème de notre temps)*, Paris, 1971.
5. Brezhnev, op. cit., p. 58.
6. Section 4, 'Science and Technology'.

Social functions of science

Günter Kröber

Corresponding Member, Academy of Sciences, German Democratic Republic

Most commentators on the processes observed nowadays mainly in the industrially developed countries, and conceptually subsumed under the term 'scientific and technological revolution', agree that this phenomenon cannot be confined to any scientific discovery, however important, to any technological disclosure or innovation, however spectacular. The impact of the multifarious effects of scientific and technological advance upon the forms and mode of the production of material and spiritual values, upon the content and character of labour, upon the people's way of life, upon interhuman relations, etc.—in short, the multifarious social consequences of the scientific and technological revolution—are more and more clearly and drastically moving into the focus of public interest.

Being conscious of our responsibility for the coming generations, we cannot be indifferent to whether the scientific and technological potential that becomes manifest today is fully unfolded, above all in the military sphere, and transformed into destructive forces which endanger not only social progress but the very existence of humanity, or whether it is purposefully and consistently made to serve social progress and the benefit of nations. We cannot be indifferent to whether the achievements of science and technology in the industrially developed countries continue to widen the social, economic and technological gap between the developed and the developing countries, or whether they are purposefully used gradually to eliminate this gap, caused by the colonialist practices of a historical epoch that has outlived itself. Finally, we cannot be indifferent to whether the present scientific and technological progress leads to changes in nature and to social transformations which cannot face up to the judgement of the coming generations, or whether the social organization and dynamics of social development can be shaped so that scientific-technological and social progress are correlated, the development of science and technology is controlled according to basic social needs, and social development is effectively advanced through the achievements of science and technology. Our question is put in such a way that every scientist or politician thinking and acting in a responsible way cannot answer it in any other sense but that of the last-mentioned alternative. This does not, however, mean that the problem has already been solved; on the contrary, if one accepts as the only possible alternative that scientific and technological advance, and social progress, must be controlled and shaped in a way enabling them to reproduce the conditions of their further advance for the benefit of mankind, the question of *how* proves to be a decisive issue. It relates not only to the natural sciences or technology, but, in the first place, to the social sciences.[1]

It would be preposterous to expect that the answers could be exhaustive—all the more so in that, in accordance with the evidence available, the social sciences in the developed Western countries are characterized by a plurality of different and even contradictory solutions, and, for this very reason, can be suspected of incompetence in dealing with the problem. Neither can anybody expect that the answers which can be given by the socialist countries will meet with the full approval of all those concerned with the problem and interested in it, because they imply preconditions which must not only be accepted but, above all, created in practical actions.

Aside from these limitations, let me broach and submit for discussion a problem which throws light upon a single aspect of our thematic area and yet, in my opinion, touches on a central problem of the relationship between scientific-technological advance and social progress: the problem of the social function of science under the conditions of the scientific and technological revolution.

It was J. D. Bernal who, in the late 1930s, was the first to move—with a still prognostic view of the scientific and technological revolution—the problem of the social function of science into the focus of public attention. Bernal's standpoint in this question was based on his conviction that science is

> 'an integral part both of the material and economic life of our times and of the ideas which guide and inspire it. Science puts into our hands the means of satisfying our material needs. It gives us also the ideas which will enable us to understand, to co-ordinate, and to satisfy our needs in the social sphere. Beyond this, science has something as important though less definite to offer: a reasonable hope in the unexplored possibilities of the future, an inspiration which is slowly but surely becoming the dominant driving force of modern thought and action'.[2]

It is precisely in this sense that the social significance of science in socialism goes undisputed. It results, on the one hand, from the various separate functions of science as a productive force, as an instrument of managing and planning social-development processes, as a factor and component of the social-reproduction process, as a means and sphere of personality development, and so forth. At the same time, the social function of science in socialism is based, above all, upon integrating socio-economic-productive potential of science. This makes the changes induced by science in the productive forces of society and in the social relations on the one hand, and the demands made by society on science on the other, converge in a direction in which scientific and technological progress, and progress in the development of productive forces and social conditions, i.e. scientific-technological and social progress, combine into a consistent stream of revolutionary changes aimed at strengthening the socialist positions. Science within socialist society is therefore not only a means and an instrument of rational action in all the areas of life of the socialist society, but represents, as such, also one of the imperishable values of that socialist society, last but not least by dint of and in compliance with its revolutionary, progress-promoting and humanistic character.[3]

On the other hand, we can see that in the capitalist countries, too, interest in the issue of the social significance of science has gained much ground, particularly in connection with the discussion of the arguments put forward by representatives of the anti-science movement and based on concern with the negative effects of scientific and technological advance and/or with disappointment arising from the discrepancy between the accomplishments of the scientific and technological advance as contrasted with great numbers of social problems and contradictions defying solution or even gaining in weight. In this connection it is sometimes considered necessary to recommend to the developing countries not to make any special efforts to establish and develop an efficient system of science and research of their own—in particular in the area of basic research—but to concentrate their endeavours on the solution of urgent economic and social problems, and, in so doing, to rely on scientific and technological achievements taken over from the developed countries.

However, it is evident that social problems cannot be automatically eliminated by the progress of knowledge—no matter whether this is brought about by the country's own basic research or by taking it over as a part of existing world knowledge—without simultaneously creating new and more complicated ones. Thus the question arises whether, in general, it is possible unambiguously to orient science so as to promote progress and humanism, or whether all attempts to obtain continuously 'products' from science with anticipated and desired use values—and, as we know, all purposive planning of science is based on this possibility—must necessarily lead to negative accidental and secondary effects that annul or overcompensate the immediate desired and achieved advantage. This question implies the fundamental significance of the term 'social function of science', if it is conceived as a historical category and not included in a static, synchronous, functionalist scheme.

At this general level, we conceive the social function of science in terms of its value for the development of human society and thus also for the future of society, and consequently regard it as an integral determination with both an axiological and a theoretical historical-developmental aspect.

At the level of categories describing the process of development of human society as a whole, external disintegration, and the attribution of values and means from without, disappears and, from the aspect of the historical-materialist conception of history, the sphere of values appears neither as primary nor as an autonomous area but as a mediating instance of the regular socio-historical process, and, consequently, is itself historicized. In this sense it applies also to science in that it is involved not only in the instrumental sphere of society, but just as much in the sphere of values; indeed we consider science to be one of the central social values.

This unity of value and means, of the determination of ways and objectives, finds its most pregnant expression in the primary task of social development in socialism. For social developments can be neither formulated nor realized without a dialectical mediation between the satisfaction of social needs and scientific technological advance, in which scientific and technological progress takes place also—and of course, primarily—as the main vehicle of increasing the productivity of labour as well as of the growth and the rising effectiveness of production; while, at the same time, the satisfaction of the growing need for creative intellectual activity increasingly manifests itself as a means of promoting scientific and technological advance.

Here I use the term 'function' in a very general and simultaneously essential manner. I also consider it right to use the attribute 'social' in its most general sense, namely to describe all interpersonal relations arising in the process of the production and reproduction of people's lives, and to encompass the totality of these relations.

From the Marxist point of view, the social evaluation of science is, in substance, positive and optimistic. This evaluation is based on the fundamental Marxist thesis that labour, the metabolism between society and nature realized with the aid of purposefully produced material mediators, has created man as a social being and continues to create and shape him in the successively achieved higher stages of the historical process. In the Marxist view, the social wealth of mankind does not result from the simultaneously mystical and ahistorical 'interior' of the personality but from a productively appropriated diversity of nature and from the history of society itself.[4] Historically, the productive force of labour has been increasing over a long period of time, mediated by an advance of knowledge which is incorporated within the immediate production process in a primarily undifferentiated way. However, from a certain period onwards—characterized perhaps by the general victory of large-scale machine production in industry—the essential part of the advance of knowledge needed for raising the social productive force of labour can be produced only in a specialized and institutionalized way. This is a knowledge-producing system whose qualities, required for the beginning of this process, were already modelled by the modern experimental natural science, constituted between the fifteenth and the seventeenth centuries. In so far as one accepts the premise that, in the final analysis, all historical progress results from the growth of the social-productive force of labour—though by no means as its automatic, immediate or inevitable consequence—one must accept the absolute indispensability of science as an instance of knowledge necessarily mediating the entire further progress of mankind.

The idea that unlimited possibilities are opened for the development of human society without a further growth of the social-productive force of labour appears to us as absolutely Utopian. The self-improvement in the sphere of the society's intellectual life, which is offered as a 'crutch of progress' (if I am allowed to use Stanislaw Lem's satirical expression), cannot go on for a long time either in the moral or in the aesthetic sphere, if the motive forces of the productive confrontation with nature slacken. The totality of the social roles actually available to society depend on the level and the intensity of the development of productive labour.

It is, of course, also unthinkable that an advance of scientific cognition could take place, as a playground for the individual's creative potential, if the productive force of labour were to stagnate. Already it is evident from the history of science that the development of the natural sciences decisively depends on the advance of production technology. Thus, for instance, although the measurement and experimental techniques can represent the best of what is possible in a given situation from the production-technological point of view, they can never go beyond the possibilities of the existing level of the productive forces. We are also aware of the fact that even so subtle a field as mathematics, which, over considerably long periods, seems to have developed from immanent motive forces alone, is forced ahead, at the decisive points of its development, by the demands made by natural sciences and social practice. The idea of an advance of knowledge, arts or morality over longer periods of time while the productive forces remain constant is totally Utopian. For this reason, science is to be regarded as an independent and irreducible, though by no means autonomous and unconditioned value.

Would it perhaps be more consistent to consider a stationary level of productive forces, and with it also a conservative, essentially complaint science which becomes innovative in a practical sense only where non-renewable resources of production must be substituted, as a realistically possible condition of the world? In such a picture, science would be largely epiphenomenal for society, and also axiologically second rate, a mere repair service for occasional defects of a stable mechanism. We shall abstain from any political and moral assessment of the idea of bringing about such a condition. But it may suffice to say that, if social forces capable of action existed, whose interests would in fact intend to bring about this situation of 'no-growth', this would be objectively inconceivable without a self-dissolution of the social reality or at least without a relapse into pre-capitalist forms of primitive natural economy. With the emergence of the capitalist mode of production, the permanent dynamic of the productive forces has become a condition of social existence:

> The bourgeoisie cannot exist without continually revolutionizing the tools of production, i.e. the production relations, i.e. the totality of social relations. The unchanged preservation of the old way of production was, however, the first condition of the existence of all previous industrial classes.[5]

This historical transition of the social-productive forces from a quasi-stationary to a stable-dynamic mode of existence, as it was stated by Marx and Engels, is irreversible.[6] This was, after all, well understood and put into drastic words by no less a personality than Norbert Wiener:

> We must not look back if we wish to survive. Our forebears tasted the fruit of the tree of knowledge and although that fruit leaves a bitter taste in our mouth, the angel with the flaming sword stands behind us. We must go on inventing and earning our bread, not only in the sweat of our face but in the exudation of our brain.[7]

Although the works which nowadays signal the possibility of an ecological catastrophe consider the ecologically negative after-effects of productively applied science, they ignore the potential of science for the stabilization and even systematic improvement of the ecological situation on a regional and global scale, accompanied by a steady increase of the productive force of labour. There is no reason to assume that there should be any cognitive limitations to the understanding of the laws of ecological motion within a progressively developing society, or that the society might be unable, for some basic reasons, to muster the material resources needed for a genuinely humanistic utilization of such knowledge. The potency of science to offer possibilities for the stabilization and improvement of the global ecological situation is no less inexhaustible than the capacity to show new ways for the productive appropriation of nature. This potency is realized in practice, whenever science becomes effective under suitable social conditions. We believe that the fundamental social function of science in relation to the world-historical process lies in its potency to secure the future of mankind which is developing on the basis of the continually growing productive force of labour.

Potency, however, is not yet reality. Between potency and its realization lies practical action, the exertion of material force connected with economic conditions, configurations of interests and value patterns of the actually existing forces, and the effectively available material and personal resources. Marxist scientific optimism has, therefore, nothing in common with any naïve belief in progress, as is occasionally maintained by the critics of Marx. The possibilities of science to secure a movement of society which, though never free of problems, makes an unlimited advance through timely diagnosis and overcoming of developmental problems, find their realization only through the mediation of the totality of the social conditions in which science is practised. This mediation can, however, be of such a nature that it dampens and blocks the realization of those historical possibilities, or even forces it to take the opposite direction:[8] 'Dampens', e.g. by investing immense means into armament which could be used to accelerate social and scientific-technological progress; 'blocks', e.g. by stopping the advancement of basic research or—following the advice of Western experts—by not initiating it in the developing countries; 'forces it to take the opposite direction', e.g. by using the achievements of science and technology as a destructive military potential.

In a world characterized by the co-existence of two types of social systems with basically different economic conditions, and the ensuing interests and value scales, it would be scientifically incorrect to speak in a generalizing way, of the social function of science without taking into account its real existence in antagonistic societies. This does not deny the existence of serious global problems concerning the future of all mankind which cannot be solved without science; however, the conditions and possibilities of reaching such a solution must be realistically evaluated.

Forecasts of a global ecological crisis imply the point of view that science, in so far as it is applied only for the purpose of continuously increasing the productive force of labour, inauspiciously engenders social dysfunctionality which overcompensates all its useful effect and leads *ad absurdum.*[9] The problem is seen in the fact that the intended, positively assessed and immediately applicable effectiveness of science makes itself felt within a short time, whereas the destructive, unintended effects appear, with considerable delay as a rule, on account of the non-linear long-term couplings within the ecosphere; when diagnosed, they can be controlled only at a disproportionally high cost. Apart from the problems resulting from a temporary lack of knowledge about certain ecological implications, the decisive issue after all, seems to be whether it would not be possible to consider, to pose, and to realize scientific growth and the long-term stabilization and improvement of the ecological situation as an integrated task; and thereby to address the demand for producing the necessary preconditions of knowledge through science as an integrated social order.

The possible answers to this question are very closely linked with dominant economic interests. From the historical point of view, the growing productive force of labour increases the general preconditions for a uniform control of the reproduction cycles of the conditions of human existence, gradually growing in range and covering longer periods of time, as long as these cyclical contexts are sufficiently known. These possibilities are realized in the measure in which all the means of effectively controlling such a reproduction context on the basis of a uniformly constituted economic interest can be concentrated. The more the interference of divergent and individually unforeseeable economic interests must be taken into consideration, the less reliable does long-term management become; and then the reproduction cycle can degenerate in such a way that the preconditions of the productive activity with which it is linked are no longer reproduced but, on the contrary, undermined. The breakdown of the total social potential into independently acting sources of capital produces a strong pressure towards giving priority to goal orientations allowing relatively short-term realization of the capital advance necessary for their implementation.

Consequently, for the global ecological dysfunctionality of productively applied science, there exist objective preconditions in the character of economic goal orientations under which it is applied, and of the economic conditions on which these goal orientations are based. It is not the science-induced growth of the productive force of labour as such but the socio-economic character of this growth that leads to global ecological danger.

The economic conditions of socialism, however, give rise to a strong urge to take long-term and global requirements into consideration in shaping the aims and ways of immediate economic action. In the early period of the socialist social relations, this impulse was frequently, though temporarily, dominated by the necessity to solve urgent social and economic problems rapidly and to remain below the ecological optimum in implementing corresponding measures. Here, however, the particularity and advantage of the socialist production relations consists in the fact that economic growth is directly transformed into increasing possibilities of long-term-oriented action; and that this growth, so long as socialist relations prevail, leads to a gradual filling of the gap between short-, medium- and long-term social interests. In so far as one ignores the difficulties arising from a lack of knowledge, the ecological dysfunctionality of productively applied science is not the result of economic growth but, on the contrary, follows from a still insufficient level; step by step, the dysfunctionality is necessarily reduced with the increasing economic effectiveness of the society.

Thus, the ecological dysfunctionality of science is not in itself a given fact but an objective tendency of capitalist production relations. In this context, we do not by any means overlook the fact that in capitalism, the opposite tendency is also sure to exist, since undermining the general conditions of human life and labour would naturally interfere with the elementary preconditions for the existence of capitalism as such. But this tendency bears the distinguishing mark, the specificity, of capitalist production relations. The active–reactive scheme is clear, in which, at first, the means for a certain 'playing against nature' would be produced for the sake of profit, and then—after the ecological 'response'—funds would, again for the sake of profit, be put on the market to make up for the damage incurred. The science applied for this purpose is, to a certain extent, divided into two practical functions, a productive-force function and ecological function, which complement each other in a certain sense. More difficult, though substantially still adequate for integrations into the profit system, are prompt and partial combinations of productive and ecological activity: productive plants and equipment which *a priori* exclude certain negative environmental consequences of their own application. In the mental preparation of such means, the productive-force function and the ecological function of science enter into a partial symbiosis.

The greatest hindrances arising from the capitalist production relations can undoubtedly be found where large territories—inclusive of the prospect of the production branches, etc., located there—are to be developed according to ecological criteria on a long-term and complex basis. However, exactly such forms of long-term effective planning are, without doubt, indispensable for bringing about a radical improvement in the global ecological situation. In activities of this order, we see one of the levels on which, nowadays, after the developed resources have become strong enough to make it possible to surpass mainly compensatory measures of environment protection and only partial measures of an environment policy, the objective advantages of socialist conditions can be demonstrated in the area of solving serious developmental problems of mankind. At the same time, favourable preconditions have arisen from the increase in the tendency towards ecological functionalization of science, which asserts itself even in capitalism. The Final Act of the Conference on Security and Co-operation in Europe (Helsinki, 1975) defines this area as a sphere of partially coinciding interests of countries with antagonistic social systems, and suggests appropriate forms of co-operation.[10] The scientific search for integrated, economically feasible solutions of problems connected with production and environment undoubtedly can, under these conditions, exercise a peace-promoting effect.

The example of the relations between economic-productive and ecological effectiveness of science demonstrates a general tendency based on modern productive forces: the tendency towards the penetration of scientific results and scientific work into all spheres of human life, and towards an increasing interdependence of these different spheres of action. From this, in fact, the possibility arises to speak of the social function of science (in the singular!) as a general phenomenon and not simply as a class of various, more or less independent ways of the applications of science. The breakdown of this integral function into components (function of the productive force, of education, of the view of the world, etc.) is possible but always relative, and continues to be an analytical decomposition of a whole. Since the functional relation of science to the entire historical process is necessarily mediated through the different social systems, their antagonisms and their relations, any closer elaboration of the functionality of science must take place within the reference frames of concrete social systems. It, therefore, seems to us advantageous to base the inquiry into the social function of science upon a comprehensive concept of socio-historical development which forms the overlapping reference frame for the analysis of functional relations in the narrower sense of the term, and, at the same time, allows adequate description of the differences in the social position, function and value of science under the conditions of the scientific and technological revolution in antagonistic social systems.

We have the question of a social-scientific theory of the development of science and of the scientific and technological advance, i.e. of a science of science, about which Bernal said that it 'must be wide-ranging: it must include the social and economic as well as the material and technical conditions for scientific advance and for the proper use of its tools'.[11]

Notes

1. This statement has not only the general sense in which Hegel already said: 'Man cannot become the master of nature unless he becomes the master of himself.' (G. W. F. Hegel, in J. Hoffmeister (ed.), *Jenenser Realphilosophie. Vorlesungsmanuskripte zur Philosophie der Natur und des Geistes von 1805–1806*, p. 273, Berlin, 1969. It rather refers to *concrete* conditions of the present-day scientific and technological revolution and to the contradictory conditions of social progress in the present world. In the latter sense, J. D. Bernal stated back in 1938: 'It is the function of science to study man as much as nature, to discover the significance and direction of social movements and social needs.' (J. D. Bernal, *The Social Function of Science*, London, 1939 (quoted from the new edition, London, Routledge & Kegan Paul, 1967, p. 411).)
2. ibid., p. 408.
3. cf. Programm der Sozialistischen Einheitspartei Deutschlands, p. 25, Berlin, 1976.

4. Marx says: 'Yet the *development of science*, of this spiritual and simultaneously practical wealth, is only one aspect, one of the forms in which the *development of the human productive power*, i.e. of wealth, manifests itself'. (K.Marx, *Grundrisse der politischen Ökonomie*, p. 439, Berlin 1953.)
5. K. Marx and F. Engels, 'Communist Party' Manifesto, *Selected Works*, Vol. I, p. 26, Berlin, 1955 (in German).
6. cf. the remark made by Marx in *Grundrisse*, op. cit., p. 440.
7. N. Wiener, *Mensch und Menschenmaschine*, p. 53, Frankfurt am Main, 1952.
8. It is interesting to note that this conclusion has already been drawn by the members of the Club of Rome, ' . . . the real limits to growth, then, are not material but political and social'. (M. Siebker, Y. Kaya, 'Towards a Global Vision of Human Problems', *Technological Forecasting and Social Change*, No. 3, 1974, p. 258.
9. 'The growing concern about pollution of air and water by radioactivity and industrial-waste substances mounted into *crescendo*. . . . Whereas at the beginning criticism had been directed towards the "users" of the products derived from scientific research, it was not long before those who did the research came equally into the focus of criticism.' (E. Shils, 'Anti-science: Observations on the Recent "Crisis" of Science', *Civilization and Science in Conflict or Collaboration?*, p. 41, Amsterdam, London, New York, 1972.
10. cf. *Neues Deutschland*, 2–3 August 1975, p. 7.
11. J. D. Bernal, 'After Twenty-five Years', *The Social Function of Science*, op. cit., p. xxv.

Some conditions for an effective science and technology policy

P. Piganiol

Commission of the French Republic for Education, Science and Culture

This article will set forth some of the conditions which must be filled in order for a science policy to reach its objectives and, in particular, to facilitate the creation of new goods and services which are useful to society.

The objectives of a science and technology policy

These are well known; nevertheless I shall restate them in order to highlight an essential difference between three interdependent elements.

A science and technology policy (STP) creates, runs, and co-ordinates the means of research, which are understood to cover, in a very broad sense, university or industrial laboratories, research departments, large and small institutions devoted to research, etc., and to include human potential as well as the equipment of the researchers and sources of funding. In a word, it is responsible for an infrastructure.

STP attempts to make possible the assimilation of world-wide knowledge—which is the common heritage of mankind—and to participate in its advancement. Although all knowledge, sooner or later, is called upon to open the way to applications, I shall stress here the cultural importance of STP, provided obviously that the notion of culture is not deprived of one of its essential components—scientific knowledge.

STP attempts to make possible the assimilation of global technologies and to improve upon them. It seeks optimal adaptation of the technologies which it creates or imports, and must know how to evaluate them in terms of their impact on present or future society.

Sometimes a fourth element intervenes: direct action on the process of 'development'. Despite its importance, I shall not consider it in what follows.

It should be noted in passing that a national or even regional STP must be placed within a broad geographical context, since STP is inseparable from an international 'science diplomacy'.

The effectiveness factors of these elements are not identical. I shall omit those which are related to the infrastructure in order to consider briefly those which influence the development of knowledge and, at greater length, those which condition the relevance with respect to men's needs, of such technical means as can be created.

The effectiveness of fundamental research

Many university-trained scientists tend to deny the existence of effectiveness factors other than the quality of the researchers and the magnitude of the means placed at their disposal. This means ignoring the role of organization and communication. But it also means ignoring the fact that the progress of knowledge, despite its uncertain character, actually very often develops from a logic within science, logic which is linked to the structure of knowledge and its own dynamic. A large part of the domains of the growth of knowledge is foreseeable if we have at our disposal that global, structural vision of science which can be called scientific conjuncture. To bring it to light and to formalize it is one of the principal research effectiveness factors. Furthermore, it is often at this level that the researcher's intuition, which has every interest in founding itself on reliable methods, comes into play.

I shall not develop this basic theme here, but simply restrict myself to pointing out that it is urgent to reconsider and develop René Leclerq's research on the logic of discovery and to recall the role of documentation. The latter must not be limited to dealing with information—which is only the means—but must also furnish a well-ordered image of knowledge in order to reveal the list of new fundamental questions for which a research effort is likely to provide the answers. This is the true mission of documentation.

The effectiveness of applied research

The aim of applied research is to provide mankind with new or improved means to satisfy its present and future needs. Therefore, the concept of effectiveness covers two totally distinct notions: the effectiveness of the research process itself, and the relevance of its results in terms of the objectives pursued. The first corresponds to the methods of creativity and the organization of work whose planning becomes increasingly rigorous as it advances. These methods are so well known and time-tested that there is no need to discuss them here.

The remainder of this study will be devoted to the relevance of the results. I shall show that the complexity of the modern world demands that every governmental, regional or international science and technology policy be complemented by two activities: technological evaluation and forecasting and planning for the future. I shall define the content of these activities and specify the localization and structure of the social organs to which they will be entrusted.

The modern world is characterized by the breadth of the impact of technology on our life, an impact not limited to the material transformation of our environment but which involves modifications in behaviour, psychology and social structures as well. An initial period of technical development, when everything that was possible was considered beneficial and necessary, is now followed by the era of choices—because of the immense and ever-broadening field of possibilities—and of reflection as to the best use of our means. It should be noted that most of the dissatisfactions of industrial society arise not because of the nature of the available means but because of the inadequacy of their mode of use. We must take into account the very powerful interactions between ends, means and the structures of society, and stop accumulating naïve recriminations and demands which run the risk of bringing misfortune to a civilization which has never held so many trumps. It is as if society did not recognize the forces that led to technical progress and industrialization as having issued from within itself. This is a very serious cultural problem which only appropriate education will resolve. But it can also be asked, and with some anxiety, whether society has attained the maturity to allow it positive choices and whether these choices will be better oriented by an embryonic clear-sightedness than by the somewhat blind energy of the pioneers of progress.

This scepticism will not prevent us from proposing structures pervaded with clear-sightedness; but it does suggest a certain prudence in the use of methods, and the refusal of solutions without nuances, without appeal, without the possibility of a return to sources.

The conditions for clear-sightedness

Since effectiveness arises essentially from the clear-sightedness which should guide progress, it is important to study the methods to aid our decisions as well as the structures which facilitate their reasoned use. These methods can be classed into two groups: those which support the 'prospective' and those used to 'evaluate' technologies. Still, this distinction is a largely arbitrary one: a prospective which ignored the evaluation of technologies as they are developed or anticipated would be doomed to failure, and a technological evaluation would be useless if it did not contribute to a definition of possible futures. Nevertheless, I shall retain this classification for the convenience of the argument.

I shall apply myself to defining these two approaches, recalling their methods, fixing their limits and indicating the most suitable mechanisms which will allow them to reach their goals. But two indispensable conditions for success must be stressed at the outset: imagination and creativity, on the one hand, openness to social innovation, on the other.

Prospective, born of the acceleration of change caused by the development of technology, has very quickly succeeded in establishing a link between a technological forecast which was originally highly materialist, and the major currents of possible social evolutions. This 'globalization' makes prospective very difficult but also gives it its letters patent of nobility; a mere mental attitude at the beginning, it has become one of the essential factors in drawing together very different disciplines and activities, and in integrating science policies into a broader human fabric.

The situation today is represented diagrammatically in Figure 1, and we shall study some of the essential points marked (1), (2) and (3) in the diagram.

Technological evaluation

First of all, from the multiplicity of solutions made possible by modern technology to the practical problems we meet. Considered in this limited

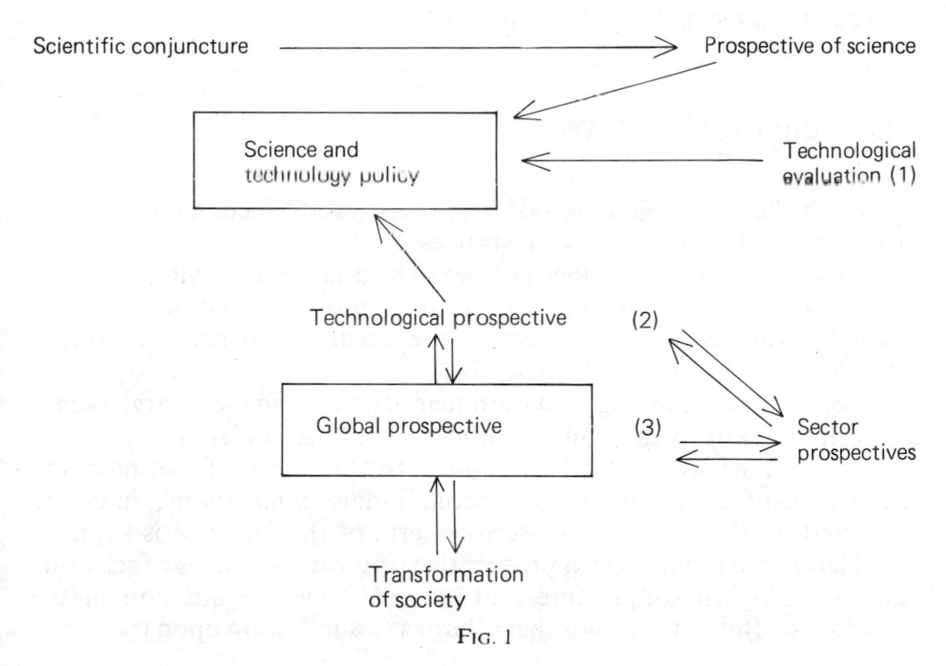

FIG. 1

sense, technological evaluation is above all an analysis of relevance and cost-benefit calculations. It is a purely technocratic evaluation.

But the extent to which the impacts of technology could make themselves felt outside the specific domain of the problem to be solved has been measured very quickly. To put it more exactly, all grand-scale technology inevitably disturbs the environment. Supplements, whose costs must be included in the cost-benefit analysis, are needed to minimize these harmful marginal effects. But there is a fundamental difference between the cost of technology exploitation, strictly speaking, and the costs of correcting these marginal effects. The latter are in effect 'elastic': they hinge upon the degree of perfection reached in achieving the elimination of nuisances. Their level depends upon social consensus or a regulatory decision. Neither could one proceed from a market-economy mechanism. On the other hand, the global cost thus defined puts technology in normal competition—provided, of course, that the rules adopted to fix the level of correction are the same for all the competing technologies and for all the countries among which goods and services are freely exchanged. (In extreme cases there would be the effects of 'dumping' on the part of countries willing to sacrifice their environment.)

Furthermore, it is clear that technology modifies our mode of life in different ways according to its mode of use. The society–technology interactions must be studied more deeply not only in the present context but also in the framework of future society. Technological evaluation then becomes technological prospective.

The technological prospective

Although the terminology is not at all fully established, let us say that prospective takes place in two instances.

First, the technology does not yet exist and we are trying to determine its chances and the date of its completion: evaluation of future possibilities as well as of the pressure society will exert in order to bring it into being: evaluation of the desirable.

Second, the technology is within reach; its cost (in the enlarged sense of 'technological evaluation') is known, but its social utility, likely impacts, and the necessary adaptations both of its mode of use and of the social structures which are to accommodate innovations, must be measured. In this instance, the second term of the first case is kept.

Hence, the technological prospective rests on the studies of scientific and technological conjuncture and on cost-benefit evaluation in the broad sense. But it transcends them thanks to a reflection upon the future society which is eventually modified by new technologies. Its originality lies in the consideration of social impact.

Technological prospective, in fact, furnishes food for thought for the different sectoral prospectives and vice versa. In a very direct way it inspires science and technology policies. It should be noted here that technology policy has always been inspired by a prospective, even when the concept of it was still unknown. But this as yet unnamed prospective was most often at the intuition stage; and of the social impact of technology it retained only its aspects of material progress.

The prospective which should guide the material progress policy must be clearly and rigorously formulated today, and for that reason its uncertainties ought to be underlined; likewise it must be joined to a global prospective without which all reflection upon the social impact of technology runs the risk of being very elementary and inadequate.

The global prospective

We have said that technological prospective must take into account the social impact of progress. However, it is not only a matter of its impact on the present society but, above all, its effects on future society. No real technological prospective exists without *scenarios* for the future society; and such scenarios are global in nature. They define the modes of use of our material means and the very evolution of these modes of use.

But here we must avoid a frequent error: there are many who naïvely believe that prospective will provide a plan for an ideal society, a coherent, non-contradictory model of society in which the functions of the different technical means will be clearly defined. This is a simplistic vision of society. To be sure, in reflecting upon society it will be appropriate to assimilate it to a 'system' and to make a structural analysis of its interactions. This is an irreplaceable working method which has considerably improved our clear-sightedness. But its limits must be recognized: assimilating society to a system inevitably means neglecting the elements of contradiction inherent in every civilization, without which there would be no fluctuations, no possibilities for change.

Social reproduction only gives way to social evolution thanks to the dynamic of the systems (supplemented with the list of its blockages), on the one hand, and thanks to the incoherent factors which allow the system to be modified (ideally before every crisis) by adding or suppressing elements or interactions, on the other.

Prospective is thus a reflection on what is or will be possible, probable or desirable in a civilization which cannot be assimilated to a social system without cutting away the essential part of its human values.

Towards more effective structures while respecting the essence of civilizations

Everything said above suggests mechanisms for reflection and study in order to make our technology more effective, i.e. to adapt it better to our needs and desires. But this also implies the necessity for knowing the limits of every social conception whose absolute character can only be contingent, and must lead us to a great deal of flexibility and prudence in our choices.

In what follows I want to propose mechanisms which are defined in terms of their functions and their localizations. Other solutions are possible: the advantage of proposing only one is to make the aims, means, and difficulties clearly apparent.

Attached to the social organism responsible for science and technology policy, there must be three research units whose work must facilitate science policy decisions.

A UNIT FOR THE STUDY OF THE SCIENTIFIC AND TECHNOLOGICAL CONJUNCTURE

This unit is responsible for a permanent analysis of the structure of scientific and technical knowledge through study of its dynamic, and proposes ways to optimize progress. The problems studied here ignore political frontiers, making it desirable that broad international co-operation enrich these studies. We can even imagine that several countries will decide to create a single agency in order to accomplish this mission. This would be a very good solution but it must not involve the total suppression of every similar body at a national level: each country must have a smaller body to serve as a link, a place for transfer, assimilation and assessment of knowledge.

A TECHNOLOGICAL EVALUATION UNIT

This is essentially a technical department conducting studies to allow choices which take into account all the secondary effects of technology. Its cost-benefit analyses must be understood in a broad sense with consideration for qualitative elements. Its most utilized tool will be multinomial analysis, or better the tree of relevance.

Working closely with the above unit, the office of science and technology prospective will attempt to evaluate social impact in the broadest sense including the psychological and cultural aspects. It serves as a liaison body with the global prospective whose conclusions it interprets with a view to establishing science and technology policy, and on which it reacts by contributing knowledge about the field of possibilities.

Naturally there must be a department of sectoral prospective attached to each agency, responsible for a sector of activity. Thus, each ministry—and more generally each large public or private national authority—should include such a department, at once guide and conscience, which furnishes elements for a global prospective and in turn reflects the latter.

This tool of global prospective, which is now indispensable to every modern country whatever its degree of industrial and agricultural development, must operate at the highest governmental level, or, at least, at the interministerial level (office of the president of the government or where not possible, the directorate of planning).

But such proximity to political power carries the risk of accentuating the normative character of the work to the detriment of opening research towards critical and philosophical reflections of an advanced exploratory character. It is therefore important to invigorate the whole of national prospective studies by facilitating very open meetings at the heart of an informal association for all who participate in them, with the association itself linked to the world prospective network.

This is the role played in France by the international association 'Futuribles' which periodically gathers French prospectivists together, introduces them to their foreign counterparts, and publishes basic studies reflecting all the diverse points of view.

Prospective thus becomes the guide for technical decisions without thereby becoming technocratic. For it must remain the clear conscience of a civilization and the ferment for its evolution.

It is through prospective that deep ties are established between the physical sciences and technology, and the human sciences. And it is prospective which necessitates modern multidisciplinary studies without which progress would have no soul.

The assessment of technology and technological policies

Karl A. Stroetmann

Vice-President and Senior research associate, Abt Associates GmbH, Federal Republic of Germany

Introduction

Supposedly we are living in a technological age. Technology is held responsible for the unprecedented economic growth in industrialized countries and the resulting increase in material welfare as well as for the destruction of the natural environment and a decrease in the 'quality of life'. Technology is godhead and devil at the same time. Technology means a possibility to make the earth lifeless as well as a challenge for more equality among nations and an opportunity for more peaceful times.

Technology is not an isolated fact in today's civilization; rather, it permeates societies in their entirety and is related to every factor in the life of modern man. A major task still ahead of mankind is to learn to manage technical progress and to direct innovative activities in such a manner that both social needs and private wants determine the rate and direction of technical change rather than mankind becoming the slave and victim of autonomous technical developments. This requires a better understanding of processes of technical change, of their impact on their interrelationships with economic and social developments and of the means and measures by which both individuals and governments can influence and direct technical progress.

Before I expand on this idea, let me briefly clarify the meaning of a few terms.

Technology, science and innovation

Let us define 'technology' as the set of instrumental rules, rules which prescribe a rational course of action to achieve a predetermined goal and which have to be evaluated by their usefulness and practical efficiency.[1]

Two important subsets of such rules should be distinguished: instrumental pre-scientific and instrumental scientific rules. Under the former we subsume an engineer's rule of thumb and rules based on experience, under the latter those based on scientific inquiry and knowledge. Even in our 'scientific age', skill and personal knowledge gained from past experience play an important role as evidenced, for example, by the enormous economies observed in the semiconductor industry.[2] Also, they constitute a major barrier to the transfer of technology, particularly to developing countries, because a direct contact of people from often quite different cultural backgrounds is involved.

Technology in the above sense may be contrasted with pure science, the aim of which is to attain fundamental and general scientific knowledge, cognitive theories, which should be judged by their truth and not according to their usefulness.[3] Pure research can then be looked upon as being performed for science's sake, whereas applied research is being pursued with the objective to be used outside the domain of science. This does not, of course, imply that the results of pure science may not be employed for practical purposes or that its findings are meaningless outside its own domain. Quite the contrary, the knowledge gathered by pure scientists has often proved to be a powerful tool for the practitioner, and problems encountered in applied research have often directed the efforts of pure scientists to new fields of inquiry.

Applied science, then, should be regarded as a proper subset of technology.

However, technological knowledge as such has only marginal effects on mankind, if any. The incorporation of this knowledge into innovations, new practices of the industrial arts, the manufacture of new artefacts, the re-organization of institutions, the creation of new services is what fundamentally affects each individual person.

The dynamics of technological development

The history of technology is the history of man's endeavour to secure the survival of his race and to improve his living conditions within an inclement and ominous environment. Being without specialized organs and possessing no natural instincts, he had to use his unique resource, intelligence, to adapt, change and transform natural resources to suit his specific needs. Man invented, for example, tools to improve upon and even substitute for his organic dexterity; man harnessed natural sources of energy to substitute for his organic energy, and developed modern information processing devices to assist his organic intelligence. This drive to create new technology, make intelligent use of innovations, and continuously improve them, may even be regarded as a characteristic of human nature.

During man's comparatively short history on this planet, these dynamics of technological development led to an enormous increase in the productivity of human labour, particularly in food production, which made the transition from precivilized to civilized societies possible, and allowed the build-up of modern industries. Today, several countries are in the midst of another great transition, that from a civilized society of the classical type to what has variously been called a post-civilized, post-industrial or information society. This transition is to a large extent due to modern developments in electronic technology in which artificial intelligent devices are substituted for organic ones. The complete automation of manufacturing processes comes within economic reach, a reduction of the industrial labour force by 50 per cent or more is envisaged in the longer run, and we may be able to spare 70 to 80 per cent of the working population for the tertiary industries.

A challenge to the social sciences

As Boulding has pointed out,[4] this new transition to a post-civilized society has many facets: 'economic, political, social, psychological, philosophical, educational, and religious. It is indeed a change which affects every area of human life and reaches far down into the human personality.' However, does this imply that man has broken through the prohibitions, taboos and rites that bound primitive man, only to become conditioned by technological civilization? Is it true that 'in the modern world, the most dangerous form of determinism is the technological phenomenon'?[5]

Probably there is considerable truth in the above statement. We are all aware of the many problems the scientific and technological revolution has brought to mankind. But I do not believe that we should, or are able to, reverse this development. Rather, we should regard it as a challenge to social scientists to contribute towards understanding these phenomena, towards managing technical change according to the needs of society, and towards transcending technology by making it the slave, not the master of progress.

In modern industrialized societies technical change does not occur in a vacuum. Rather, science, technology, economy, society and the State constitute a closely interrelated dynamic system subject to permanent change originating in different subsystems and forcing the other elements to adjust.

Simplifying, we may distinguish only two subsystems: society and technology. In society, individuals act whose needs and perceptions influence attitudes, goals and values; and institutions exist which influence and to a large extent even determine how we organize, relate and make social choices.

Society makes choices—explicitly and implicitly—of particular technologies in accordance with the dictates of its values and institutional structures. From the application of such technologies, in turn, derive important consequences that make impact on these values and structures.[6]

Also, the technological subsystem, by offering certain solutions to certain individual and social needs, may directly influence the choices made. It is the nature of these interactions which need to be explored and understood to help us to avoid the well-publicized negative side effects of modern technological developments. By identifying and analysing the crucial relationships in the technology–society system social sciences can contribute towards overcoming the 'failure of mankind to match technological with social ability'.[7]

The need for an assessment of technology

If society is to make rational choices of technological research to be supported, technics to be developed, and innovations to be accepted, an adequate perception of problems and sufficient information to arrive at consistent decisions are needed. I believe that technology assessment—the systematic study of the potential effects on society of the introduction, extension, or modification of a technology, with special attention to unintended, indirect, or delayed effects—could be an important tool in this respect. Here, technology assessment should be understood in a rather broad sense, as an approach towards analysing the impact of technology on a small group of workers as well as a means to identify potential effects on a global scale. Not only technical and economic aspects need to be considered but also social, cultural and individual impacts. It should not only be applied to technologies being developed or already in existence but also to alternative technologies being envisaged to satisfy existing or foreseeable new needs of society. Not only new so-called high technologies but also existing technologies being considered for transfer to developing countries and new so-called intermediate or appropriate technics need to be assessed.

So far, technology assessment stands for an idea rather than a proven approach or method. In view of the urgent problems posed by technical change, international efforts to develop research strategies and apply them to concrete examples to prove their feasibility seem warranted. A broad range of multi-disciplinary activities would be needed.

The successful implementation of technology assessment

may require new attitudes towards science and technology on the part of government, industry, scientists and engineers. Policy-makers may find it appropriate to clarify—in a much broader consultation with the public—the goals of society towards which technology would be aimed and evolve suitable methods of assessment, incentives and deterrents. Industry and socio-economic groups should accept a higher degree of social responsibility and co-operate with governments in the development and application of these new approaches. Science can develop more knowledge both of the individual disciplines and the total systems involved and technology can devise new technological solutions.[8]

However, technology asessment is not a procedure that will help us solve all future problems of managing technical change. Rather, its inherent limitations and problems encountered when implementing it need to be perceived and studied as well.[9]

The scope and precision of the results obtained depend on the availability of: (a) a methodology to translate the statement of a complex problem into a representative structure or model; (b) sufficient quantitative and qualitative information; and (c) sensitive methods for anticipating the 'essential' impacts of the assessed technology and for identifying critical factors and relationships.

Major problems are usually encountered: (a) when attempting to adequately quantify social variables, to identify their causal relationships and to model these; (b) when trying to unveil the often contradictory value systems and interests of affected groups; (c) because consciously or unconsciously the value system of the research team, which need not be identical with that of the user of research results, will influence the outcome of a technology assessment study.

Principal limitations result from (a) our understanding of structural relationships being conditioned by and depending on present and past experience and values; (b) our virtual inability to forecast future value changes and related changes in individual and social needs: and (c) an inability to forecast stochastic events of exogenously determined variables (e.g. basic innovations or political price changes).

The need for an assessment of technology policies

The assessment of technologies should lead to an evaluation of benefits as well as of costs, both in social and economic terms, and to an identification of potentially beneficial or detrimental effects, soon enough for supportive or remedial measures to be implemented by policy-makers. This indicates that technology assessment can only be a first step towards coping with today's problems posed by the development of technology. In a second step, society, its groups and institutions, has to choose amongst alternatives and to implement appropriate technology policy measures. Descision-making is a political task; scientists can assist here by providing information on the probable impact and side-effects different policy measures would have.

The need for a comprehensive assessment of existing or newly devised technology policy measures seems obvious. Most present decision and evaluation rules related to technology policy measures, and the quality of the available information on the impact of governmental activities related to technical, economic, and social developments, hardly provide a comprehensive basis for rational choice. Methods for a systematic assessment of policy measures encompassing the whole field of determinants and effects of such measures need to be developed. Research in this area must go beyond most traditional innovation (policy) research which has mainly focused on an analysis of the determinants of basic innovations and has largely neglected factors affecting the overall process of technical change which is characterized by incremental rather than by fundamental change. A systematic assessment implies, too, that simultaneously operating government measures—including those that are not directly related to technology policies—which affect the process of technical development should be considered. In addition, the fact that actual decisions do take place in a political context, characterized by (a) political restrictions (like conflicts of interests among governments, parties, labour and capital institutions, federal and regional authorities, etc, and the need to achieve short-term political consensus due to periodical elections), and (b) economic restrictions (like principles of economic conduct and changing availability of governmental funds due to the overall situation in the economy) needs to be taken into account.

Notes

1. Mario Bunge, *Scientific Research*, Vol. II, p. 132, New York, Springer Verlag, 1967.
2. John E. Tilton, *International Diffusion of Technology: The Case of Semiconductors*, p. 85–7, Washington, D. C., Brookings Institution, 1971.
3. Richard Mattessich, *Instrumental Reasoning and Decision Systems* (in press).
4. Kenneth E. Boulding, 'Research and Development for the Emerging Nations', in Richard A Tybout (ed.), *Economics of Research and Development*, p. 422–3, Ohio State University Press, 1965.
5. Jacques Ellul, *The Technological Society*, p. xxiii, New York, Knopf, 1967.
6. Roy Amara, 'Some Observations on the Interaction of Technology and Society', *Futures*, Vol. 7, No. 6, December 1975, p. 515.
7. Charles F. Carter, 'Future Darkness 2', *Science & Public Policy*, Vol. 1, No. 6, June 1974, p. 99–100.
8. OECD, *Methodological Guidelines for Social Assessment of Technology*, p. 6, Paris, 1975.
9. OECD, op. cit.; Eberhard Jochem, *Die Motorisierung und ihre Auswirkungen*, p. 8–9, Göttingen, Verlag O. Schwartz, 1976.

Part Four

Science, technology and development

Conditions in developing countries

Alassane M. Cìssé

Professor, École Normale Supérieure, Bamako, Mali

We find ourselves somewhat bewildered when we want to express ourselves on so vast a subject as the scientific and technological revolution and the social sciences. The subject embraces all the different aspects of man's life. A choice must be made. I shall not talk about the evolution of Western European societies during the period of the scientific and technological revolution. I shall limit myself to the case of countries which until recently were still colonized and which now live under a neo-colonialist system (I have in mind above all the countries of West Africa). The evolution of science and technology in industrialized countries, their multifaceted manifestations, their impact on the social structures and on man will certainly be masterfully expounded by other participants.

The colonialization of the countries of Asia and Africa by the Western powers during the nineteenth century was one of the consequences of the industrial development of Western Europe. The numerous scientific discoveries and the gigantic progress of technology gave an unprecedented stimulus to economic growth. Very quickly, national frontiers appeared too confining. Capitalism rushed towards the conquest of other sources of raw materials and other workers to exploit. Under the cover of generous principles, Western European countries undertook the conquest and systematic division of the less-developed regions of the world.

What consequences did the importation of science, of technology, and, above all, of Western civilization have on the human beings and on the social structures of the former colonial countries? This is what I am going to explain briefly.

On the level of science and technology, Africa's backwardness was considerable. The means of production were not developed. Societies lived primarily from agriculture, raising stock, fishing, and hunting in a system of local or auto-consumption. Nevertheless, in certain regions, production was very important, so that commercial channels were established on a large scale, e.g. the African Sudan. The social organization then could assume the forms of kingdom and empire, and was spoken of here as a feudal system. In other instances the social system remained more primitive. In all cases, these traditional societies were perfectly balanced with the role of the individual well defined. There was a harmony between nature, society and man.

Science and technology brought in their wake the rupture of this traditional framework, the breakdown of the equilibrium between man and society, and a profound change in the environment.

Before going further, I must state clearly that I am not directly calling science and technology into question. It is absolutely necessary to continue developing science and technology in order to expand knowledge and to improve continually the possibilities for action offered to man. The misdeeds I am going to denounce here result from the fact that science and technology serve big capital which is exclusively after profits, and is not concerned about man, society or the environment. Having made this clear, let us return to the impact of the scientific and technological revolution on traditional societies.

There has been a profound alteration of family ties. The cohesion of the family has very much decreased, above all in urban areas, which are the most affected by Western civilization. The authority of the head of the family, the guarantor of family stability, if not openly contested, is not always respected.

The individual has been freed from the family and social group. He has escaped from family control and is no longer subject to the beneficent influence of the social group. Traditional values are no longer respected. The individual is drawn to the mirage of the new mode of life, the new life-style offered to him by a blatant and encroaching modernism.

Man is increasingly depersonalized. More and more, social groups lose their identity and drown in an amorphous mass without originality and without culture. Some begin to feel that everything not inspired by science and technology is backward, obsolete and, consequently, must be abandoned. Man lives on without ideals, with a sole objective of acquiring the material goods made possible through science and technology. The result of this situation is a powerful moral degradation.

The individual lived in equilibrium with society and nature as an artisan, peasant or herdsman. The setting up of factories, the development of commercial and administrative sectors created a need for manpower. The country peasant and artisan became city workers. The urban centres grew much too quickly and shanty-towns, in which miserable men led miserable lives, sprang up.

Everything I have just said is equally true of Europe. But there is a difference in intensity which is worth emphasizing. Science and technology in Europe developed gradually over the course of the last centuries even if the most spectacular results have been obtained during the contemporary period. Science and technology are the products of European society, and these societies, in turn, have been modified under the effects of the evolution of science and technology. In Africa, we have witnessed a sudden introduction of science, technology, and the so-called 'technician' civilization into a society at another stage of development. This introduction was directed by a colonial system which had every interest in shattering the earlier social organization. It is, therefore, easy to understand that the transformations at the level of social structures and man are more violent.

There has been no real introduction of science and technology into developing countries. Western capitalism, which exploits these countries, has no interest in permitting a genuine development of science and technology in the countries of the Third World, and would certainly condemn itself if it did so. These countries, then, suffer all the disadvantages of the scientific and technological revolution without really enjoying its advantages.

The progress of science and technology is extremely rapid in developed countries. The acquisition of science and technology by developing countries is so slow that the gap widens from day to day between the two groups of countries. This backwardness contributes to maintaining and even to aggravating the dependence of developing countries upon developed countries. This creates conditions of permanent assistance and the continuation of neocolonialism.

Faced by this situation, Western social sciences have not been able to provide an accurate analysis which would allow a definition of clear perspectives for the future. The societies are considered *en bloc* without an analysis of contradictions at the heart of these societies. They speak of economic development and increasing income, while using every means to delay this development. They call for the limitation of population growth because it cancels out the effects of economic growth. They talk of the need for the development of education, etc.

Just as they have not been able to analyse the fundamental causes of the crisis provoked by the scientific and technological revolution in Europe, the Western social sciences, as far as the developing countries are concerned, concentrate only on certain apparent phenomena which, in fact, hide a profound reality.

The nationalists in Third World countries who struggled for political independence are now confronted with the problem of scientific and technical development. Scientific research must be undertaken in order to create a national base capable of receiving scientific data from the outside. The development of science and technology allows the consolidation of a national identity when the colonial period is over. Man must preserve his personality and the societies their originality, while remaining receptive to scientific and technological progress. The introduction of new technologies must give a predominant place to man. The situation currently prevailing in the West of man dominated by science and technology absolutely must be avoided.

Given the close dependence of developing countries on industrialized nations, the crisis of perspectives in developed societies is the same as that of the developing countries.

The experience of the Western countries has shown the ineffectiveness of the solutions proposed by the social sciences to control the scientific and technological revolution and to avoid, for the society, the profound crisis now in progress.

In developing countries solutions for the crisis have been sought. Thus, in sub-Saharan Africa, people have spoken of authenticity, return to origins, Africanicity, etc. All these solutions have remained ineffectual because they do not rest on a valid theoretical base.

It is necessary to install new productive forces in the developing countries which are more powerful than those of the traditional societies. This is the condition for their survival. The fundamental question to be posed then is the following: How must the installation of these new productive forces be done?

In Third World countries, the introduction, development and control of science and technology can be realized only if the social conditions are met, i.e. if there is a mobilization of subjective social forces. Only these social forces can lead the countries along the path to a harmonious social and economic development. The European experience is sufficiently revealing so that the developing countries may avoid repeating the same mistakes. It is absolutely necessary not to take the capitalist road, which will simply contribute to the maintenance of underdevelopment and dependence on the outside world.

The subjective social forces will be able to effectively and rapidly restrain sudden change of traditional societies, realize a new equilibrium, and determine clear perspectives.

The scientific and technological revolution in the developing countries can be managed only within the framework of close collaboration with the developed ones. But the developed countries must be sincerely concerned to aid in the development of the Third World countries. They must recognize, as a historical necessity, their contribution to the development of the Third World.

This will help defuse the confrontation now looming between developed and developing countries. But we must not cherish any illusions. The aid from the developed countries of Western Europe will be effective only if profound social transformations are produced in these same countries. As long as scientific and technological progress is exploited by capital, the Western countries will try to keep the Third World countries in their state of underdevelopment and continue to use them as suppliers of raw materials.

The developing countries must look for a correct theoretical base for their scientific and technological revolution, they must use the subjective conditions of this revolution, and finally they must strive at a more general level for the establishment of a new world economic order.

By way of a conclusion I wish to call attention to the following points: (a) colonization is one of the consequences of the economic growth of the European countries; (b) the profound alteration of the social structures, and of the meaning of man's existence, stems from the scientific and technological revolution; (c) the scientific and technological revolution has made the modification of the old social structures absolutely necessary in order to adapt them to the new situation and avoid more serious crises; and (d) the scientific and technological revolution has made co-operation between developed and developing countries necessary in order to accelerate the growth of the latter towards higher living standards and to establish a new economic order.

A view from the Third World

Ezzat Hegazy

Sana'a University,
Yemen Arab Republic

Throughout the entire history of mankind, man has endeavoured to secure and improve his living conditions through better understanding of the world around him and by more effective control of it, i.e. through ever-developing science and technology.

The present scientific and technological revolution, however, started with the beginning of the modern era in Western Europe, whence it spread to the New World and other European settlements overseas. Significant among the factors responsible for this historical development were the crisis that befell the feudal regime, the Enlightenment of the seventeenth and eighteenth centuries and geographic explorations and the resulting waves of European colonization which were made possible, and at the same time contributed to, scientific and technological advances.

Some sixty years ago, the countries of the 'second world'—the Union of Soviet Socialist Republics and, later, other socialist countries—embarked on a similar revolution. Thanks to strong commitment and devotion to socialist ideals, and a number of other factors, some of these countries could accomplish in little more than half a century what the countries of Western Europe needed about three centuries to achieve.

Most of the countries of the Third World remained under European domination until after the end of the Second World War. Contrary to their professed goals, European colonization retarded the development of the conquered territories, to the extent that some of these countries to this day live a life very similar to that of their distant ancestors.

Until the disintegration of the colonial regime, most of the leaders of national liberation movements in the colonies were convinced that the real challenge that faced them was political independence. The huge gap that separated their countries from the industrialized world represented no serious problem to them, for it was thought that the introduction of modern science and technology was possible and could close that gap.

The experience of the third quarter of the present century, however, has proved that that attitude was naïve, even faulty, and that the gap is widening rather than closing.

This calls for a fresh analysis of the major barriers to the transfer of science and technology, the efforts these countries have been able to make in both fields, and their impact on the social situation, and their interrelationships with local and world economic and social developments.

An attempt is thus made here to highlight the conditions responsible for the developing countries' failure to adopt the scientific and technological revolution, in terms of two factors: the international economic system, on the one hand, and the peculiar social structure of these countries, on the other.

There is no doubt that the quantitative additions as well as the qualitative leaps achieved in the fields of science and technology during the last three centuries far surpass all that had been accomplished throughout all the history of mankind until the beginning of the modern era. What has been achieved in the various branches of science makes the wealth of human knowledge at the end of the Middle Ages look insignificant. On the applied side, man has experienced the change from manual labour in which he utilized his muscles and animal energy, to the stage of the tool acting as a machine controlled by man, to automation, and finally to the stage of cybernetics, a change that has made a substantial increase possible in the productivity of human labour.

It is because of this, and the fast pace—sometimes the mutations—with which progress occurs, that it is called a revolution. And if this is true of the situation since the beginning of the modern era, it is all the more typical of what has been happening since the start of the present century.

The growing sophistication of the scientific method has made social forecasting possible, something rendered necessary by the accelerating pace of social change.

The scientific and technological revolution has even gone so far as to influence man's mentality too. Man is no longer captive of past and present conditions, but is now future-oriented, attempting to foresee developments to come and to influence them.[1]

Thanks to the breakthrough made in science and technology in the past decades—the greatest revolution man has ever made—science is no longer simply one of several elements making up the productive forces, but has become a key factor in social development, penetrating more and more into the various fields of life.

However, since the early phases of the modern era, much more time, effort and money has been devoted to the development of the natural sciences, a trend that has been accentuated by the so-called military-industrial complexes in the richest and most advanced countries. The social sciences have drawn very little attention, and that only recently. As a result, there exists a serious imbalance between the two major fields of human knowledge, with the social sciences progressing much more slowly than, and still lagging behind, the natural sciences.

This, however, is not the cause of the world crises, but rather one of their manifestations; and yet the development of the social sciences can contribute to a better understanding of the crises and more effective handling of the problems they pose.

What is needed is neither the clarification of some issues nor the accumulation of more data, but rather the development of a new perspective upon, and comprehensive plan of, societal and global actions capable of better handling such world problems as security, development, diminishing natural resources, and the like. This, unfortunately, is something that the social sciences, which have reached some kind of impasse with regard to theory and method, cannot provide, nor the present world situation can welcome. And herein lies the crisis.

Europe's colonization of the 'New World', Africa and parts of Asia was—for the colonized—a devastating experience. The violence that accompanied it did away with the relative stability that characterized life in many of these areas, and abruptly changed social conditions. Since most of the colonized were inferior to the Europeans in terms of science and technology, they could not but resign themselves to the fate that befell them, for some time at least.

The old colonial policy guaranteed a regular flow from the colonies of cheap raw materials needed by the new and expanding European industry, and opened up the colonies to the manufactured goods of Europe at rewarding prices. Achievement of these goals demanded that the colonies' self-development had to be slowed down, if not stopped.[2]

Nevertheless, in order to accomplish these goals, the colonial powers could not avoid introducing some 'reforms' into the economy of the colonies—new cash crops, modern irrigation systems, up-to-date transport and communications networks, new administrative apparatus, etc.

The introduction of modern science and technology on a large scale into the colonies was discouraged by strengthening numerous survivals of archaic social relations. Also, the investment of colonial monopolies in local manufacturing industry, the sales of licences and patents, technical 'co-operation' and the trade policy, all posed serious obstacles in the developing countries' way to utilize modern scientific and technological innovations.

Thus the science and technology that was introduced into the colonies, little though they were, constituted a tiny enclave of modernity in a vast world of backwardness. What is more important is the fact that the coexistence of the tiny modern sector with the traditional, even primitive one led to a kind of technological dualism, but one that caused no concern on the part of the colonial administration. The resulting social and cultural repercussions were ignored, and the harmony or 'fit' between technical 'reforms' and social and cultural conditions were overlooked.

These and other developments, nevertheless, gave birth to important structural changes. Significant among them was the rise of two new social strata—the middle and the working classes—different in many respects from their European counterparts—the bourgeoisie and the proletariat.

The colonial administration, however, intervened to impede the natural growth of these social strata, and thus prevented them from playing a role similar to that played by their European counterparts in the modernization of Europe and the rise of modern civilization. This policy was pursued by the colonial administration through the linking of the interests of the new social strata to its own, and by the creation of a conflict of interests between them and the masses in their countries.[3]

The 'harmony', in a sense, between the interests of the colonial administration and those of the local middle classes in the colonies did not last long, nevertheless. For the more these middle classes gained in wealth and power, the more their ambitions outgrew the alliance with the colonial administration. Hence national liberation movements developed and a nationalist consciousness too, both of which were related to—if not connected with—the growing ambitions of the middle classes in the colonies.

The countries of the Third World have, no doubt, experienced a series of very important transformations—the emergence of new nations and the rise of modern native élites replacing the traditional ones, the establishment of modern factories and service facilities, the opening of schools at various levels, etc.; nevertheless, the end of foreign military occupation did not end backwardness and exploitation. For though the new ruling élites in most of the developing countries raised slogans different from the goals of the colonial administration, they did not behave very differently.

A number of reliable indicators—such as labour productivity, gross national product, the standard of living, literacy rates, health conditions, and the like—show that what the Third World countries have achieved in their attempts to overcome backwardness remains very modest. There have been some major transformations, but these have not radically changed the multiform nature of the economic structure of their countries. For economic development proceeded very slowly, and mainly in the sphere of service and small industry, while the development of large-scale industry and the organization of economic development on a nation-wide scale have been largely ignored.

When the State-owned sector began to take part in the task of modernization, the introduction of modern science and technology fell victim to the 'demonstration effect', something which was very costly and left its impact on consumption patterns and led to changes in the structure of the domestic market that proved to be in conflict with the needs of further growth.[4]

Modern factories, farms, mines and banks and other service establishments are mostly overstaffed with under- and sometimes unqualified personnel working under obsolete systems of administration, and therefore producing well below the level of marginality.[5]

Despite the opening of many schools at various levels, illiteracy rates are still very high, and there is still a need for an up-to-date educational system that serves modernization by enhancing creativity, and by allowing for the accumulation of experience.[6]

Research, vital for scientific and technological progress, either does not exist or is of a primitive nature. Research organizations, when they do exist, are nothing more than 'prestige projects', that cost much, but function at a very low level of efficiency.

These and other conditions lead to the 'brain drain', the migration of the cream of the professionals from where they are badly needed but are mostly unemployed and, in a sense, alienated, to the advanced countries of North America and Western Europe, where there are many of them, but where work conditions and incentives are irresistible.

So far the developing countries have been very slow, and only partially successful, in adopting and utilizing the achievements of science and technology. The transplantation of modern science and technology from where they have been invented to the developing countries, a completely different environment, seems to have been a very painful and costly experience.

But why has this been so?

Though it is true that the scientific and technological revolution proceeds faster in countries with a higher economic potential, more mature forms of social division of labour, larger markets, more skilled labour force, etc., the argument that scientific and technological progress assumes forms that are consistent only with the needs and conditions of the advanced countries does not withstand serious critical consideration.

The explanation of the slow progress achieved by most of the developing countries by what is called 'resistance to change' both needs specification and leaves many problems to be explained. For even the most primitive peoples known to researchers have welcomed some kinds of change under certain circumstances, while in some of the most advanced countries there is resistance to change—such as labour unions' rejection of technological innovations leading to the lay-off of some workers, students' protests of curriculum changes that do not fit in their conception of what suits them, etc.[7]

The followers of the 'pessimist' branch of the school of technological determinism put the blame on the progress of science and technology as such, deliberately ignoring the fact that development is such that the adoption and conversion of science and technology into productive forces are dependent not on progress itself but on the type of social system.

The ruling élites in most of the Third World countries blame the crises in which they find themselves and their countries on the colonial policy and conceive of their problems as survivals from colonial times. This is not far from true. We can even add that the disintegration of the colonial regime did not end the colonial strategy's adverse impact on the development of the former colonies, a fate very difficult to avoid in as much as the developing countries exist not as isolated entities but as part of an economic order which is dominated by the former colonial powers.

The developing countries still now, and will for some time, have to import most of their needs of the new products of science and technology from the advanced countries. They are sold 'intermediate', not advanced technology, for which they have to pay a much higher price as compared with preceding periods.

Even when the needed capital is available, the developing countries have to make a very difficult choice. For the adoption of modern capital intensive technology does not create real possibilities for resolving the problem of unemployment.

Besides, these countries are extremely short of skilled personnel, lack experience in operating large enterprises and, what is more important, suffer from a socially and technically defective State apparatus.

But if the cause of difficulties stemmed only from the colonial heritage, independence and the severing of ties with the old colonial powers could have led to a gradual easing of the difficulties and an acceleration of the process of development.

In contrast to Western Europe, where modernization proceeded as a result of the disintegration of the feudal regime and the gradual emergence of the bourgeoisie—an ambitious, resourceful and hard-working class—the modernization of the Third World countries is going on under outside influences. It is true that national liberation movements have abolished the colonial administration and destroyed most of the obsolete social structures connected with it. But the modernizing élites in most of the developing countries lack the initiative and the readiness to take risks of the European bourgeoisie. Rural in origin and still strongly attached to the country, a large portion of these élites are still tradition-oriented and though motivated for economic development are unequivocally against change in the value system or the system of social relations. These are among the reasons why they have failed to create new social structures favourable to modernization and have failed to appreciate the far-reaching implications of the fact that 'the transformation of man's nature and living conditions represents not only wide-ranging technological tasks but also grave social and human problems'.[8]

Most of the developing countries carry out the task of modernization under no clear ideology. Their development is not served by their attempts to create a replica of one or the other of the present-day economic structures of the advanced countries. Without their own ideology and base of development planning, it will be practically impossible for them even to apply the inventions at hand.

Finally, the developing countries are unlucky not to have had something like the European Enlightenment which habituated men to the idea that both the physical and social worlds obeyed natural laws, that human intelligence was the most effective instrument for the mastery of nature and social reform, and that under the guidance of human reason development could proceed toward unlimited frontiers.[9]

The crises encountered in the transfer and utilization of the immense advances of science and technology are not inherent in the scientific and technological revolution. The deformed course of the revolution, the stagnating social conditions and the resulting lack of faith in science and technology in most of the countries of the Third World are all traceable to the defective system of production and social relations prevalent in these countries. The scientific and technological revolution only makes potentialities the realization of which is dependent on the social system.

As Radovan Richta puts it:

> The crucial issue of the present is not the halting of the process of transforming nature, the taming of the scientific and technological revolution or its reduction to the level of the possibilities of the old social order . . . [On the contrary, society's] vital interest is to develop science and its application into new dimensions, capturing the entire complex of the transformation of nature and society . . . A society which is to master the conditions of the scientific and technological revolution can only be a 'controllable society'—a society which controls the entire complex of people's living conditions, affecting their activities, interests and stimuli. . . .[10]

Since scientific and technological progress requires profound departures from previous techniques rooted in cultural traditions, the developing countries have to undergo progressive socio-economic changes, update their scientific and technological infrastructures—something that is vital for the transition from the state of dependency to one of independence—and fight for a better position in a new international economic system.[11]

Notes

1. To appreciate the important implications of these developments, see Mihai Rotes, 'The Revolution of Science and Technology and the Future of Mankind', *The Revolution in Science and Technology and Contemporary Social Development*, p. 231–43, Bucharest, The Academy of the Socialist Republic of Romania, 1974; and Walter Roman, 'Controversial Theoretical Problems of the Revolution in Science and Technology', *The Social Future* (Bucharest), 19–24 August 1974, p. 35–41. (Special Issue for the 8th World Congress of Sociology, Toronto).
2. George Balandier, 'The Colonial Situation: A Theoretical Approach', in Immanuel Wallerstein (ed.), *Social Change: The Colonial Situation*, p. 34–61, New York, John Wiley, 1966.
3. William Morris Carson, 'The Social History of an Egyptian Factory', in Abdulla M. Lutfiyya and Charles W. Churchill (eds.), *Readings in Arab Middle Eastern Societies and Cultures*, p. 365–74, The Hague, Mouton, 1970.
4. A. Elyanov, 'The Scientific and Technological Revolution and Socio-economic Problems of the Developing Countries', *Social Sciences* (Moscow), Vol. 4, No. 2, 1973, p. 133–44.
5. Carson, op. cit.
6. Gall Erno, 'Intellectual Creativity and the Revolution in Science and Technology', *The Social Future*, op. cit., p. 35–41.
7. The frequently observed lack of discipline, feeling of fatigue and work injuries on the part of many workers in modern factories in the developing countries are, in most cases, manifestations of the resistance to change that does not help the workers attain their goals, or so abrupt that they cannot make the necessary adjustments to it or meet its requisites, or implies certain transformations they are not ready to accept.
8. Radovan Richta, 'The Role of the Social Sciences', Paper for Unesco conference on 'The Scientific and Technological Revolution and the Social Sciences', p. 37, Prague, 6–10 September 1976. (This vol. p. 63.)
9. Raj P. Mohan and Don Martindale, 'Introduction to Part 1', in Raj P. Mohan and Don Martindale (eds.), *Handbook of Contemporary Developments in World Sociology*, p. 3, Westport, Conn., Greenwood Press, 1975.
10. Richta, op. cit., p. 33 and 54. (This vol. p. 60 and 74.).
11. Herein lies the solution of the main issues.

The Latin-American experience

Philip Maxwell

The IDB-ECLA/Regional Programme of Research on Science and Technology, United Kingdom

The last ten years have seen a remarkable upsurge of interest among Latin-American intellectuals and in several Latin-American governments in understanding, criticizing and changing the role played by science and technology in their countries' development.

One outstanding result of this upsurge has been the passage of significant new legislation on foreign investment and technology in the Andean Pact countries,[1] in Brazil, in Argentina,[2] and in Mexico.[3] Furthermore, the outpouring of new studies, new legislation and new experience from the Latin-American countries in the past few years in the technology field has given a powerful spur to parallel ongoing international efforts in such fields as draft codes of conduct for the transfer of technology by multinational companies,[4] the reform of the Paris convention on the recognition and exploitation of invention patents,[5] etc.—which initiatives are clearly in the spirit of the 'new international economic order'[6] in that they imply non-trivial changes in some of the 'rules of the game' under which international investment, trade, and competition take place as between the 'advanced' and the 'developing' countries.

At this point there is an interesting historical parallel worth mentioning. In the 1950s and early 1960s, it was Raul Prebisch and the Economic Commission for Latin America who helped to launch the issue of fair terms for raw-material exporters on to the world negotiating stage, via the creation of the United Nations Conference on Trade and Development (UNCTAD). Ten years later, it has been Constantine Vaitsos, and the Andean Pact countries (Colombia, Venezuela, Peru, Chile, Bolivia and Ecuador) who have done most to project the issues of 'transfer of technology' and of the need to control the restrictive practices of multinational corporations on to the world forum. Therefore, once again, Latin-American experience can be seen to have acted as a kind of unofficial 'laboratory' for ideas now taken up by many Third World countries and adding to the growing package of 'areas of dispute' between the dominant and relatively weaker countries in the present-day world.'[7]

This article now proposes to trace out four broad lines of approach along which Latin-American writers have analysed Latin-American experience in the science-technology-development field. These four broad lines of approach are outlined below.

First, the 'technological-dependence' approach, which seeks to identify the adverse consequences to Latin-American economies of its broad dependence on sources of technology from the advanced Western countries. Here, after briefly introducing the 'dependency' theories advanced by Latin-American authors such as Furtado, Cardoso and Sunkel in the 1950s and 1960s, the article goes on to present a brief selection of the notable discoveries and advances made in the 1970s by Latin-American authors like Constantine Vaitsos, Fernando Fajnzylber and Miguel Wionszek in the field of the 'transfer of technology' to Latin America, and the role played by multinational corporations as agents of technology transfer. A summary is then given of the principal features of the detailed new diagnosis of 'technological dependence', which has emerged from these recent studies, and of some of the 'counter-measures' to combat this dependence which are contained in the Andean Pact legislation on foreign investment and technology, and to a lesser extent in the Brazilian, Mexican and Argentine legislation.

Second, the 'autonomy-promotion' approach, which is concerned with boosting the build-up of local science and technology so as to harness it in favour of national autonomy. Here the effort is made to briefly review Latin-American experience and thinking both with regard to the build-up of local scientific and technological institutions and with regard to the efforts made to promote and plan national (and regional) science and technology policy. The authors cited here include Jorge Sábato, Máximo Halty and Francisco Sagasti—and the review seeks to emphasize the new ideas and original approaches to science and technology policy which these and other authors have developed and promoted in response to Latin American conditions.

Third, the 'micro-economic-historical' approach, which is concerned with detailed hindsight studies of how local technological capacity has actually developed in Latin-American public, private and foreign-owned enterprises, and in specific manufacturing sectors such as pharmaceuticals, automobile manufacturing, petrochemicals, etc. Here the effort is made to review some of the interesting findings made by a leading exponent of this approach, Jorge Katz.

Fourth, the 'appropriate technology' approach, which is chiefly concerned with a critical attack on the type of consumer products which modern science and technology is used to generating on the grounds that these products are often highly 'inappropriate' to the most basic needs of Latin-American societies with at present low per capita incomes and serious unemployment problems—and on the grounds that development based on these products helps to generate and perpetuate an extremely skewed pattern of income distribution. Here the review mentions the interesting polemic views of Ivan Illich on the subject along with a number of reservations and points in relation to the application of the appropriate technology concept in the Latin-American context.

Of course, as the reader is bound to note, this is quite a rough-and-ready classification. First, many of the authors mentioned combine in their thought some distinct elements from all four 'approaches'. Second, there are strong complementarities (as well as antagonisms) between certain of the views held by the different authors represented as belonging within the different approaches. But this classification is not meant to be watertight. It is simply a convenient way to organize this brief and inevitably incomplete 'overview' of the fund of Latin-American experience which exists in this critically important field. (In this particular article Cuba's experience is not considered.)

Having first explained briefly the four different approaches to be reviewed in the article, we will now proceed to examine each of these four approaches in turn.

The technological-dependence approach

In introducing this first approach it is useful to refer briefly back to the framework of 'dependency theories' which were already being developed by prominent Latin-American economists and sociologists throughout the 1950s and 1960s. In these theories, which are associated with authors such as Celso Furtado, Fernando Cardoso, Osvaldo Sunkel and others,[8] the starting point was the explicit recognition of the historical 'specificity' of the type of development which has been taking place in Latin America from colonial times until the present day.

These authors showed that Latin-American economies were not likely to follow a similar sort of path to the one taken until now by the advanced Western economies. This was because the interaction between the gradual growth of local Latin interests and the always very powerful foreign economic interests of the advanced capitalist world in Latin America had produced a set of 'hybrid' types of economies which have been evolving according to a substantially different set of rules, conflicts and constraints from the ones which applied at earlier stages in the history of the advanced Western countries.

Furthermore, in view of the preponderant and continuing historical role of these external economic forces in Latin America, the name given by these theorists to the type of development experienced in Latin-American countries was the term 'dependent' development—where by this term they understood not a fixed unchanging relationship of dependence between Latin-American economies and the advanced capitalist world but an *evolving* relationship in which new forms of dependence tended to evolve from the previous ones. In Celso Furtado's words, 'underdevelopment is linked to the complex relations of domination-dependence between peoples, which tend to perpetuate themselves under changing forms'.[9]

As will be seen shortly, this particular way of focusing on Latin-American development problems gets quite a lot of reinforcement from recent Latin-American research findings on the subject of the 'transfer of technology' to Latin America by multinational corporations, particularly in the field of manufactured products. This research shows convincingly how the mastery and control by multinational corporations of technology (both manufacturing technology and marketing technology) has become—since the end of the Second World War—an extremely important new factor in reinforcing the build up of foreign ownership in the Latin-American economies.

In other words, whilst in prior periods, it was factors such as the outside powers' military threat, or else their control over the export markets for Latin America's primary products, or else their control over critically needed loanable resources, which made up the external 'clout' which supported foreign interests in Latin America, research now showed that an important new factor had come into play: this new factor, of clearly increasing importance throughout the 1960s and 1970s, and contributing greatly to the external 'clout', is the mastery and control by multinational corporations over technology.[10]

What evidence is there to substantiate this statement? What does it mean to talk about 'control over' or 'transfer of' technology? What consequences is this said to have for Latin-American development? These are the kinds of questions with which recent Latin-American writers within the 'technological-dependence' approach have been concerned. To look at these questions in more detail, it is helpful to start first with a brief historical outline.

Whereas up until 1930, foreign investment in Latin America had steered pretty much clear of the manufacturing sector and had concentrated its efforts in such export activities as mining, petroleum drilling and food production plus railway construction and infrastructure activities, the picture after the Second World War began to change quickly. Encouraged by high Latin-American tariff structures and 'open-arms' import substituting policies combined with attractive fiscal incentives, very many of the leading United States multinational enterprises—and, later, European and Japanese ones as well—began to set up manufacturing branches in Brazil, Argentina, Mexico and to a lesser extent in Venezuela, Peru, Colombia and Chile, with the aim of producing and selling manufactured goods such as automobiles, pharmaceuticals and home electrical appliances to these countries' own internal markets (and in some cases to the regional markets established by trade agreements). The magnitude of this 'change of pattern' was such that since 1960 manufacturing has made up around 50 per cent of the increments to United States investments in Latin America, and by 1973 manufacturing investments accounted for around 35 per cent of the total book-value of all United States direct investments in Latin America—themselves more than half of all foreign investments.[11] Table 1 shows still further data on this 'change of pattern'.

The major implication of this striking new pattern of manufacturing investments in Latin America was that it necessarily involved the multinational firms in (a) 'transferring' to Latin America very significant quantities of their manufacturing technology and know-how (so as to get their Latin-American plants working properly and turning out automobiles, antibiotics, refrigerators, etc., similar in quality to these models or products produced by the companies in their advanced country plants), and in (b) simultaneously transferring to Latin-America very significant quantities of their marketing technology and know-how (e.g. in particular, the brand names and trade marks like Kodacolor, Kent, Ford Falcon, Electrolux, etc., plus much of the associated advertising and publicity techniques linked to reinforcing the international prestige image already associated with these products prior to their production in Latin America).

In other words, the new pattern of foreign investments which aimed at producing directly for the internal markets of Latin-American countries necessarily involved a striking increase in the transfer of technology carried out by multinational firms to Latin America.

However—and this is the key point—in the case which we are examining here of foreign direct investments which involved the multinational companies in setting up wholly-owned subsidiaries in Latin-American countries, the relevant technology has not been transferred to 'Latin America' but has simply been transferred to a new Latin-American

TABLE 1. Participation of sectors in increments to United States direct investment in Latin America 1950–73

	1950–60	1960–67	1967–73
Mining and smelting	14.5	2.3	6.2
Petroleum	42.1	6.4	14.4
Manufacturing	21.3	64.8	44.9
Other	22.2	26.5	34.5

branch (a subsidiary) of the same multinational company. Obviously in such transactions, the technology which is the subject of transfer to the Latin-American subsidiary does not pass out of the confines of the multinational company as a whole, and—equally obviously—such transactions in technology are not 'arms-length' transactions, but constitute a special case of 'intra-firm trading', i.e. of transactions carried out between divisions of the same firm. This means that the actual amount which a Latin-American subsidiary branch has to pay to the rest of the multinational corporation for the technology which has been transferred to it is to a significant extent arbitrary and is greatly influenced by the policies of the corporation's head office as to what distribution of income as between its various foreign branches is most convenient bearing in mind the different tax rates and legal restrictions and obligations in different countries and the corporation's global maximization strategy faced by these differences.

To recap the foregoing very briefly, one can say that it has been the multinational corporation's mastery of the technology for producing advanced consumer goods and consumer durables plus their willingness to transfer this technology via the direct investment route to new Latin-American subsidiaries, that has won for these companies their immensely successful entry and expansion in Latin-American internal markets.

However, the very fact that these transfers of technology have taken place via intra-firm (i.e. non-arm's-length) trading immediately raises a sharp question as to what actual terms have governed these transfers of technology, and whether these terms have, on balance, corresponded to the Latin-American host countries' interests and expectations.

It is precisely with regard to this 'sharp question' that Constantine Vaitsos and his collaborators, with the co-operation first of the Colombian Government and then of the Andean Pact Governments began to create—in the late 1960s and early 1970s—a most almighty stir on the 'multinational waters' by showing (a) that the transfer to their Latin-American branches of patents, trademarks and other technological assets controlled by multinational companies was being carried through via licensing contracts which contained an extraordinary quantity of restrictive clauses many of which would be illegal in the parent firm's country of origin; and (b) that the imposition of such clauses (and their acceptance as legal by the host government) were instrumental factors in permitting the systematic 'cover up' by the foreign corporations of the real magnitude of the profits realized from their affiliates in the Andean Pact countries.

For example, Vaitsos was able to show that in a representative sample of the Colombian pharmaceutical sector, which is 80 per cent foreign owned, the average degree of overpricing of the intermediate products imported by the foreign affiliates in the sector in 1968/69 was 155 per cent by reference to internationally prevailing prices for the same ingredients. So for these foreign affiliates, this meant that their declared profits in Colombia were only a fraction of the 'true' profits they made—the 'missing' part of the profit had been collected by the parent companies in the form of overcharging their affiliates in Colombia for the needed intermediates. Indeed Vaitsos estimated that the reported profits of these affiliates constituted only 3.4 per cent of the effective returns to parent companies, with royalty payments accounting for 14 per cent and overpricing for 82.6 per cent of effective returns![12]

Similar, though not quite so blatant, cases of overpricing of intermediate goods were found in the rubber, electronics and basic chemicals sectors researched by Vaitsos and his group. Furthermore, significantly inflated intermediate product prices were often found to have occurred even when the technology recipient was an independent nationally owned firm or joint venture rather than a foreign wholly owned subsidiary.

So the first important result of Vaitsos's work was that it called dramatic attention to the possible widespread occurrence in Latin American countries of inflated undeclared profits resulting from overpricing of tied inputs linked to transfer of technology contracts.[13]

The second important result of this research by Vaitsos and his co-workers in the Andean Pact countries, and of parallel investigations into the terms of transfer of technology contracts carried out by other researchers such as Miguel Wionszek in Mexico, Francisco Sercovich in Argentina, Fernando Fajnzylber in Brazil, etc., has been to demonstrate how some very frequently found characteristics of such contracts—in particular their 'package-deal' nature and the host of attached restrictive clauses and conditions—produce some deeply inhibiting effects on local technological creation and autonomy.

Specifically what these researchers found—both in the analysis of the 451 licensing contracts in the Andean Pact countries and in studies of hundreds of other contracts in other Latin-American countries[14]—was that foreign technology suppliers had very often been able to limit the freedom of the Latin-American subsidiary firms,[15] licensees[16] or plant purchasers[17] by not permitting them to buy or develop alternative product or process technologies to those which the suppliers himself provided or specified. The result is to keep Latin-American technology recipients broadly 'dependent' on their foreign suppliers for much longer than necessary, and over a far wider spectrum of activities than necessary.

So, in effect, the existing methods of technology transfer not only create the opportunities for foreign suppliers to secure 'supernormal' profits on their technology by the overpricing of chosen inputs linked to the use of the main technological assets included in the 'package deal', but these methods also help to keep Latin-American enterprises in a state of permanent technological dependence on their foreign suppliers. This—in a nutshell—was the substance of the Andean Pact diagnosis of how the 'technology factor' has tended to work in the Andean context.

In view of this diagnosis of 'technological dependence', which was confirmed by many further studies beyond those mentioned above,[18] the Andean Pact countries, and other Latin-American countries too, became interested in the late 1960s and early 1970s in implementing legal and other governmental measures designed to combat the many abuses, restrictive practices and balance of payments losses implied by the existing pattern of technology transfers.

The studies all pointed to four key areas where counter-measures could be of real value:

In the area of a much more systematic legislation by Latin American governments so as to establish the rules applying to foreign investments and to define and control in a comprehensive way the permitted channels and agreed levels of the remittances which foreign investors should be allowed to make abroad with regard to the imports of goods and services, the repayment of loans and the payment of royalties and profits.

In the area of systematic government intermediation in the process of the negotiating and contracting of licence agreements, with the aims of forcing more disclosure of information from technology suppliers, bolstering the negotiating position of local firms, and giving the government veto power over contracts embodying tie-in clauses and restrictive business practices.

In the area of new national legislation on industrial property privilege, so that the granting of monopoly privilege to firms in the field of patent recognition, trade mark registration, etc., should be made conditional on the active use and/or subsequent diffusion by such firms of the new technology for which their privileges were granted.

In the area of promoting the build up of local technology creation in critical fields, so as to increasingly substitute for many of the items currently being included in the 'packages' of inputs being supplied from abroad.

Furthermore, these studies were not just academic efforts—they raised significant political interest, and helped to spawn the following sequence of measures: December 1970: Decision No. 24 of the Andean Pact on the 'Common Treatment for Foreign Capital and Trademarks, Patents, Licensing Agreements and Royalties'; September 1971: Argentina passes law 19321 regulating the transfer of technology; December 1971: Brazil institutes, via law 5772, a new Code of Industrial Property; December 1972: Mexico passes a law on the registration of the transfer of technology and the use and working of patents, trade names and trade marks; May/June 1974: Decisions 84 and 85 of the Andean Pact set out the bases for a comprehensive 'Subregional Technology Policy' and regulate the application of the rules concerning industrial property; February 1976: Mexico passes new 'Law on Inventions and Trademarks'.

As could be expected, none of these measures has produced overnight transformations in the respective country's technological capacity and independence *vis-à-vis* foreign investment, and some of these new laws have already begun to be watered down in the face of foreign and internal pressures.[19] However, there is no question but that laws such as these have helped to strengthen the negotiating power of the local authorities as well as of local firms and have launched the issue of technological dependence irreversibly onto the statute books as well as into the 'consciousness' of many thousands of public officials and private and public entrepreneurs all over Latin America.

The 'autonomy-promotion' approach

In contrast to the first approach which is concerned with technological dependence, and whose point of departure is the analysis of the influx into Latin-American economies of imported external technology, the 'autonomy-promotion' approach takes as its point of departure the 'promotional' goal of building up Latin-American local capacity in science and technology. Thus, whilst the technological dependence approach is largely about what has already happened or is actually happening in Latin-American countries with regard to incoming technology, the 'autonomy-promotion' approach is largely concerned with what ought to be happening in the area of the planning, organization and implementation of locally based scientific and technological efforts. In other words, it is a 'normative' approach.

The spirit of this approach can be well captured in the following lines written by Jorge Sábato and Natalio Botana in 1968.

> In every case, and whatever are the chosen roads, the access to a modern society—which is one of the objectives intended to be achieved by development—necessarily supposes decisive action in the field of scientific and technological investigation. . . . Latin America, with scarce intervention in the past and present of scientific and technological progress, must change its passive role of spectator for the active role of protagonist.[20]

The extent to which Latin America had indeed adopted a passive role in the twentieth century with regard to science can be gauged from the very primitive development of the natural sciences and of scientific research in Latin-American universities right up until the 1950s.

This was partly due to the very newness of the university idea itself so far as Brazil is concerned—this country's first real university (São Paulo) only started as recently as 1934—however, in the countries of Spanish America, with a much longer university tradition dating back to the colonial period, the lack of support and resources for science and scientific research throughout virtually the whole period from 1900 to about 1950 (with isolated exceptions) was due to the concentration of the universities on turning out graduates for the professions—especially in medicine, law and engineering. Natural sciences had to take a 'back seat' as minor wings or offshoots of the engineering faculties, whose biggest concern, in any case, was civil engineering—and the social sciences were usually organized as subsections of the law faculties.[21] Probably the only scientific activity which enjoyed genuinely high prestige and significant government support in Latin America of more than a purely fortuitous kind before the 1950s was medicine.[22] And even in this field it was research institutes—such as the famous Osvaldo Cruz Institute in Brazil—rather than universities, which led the way.[23]

Coming, next, to technological innovation, it was also clear that Latin America had historically played a mostly passive role in this field too. This is because the strategies followed in the industrialization of Latin America—i.e. import substitution policies in the leading Latin-American countries from 1930 onwards, followed by the new wave of industrial development associated with the post-1950s entry of multinational firms into the manufacturing sector—had been very largely based on imported capital goods, in the first case, and on a mixture of imported capital goods and imported technology, in the second.

So although Latin-American industrialization had in fact given rise to some official efforts at local technological development—e.g. some standards-making and quality-control institutes, and productivity centres—and although Latin-American public and private enterprises had often had to engage in adaptive research to adapt output and equipment to local characteristics, it seemed rather clear (at least until recently) that there was relatively little original design of new products, little innovative R&D, and a basic lack of State support for the development of new industrial technology in Latin America[24]—in other words, as in science a highly passive situation.

Hence, those Latin Americans wishing to build up the science and technology systems of their countries in the post-war decades certainly could point to the observable embryonic and 'retarded' state of local science and technology as needing official attention, and, of course, resources.

If one next turns to look at the historical record of the science and technology build-up in Latin America since the war one can see immediately that there seems to have been a notable shift in emphasis both in the arguments used to promote the build-up and in the type of official response to these arguments in Latin America if one compares the 1950s and the 1960s, on the one hand, and the 1970s on the other.

Thus the most publicized government-supported developments throughout the 1950s and 1960s were (a) the setting up of national councils of scientific research by many Latin-American governments,[25] and (b) the setting up a national commissions of atomic energy.[26]

These initiatives seem to have been stimulated by the enormous post-war prestige which scientific activities and research acquired in the wake of Hiroshima, and by the effective advocacy of some of the politically oriented members of Latin America's small science communities, who managed, as J. A. Sabato has pointed out, to 'create a climate' which put the fundamental emphasis on

> the importance of science, the benefits which a country would receive upon implanting it and developing it, and on the prejudicial effects that would be occasioned by not doing it. In [the published papers of these advocates] it was proposed that once the 'productive machine' of science was set to work, this would function in a continuous manner and would incorporate itself into reality, which was anxiously awaiting it, without major difficulties. The idea was something like 'Do science, then everything will go better'.[27]

However, it became clear in many Latin-American countries throughout the 1960s that scientific research by itself could not necessarily be expected to have any significant impact in accelerating industrial and technological development because of the irrelevance of many of the funded research topics to the ongoing economic activities and pressing needs of Latin-American countries, and because of the acute lack of 'coupling' between the small scientific communities in each Latin-American country and the rest of their society, in particular its productive structure. This critical view of the value to their societies of those achievements that actually had managed to be reached by Latin-American science became widely known as the view that science in Latin America is 'marginalized'—i.e. divorced from real productive and social needs—in other words, that money spent on science had been a 'consumption item' in Latin-American economies rather than an 'investment item'.[28]

It was because of this context of greater realism about the actual marginal role of much scientific research in Latin America that one can observe, from the end of the 1960s onward, a definite switch in the preoccupations of both the 'promoters' of local science and technology in Latin America and of national governments in Latin America—away from their earlier nearly exclusive emphasis on national science policy, and towards a new (and much more realistic) emphasis on national technology policy. Then, given this new emphasis on how policy might be used to promote the generation and practical application of relevant research and development results in the Latin-American context, it became essential for those concerned to devise an adequate and convincing analytical framework for the application of technology policy, i.e. to develop the 'theory' of technology policy in the Latin-American context.

As a result, research work aimed at providing a reasonable framework for national technology policy began to be supported towards the end of the 1960s both by national governments (e.g. in the Andean Pact, Mexico, Brazil, etc., simultaneous with their interest in analysing technology transfer) and also by the Organization of American States via their Department of Scientific Affairs in Washington and by other interested international agencies such as UNCTAD in Geneva and the International Development Research Centre in Canada.

Some of the most interesting outcomes of the substantial flow of new work funded by these organizations have been:

The explicit recognition, in most of the key official research documents of the many negative effects inherent in the existing methods of technology transfer—leading in turn to recognition of what Maximo Halty of the Organization of American State called 'the imperious necessity to control the process of importation of technology'.[29]

The adoption, as an important guiding lemma for national technology policy of the goal of 'autonomy of decision' in the field of technical progress, which means not the goal of autarky but the goal of being able to decide explicitly and with accurate knowledge about the relative merits and demerits of different ways of absorbing technical progress into the economy. Specifically, such autonomy involves being able to decide from the viewpoint of the national interest on the relative merits of 'packaged' versus 'unpackaged' ways of acquiring technology from foreign sources. Therefore the 'breaking up of imported technological packages', both for the purposes of negotiating more effectively their terms of acquisition and for exploring intensively the possibilities for the local substitution of some of the inputs involved in the package, has become a very widely accepted normative goal of technology policy.[30]

The recognition, in so far as effective national policy in technology is concerned, that one of the most crucial areas for government action is in the field of stimulating and requiring an active local technological effort from the public-sector enterprises already controlled by each Latin-American government.[31]

The working out and implementation of innovative methods of planning the growth of national scientific and technological 'systems' in the Latin-American context starting out from a frankly 'dependent' initial situation, and bearing in mind the priority need explicitly to 'couple' the various elements in the growing system.[32]

The advocacy of and experimentation with various forms of joint ventures funded by two or more Latin-American governments to promote needed technological developments in specific sectorial fields—on the basis that joint international efforts will help Latin American countries to achieve the 'critical mass' of R&D effort required in these fields.[33]

The development, on a theoretical basis, of an innovative new approach to planning technology policy on a sector by sector basis, 'from the bottom up', rather than from 'the top down'.[34]

The implementation of explicit new policies and financial schemes to assist in the build up of Latin American consultancy and engineering firms.[35]

Obviously, this brief *résumé* has only hinted at the great number of research documents and new government-sponsored activities which have begun to spring up in the technology policy and technology promotion field in the past few years, particularly in Brazil, Mexico, Argentina and in the Andean Pact countries.

However, whilst the 'volume of activity' in this field is much greater than it used to be, and suggests that some non-trivial changes might now be in the process of introduction into the overall picture of 'technological dependence' in Latin America, it is still by no means easy to discern the likely long-term impact of these efforts.

In very crude terms, one might say that the 'technological dependence' analysts in Latin America have proved the existence of mighty—and dynamic—interests blocking the path towards the 'autonomy of decision' aimed at by the technology promoters—whilst the technology promoters for their part have not ceased to propose a stream of new policies, new measures and new approaches aimed at maintaining the build-up of local science and technology with the stated aim of using this build-up to help win more 'autonomy of decision' for Latin-American governments and enterprises than they currently possess.

The 'micro-economic-historical' approach

This now brings us to the third approach to be examined here. This we have termed the 'micro-historical' approach. Its central focus is—like the second approach—on the issue of the internal, i.e. local generation of technology. However, it differs from the 'autonomy-promotion' approach, because whilst this latter is principally concerned with what ought to be happening to local science and technology (and thus starts out as a normative, future-oriented approach) the micro-historical approach is concerned with the detailed exploration of how local technology capability has actually grown historically in Latin America. Its goal is to try to trace out and understand the history and microeconomics of the growth of local inventive and innovative capability in specific enterprises and sectors in Latin-American countries from their beginnings up to the present day, and to explore the consequences produced by this growth in inventive and innovative capability.

The leading exponent of this approach in Latin America is the Argentine economist Jorge Katz, whose studies over the past few years have opened up some important new angles on the question of local inventive and innovative capability in Latin America. To begin with, in marked contrast to the conventional view that local innovative efforts in Latin-American economies are virtually non-existent, or in any case insignificant, Katz has demonstrated that, in the Argentine manufacturing sector:

Firms spend between U.S.$30 and 35 million per year in technical activities which could be characterized as research and development work addressed at product and/or process adaptation and improvement. . . . [Also] the accumulated flow of these technical activities is strongly correlated to the observed productivity performance of each firm, as well as to the increased manufacturing export capacity shown by the Argentine economy. . . . In many cases, the reception of foreign technology triggered initially local adaptation, which was followed by increased production and the emergence of the companies as exporting firms. In thinking about the Latin-American situation, it becomes clear that Brazil, Mexico and Argentina are writing a different technological story than is the rest of Latin America; the gap between these three countries and the rest of Latin America is a significant one, and is growing more important as the three gradually evolve into regional suppliers of technology, which has been happening more and more over recent years.[36]

A significant point to be made about these results of Katz's is that they show how the walls of 'technological dependence' are—at least in some specific fields—more permeable and subject to filtrations than might have been imagined. This should not, perhaps, have come as a complete surprise given that although much technological knowledge is patented, or is otherwise kept restricted in the interests of its value as a business asset to its owners, there is also a great deal of technological knowledge which is not restricted in this way—and even the knowledge that is restricted is by no means always perfectly appropriable. Production experience, copying, reinvention, parallel development, attraction of key staff away from competitors, or purchase of the relevant know-how from more willing suppliers are just some of the many ways in which formerly 'proprietary' knowledge can start to acquire wider diffusion.

Furthermore, as Katz's results suggest, there seem to exist some significant technological 'learning sequences' which can work even in dependent economies, whereby at least some firms gradually acquire sufficient mastery over technological adaptation and innovation as to be able to trace out new patterns of comparative advantages for themselves across time. Even though initially strongly dependent on foreign technology, such firms are clearly able to use foreign technology as a 'springboard' to launch their own technological efforts and from thenceforth to establish a synergistic rather than antagonistic relationship between external and internal technological inputs.

The interesting questions of course are: When and how often does this happen? In what sort of firms? Under the impact of what kinds of stimuli? In which sectors and with what economic results? With what macroeconomic consequences? These are some of the kinds of questions which Jorge Katz and his co-workers are now trying to explore in detail in several Latin-American countries in basic industries such as petrochemicals, steel, construction and cement as well as in consumer industries such as tobacco and pharmaceuticals.[37]

One interesting existing finding from this line of research deserves particular mention, prior to closing off the remarks on it—this concerns the discovery that the firms which account for the majority of this 'adaptive R&D' spending identified in Argentina, were foreign-owned multinational subsidiaries. Contrary to much conventional theory which claimed that the multinationals concentrate all their R&D in the central countries, this finding suggests that the multinationals have 'hit on' and are making use of new local comparative advantages in adaptation and redesign of products (and processes) to fit both national and also regional Latin-American markets.

But a question then arises—if some Latin-American countries are now demonstrating clearly that they have sufficient engineering skills, as a result of the industrialization process, so as to be able to generate significant quantities of new technology endogenously (the very thing that was taken, fifteen to twenty years ago, to be the characteristic defining the advanced countries in contradistinction to the dependent ones)—then what steps are Latin-American governments going to take to ensure that a reasonable slice of the returns to this endogenous inventive activity will accrue back to the local community rather than being transferred abroad via the remittances made to foreign parent corporations by their Latin-American branches?

Equally, although the growth of 'endogenous' innovative capacity makes it theoretically possible for a country to conceive its economic future in more autonomous terms by virtue of having much greater possibilities to 'migrate' from its existing bases of economic comparative advantages to superior ones—i.e. what John Roberts has called 'the strategic advantage of having the means to alter the structure of one's own comparative advantage'[38]—this possibility can hardly become fully used if the majority of the country's endogenous innovative talent continues to respond exclusively to the multinational firms, since these latter will only develop a given country's comparative advantages in so far as such development would not interfere with their own world wide production, sales and profit strategies.

These remarks suggest that the 'permeability' of technological dependence, whilst absolutely real, does not warrant exaggeration at this point in Latin-American economic development. What is clearly encouraging is that a significant new element of local technological creation is now introduced into the picture that was previously dominated exclusively by external sources of technical advance. However, further experience is obviously needed before the longer term impact of this important development can be realistically assessed.

The 'appropriate-technology' approach

The final approach to be reviewed here is concerned with the issues of what has come to be called 'appropriate technology'—although this is indeed a 'catch-all' concept embodying many diverse concerns.[39]

Obviously, there must exist degrees of appropriateness rather than just an either/or category—but the basic idea is in any case clear enough. To see its relevance in the Latin-American context one might, as a spur to thought, imagine oneself to be a passenger in a Concorde airliner just approaching Rio de Janeiro's Galeou airport on one of its regular flights from Paris. After enjoying an immaculate air-conditioned flight in one of the world's most expensive technological investments, which was paid for by British and French taxpayers, you make the approach run to the Galeou tarmac over some of the world's most appalling water-side slums. As you pass through airport formalities you will not fail to notice how extraordinarily modern the airport is, with visual display boards, computer terminals, and all the well-designed systems that money can buy—for the Brazilians have built a showpiece in Galeou with money from their taxpayers.

Next, on the approach road into the centre of Rio de Janeiro, if you are not in one of the airport's air-conditioned buses or taxis, you might catch the smell of the extensively polluted waters next to which the many families who dwell in those water-front shacks have to live. It might then cross your mind that to have invested Brazilian tax money in the luxurious new airport is hardly an 'appropriate technological choice' from the viewpoint of these families, or even of the majority of the Brazilian tax-payers who forked out for it—any more than it was 'appropriate' to have poured so much irretrievable tax money from the unconsulted British and French publics into the Concorde airliner in which so very few will ever eventually travel.

So this simple example taken from the air-transport field already identifies a number of characteristics which are often viewed as typical of 'inappropriate' technology—namely the large scale of the resources which must be collected and concentrated in order to develop and operate such technologies, and the extremely limited or zero access which most of the public has to their benefits.

An absolutely similar point has been made by Illich in the context of road transportation in Mexico—as part of his much broader polemic attack on the whole development Gestalt of modern technology:

> Under President Cardenas in the early thirties, Mexico developed a modern system of transportation. Within a few years about 80 per cent of the population had gained access to the advantages of the automobile. Most important villages had been connected by dirt roads or tracks. Heavy, simple, and tough trucks travelled over them every now and then, moving at speeds far below twenty miles an hour. . . . Any Mexican could now reach any point in his country in a few days.
>
> Since 1945 the money spent on roads has increased every year. It has been used to build highways between a few major centres. . . . The old, all-purpose tramp truck has been pushed back into the mountains or swamps. . . . In exchange for an occasional ride on an upholstered seat in an air-conditioned bus, the common man has lost much of the mobility the old system gave him, without gaining any new freedom. Research done in two typical large states in Mexico [showed that] less than one per cent of the population in either state travelled a distance of over fifteen miles in any one hour during 1970. More appropriate pushcarts and bicycles, both motorized when needed, would have presented a technologically much more efficient solution for 99 per cent of the population than the vaunted highway development.[40]

However, a great problem with Illich's approach—and with his objective of seeking to show that 'two-thirds of mankind still can avoid passing through the industrial age—is that in Latin America the existing level of industrialization has already introduced irreversible changes into the picture. In particular it has helped to produce what Sven Lundquist has called 'one of the greatest migrations in human history—the Latin-American flight from the countryside'.[41] The result is that the term 'appropriate technology' will have to be given meaning in the context of a basically urban-industrial Latin America, and to a great extent in megalopolis cities which still, in Jorge Katz's words, 'basically lack the technological infrastructure with which to handle the package of social services demanded by populations of 10 to 15 million people'.

It is certainly rather clear that Latin-American governments, economists, and social scientists in general are not yet devoting significant resources to this appropriate-technology question, even though the term 'social-technologies' has now been floated and is beginning to circulate in some United Nations and official Latin-American reports.

The basic focus of Latin-American governments in the technology field still seems to be very much on the autonomy-dependence question—and there has not yet been any serious official questioning of the appropriateness, from a social point of view, of what Oscar Varsavsky has called the 'style' of technological development which Latin-American enterprises of all kinds are aiming to pursue.[42]

Nevertheless, it is important to underline that significant degree of adaptation of foreign technology to local Latin-American raw materials, labour availability, market size and tastes does already occur, and this can be considered as giving foreign technology a somewhat greater 'degree of appropriateness' to Latin-American conditions than it initially possessed.[43]

Also it should not be forgotten—in favour of the drive for autonomy—that one of the most important ways in which technological development can be 'inappropriate' is if it is very largely outside the control of a country's own nationals.

Notes

1. See the publication, Junta del Acuerdo de Cartagena, *Andean Pact Technology Policies*, published in English by the International Development Research Centre, Ottawa, Canada, 1976, which contains an excellent introduction and explanation of the Pact's policies, as well as Appendixes containing selected articles from the historic Decision No. 24 on the Common Treatment for Foreign Capital and Trademarks, Patents, Licensing Agreements and Royalties, passed in December 1970, as well as the important Decision No. 84 outlining the bases for a Subregional Technology Policy, passed in May/June 1974, and Decision No. 85, of the same date, concerned with the Regulations for the Application of Rules concerning Industrial Property. See also Junta del Acuerdo de Cartagena, *Technology Policy and Economic Development*, Ottawa, Canada, International Development Research Centre, 1976.
2. For a useful brief comparative analysis of Brazilian, Argentine and Andean legislation, see Jorge Katz, *Importación de Tecnología, Aprendizaje e Industrialización Dependiente* [The Import of Technology, Learning and Dependent industrialization], Mexico, p. 206–10, Fondo de Cultura Económica, 1976.
3. See Consejo Nacional de Ciencia y Tecnología, *Plan Nacional Indicativo de Ciencia y Tecnología*, Mexico, CONACYT, 1976.
4. The first widely publicized draft Code of Conduct on the Transfer of Technology was published at the initiative of the Pugwash Movement in April 1974—see 'Anteproyecto de Código de Conducta sobre Transferencia de Tecnología', *Comercio Exterior*, Mexico, May 1974—and the leading role has since been played by UNCTAD in seeking to obtain international agreement on such a code.
5. See for example UNCTAD, *Informe del Grupo de Expertos Gubernamentales sobre la Funcion del Sistema de Patentes en la Transferencia de Tecnología* [Report of the Group of Government Experts on the Function of the Patent System in the Transfer of Technology], 21 October 1975 (UNCTAD TD/B/C.6/8).
6. It was the April 1974 Special Session of the General Assembly of the United Nations that approved 'by consensus' the 'Declaration on the Establishment of a New

International Economic Order'. It should also be recalled however that even then, the developed nations filed 230 pages of 'reservations' on the declaration.

7. Of course, the fact that 'science and technology' have now achieved the same sacrosanct status as 'aid', 'trade', 'population', 'environment', 'peace' and 'human settlements' as subjects for world-wide conferences sponsored by the United Nations (United Nations Science, Technology and Development Conference, 1979) only proves that a new issue of dispute has been officially 'noticed', rather than that anything much is likely to be done about it. In the same way, the fact that new legislation, new codes of conduct and new declarations about technology are being drafted or have actually been approved—in Latin America or elsewhere—is not by itself a cause for rejoicing. One only needs to recall the sad fate of the United Nations Universal Declaration of Human Rights (adopted 10 December 1948, forgotten 11 December 1948) so as to remind oneself of the difference between laws which are binding and laws which are bent.
8. See for example Celso Furtado, *Development and Under-development*, Berkeley, Calif., University of California Press, 1964; F. H. Cardoso and Enzo Faletto, *Desarrollo y Dependencia en América Latina* [Development and Dependence in Latin America] Mexico City, Siglo XXI, 1969; Osvaldo Sunkel, *Capitalismo Transnacional y Desintegración Nacional* [Transnational Capitalism and National Disintegration], Buenos Aires, Nueva Visión, 1972; see also Cardoso's recent article 'El Consumo de la Teoría sobre Dependencia en los Estados Unidos' [The Consumption of Dependence Theory in the United States] *El Trimes tre Económica* (Mexico City), No. 173, January-March, 1977.
9. From Celso Furtado, *Teoría y Política del Desarrollo Económico* [The Theory and Politics of Economic Development], 5th ed., Buenos Aires, Siglo XXI, 1974.
10. Of course, this does not mean that the other factors have ceased to be important.
11. G. K. Helleiner, 'Transnational Enterprises in the Manufacturing Sector of the Less Developed Countries', prepared for the International Seminar on the Foreign Investment Code in the Andean Pact, Caracas, March, 1975, published in *World Development*, Vol. 3, No. 9, September 1975.
12. The precise methodology and results of these investigations are reported in Constantine Vaitsos, *Intercountry Income Distribution and Transnational Enterprises*, Oxford University Press, 1974.
13. According to Helleiner, 'Vaitsos' findings with respect to a relatively few transfer-priced goods in Colombia are sufficiently well known in this part of the world [the Andean Pact] not to require elaboration but they are still under-reported in other parts of the less developed world.' Helleiner, op. cit.
14. The methodology and findings of the analysis of these 451 licensing contracts can be found in Vaitsos, op. cit., p. 35, 36 and 42–5, and also in Junta de Acuerdo de Cartagena, *Technology Policy and Economic Development*, op. cit., p. 60–1. A valuable analysis of 70 Argentine technology licensing contracts and a comparison with the findings of the 451 Andean licensing contracts can be found in Katz, op. cit., p. 31–51. A further useful source is UNCTAD, *Control of Restrictive Business Practices in Latin America*, Chap. V and VI, 1975 (UNCTAD/ST/MD/4).
15. Limits on the freedom of local branches naturally reflect the multinational firm's policy regarding the most suitable overall technological 'division of labour' amongst its various branches from the viewpoint of head office and the firm's global maximization strategy. In practice, a frequent policy of multinational firms is to ensure that their foreign branches in Latin America will continue to depend on high dollar volumes of inputs of technology, capital goods and intermediate goods drawn from other foreign branches of the firm, so that adequate channels will then exist for transferring income via transfer pricing from the given Latin-American branch to

other branches or to head office. In such cases, the official strategy of many Latin-American nations which aims to proceed beyond the first stage of import substitution of end products and increasingly into the next stage of import substitution (of technology, intermediate goods and capital goods) is likely to run counter to the strategy of the multinational firms—which will seek to avoid developing local technological capacity in directions which would permit local inputs to largely substitute for those being imported into the Latin-American branches via intra-firm trading. This point is well discussed in Guillermo O'Donnell and Delfina Linck, *Dependencia y Autonomía* [Dependence and Autonomy], Buenos Aires, Amorrortu, 1973.

16. See for example F. C. Sercovich, *Foreign Technology and Control in Argentine Industry*, Appendix B, p. 1–80, University of Sussex, 1974 (Ph.D. thesis), for an outstanding detailed analysis of nineteen licensing agreements signed by Argentine firms with foreign suppliers, including seven agreements signed by locally owned firms and four signed as part of joint ventures. These agreements display many examples of restrictions on the freedom of the licensees with regard to making process or product modifications.
17. See Charles Cooper and Philip Maxwell, *Machinery Suppliers Packaging and the Transfer of Technology to Latin America*, Science Policy Research Unit, University of Sussex, 1975 (mimeo.), for an analysis which shows that even in straightforward plant and machinery sales to independent Latin-American enterprises, where neither foreign investment nor licence agreements are involved, the recipient enterprises may still be inhibited from introducing or insisting on modifications to the technology that has been supplied owing to such factors as: (a) dependence on supplier financing; (b) dependence on supplier guarantees of plant performance; (c) dependence on the supplier as a continuing source of technical assistance and spare parts, etc.
18. For example, other practices associated with technological dependence which have been revealed by various studies include: the overpricing of capital goods imported via intra-firm trading, the underpricing of exports sold abroad via intra-firm trading, the capitalization of and claiming of depreciation charges on intangible assests, the preventive taking out of patents in Latin America to stall market-entry by local firms or inter-national competitors, etc.
19. For example the Andean Pact commission issued Decision No. 103 in November 1976, modifying Decision No. 24 so as to relax a number of the restrictions on foreign firms (e.g. increasing the annual limit of profit remittances, permitting foreign firms to have access to the short- and medium-term local capital markets, extending the start-up date for calculating the period required for foreign firms to convert themselves into mixed or national concerns, etc.). See R. Chapman and J. Duhart 'Relaxation of Laws on Foreign Capital', *The Times* (London), Friday, 17 December 1976. Also it should be recalled that Chile withdrew from the Pact in October 1976 and now operates its own foreign investment code.
20. Jorge A. Sábato and Natalio Botana 'La Ciencia y la Tecnologia en el Desarrollo Futuro de América Latina' [Science and Technology in the Future Development of Latin America], *INTAL Revista de la Integración* (Buenos Aires), No. 3, November 1968.
21. See Alberto Sánchez Crespo, 'Esbozo del Desarrollo Industrial de América Latina y de sus Principales Implicaciones sobre el Sistema Científico y Tecnológico' [Review of the Industrial Development of Latin America and of its Principal Implications for the Science and Technology System], in F. Suárez and H. Ciapuscio (eds.), *Autonomía Nacional o Dependencia: La Política Científico Tecnológica*, p. 69–71 and 99–104, Buenos Aires, Paidós, 1975.
22. See Sánchez Crespo, op. cit., p. 70–1.

23. See Nancy Stepan, *Beginnings of Brazilian Science, Osvaldo Cruz, Medical Research and policy, 1890–1920*; New York, Science History Publications, 1976 (also in Portuguese in a Brazilian edition, Editoria Artenova, Rio de Janeivo, 1976).
24. These points are all made, with some brief documentation and interpretation, in Sánchez Crespo, op. cit., p. 72–82 and 93–9.
25. The foundation dates of these science research councils were: 1951, Brazil; 1958, Argentina; 1960, Mexico; 1960, Jamaica; 1961, Uruguay; 1963–66, Panama; 1966, Chile; 1967, Venezuela; 1969, Colombia; 1969, Peru; see Sánchez Crespo, op. cit., p. 100.
26. The foundation dates of these national atomic energy commissions were: 1950, Argentina; 1955, Uruguay; 1955, Peru; 1955, Mexico; 1956, Brazil; 1956, Colombia; 1958, Ecuador; 1960, Bolivia; 1964, Chile; 1967, Costa Rica; see Sánchez Crespo, op. cit., p. 84.
27. Translated from Jorge Sábato (ed.), *El Pensamiento Latinoamericano en la Problemática Ciencia-Tecnología-Desarrollo-Dependencia* [Latin American Thought on the Problem of Science-Technology-Development-Dependence], p. 13, Buenos Aires, Paidós, 1975.
28. Furthermore, in explaining this situation many proponents of the 'Marginalization of Science' view—such as Argentine geologist Amilcar Herrera—point out that marginalization is a 'structural' problem. On this view, the inbuilt economic, technical and cultural dependence of Latin-American societies has the effect of systematically orienting the local demand for technologically relevant science and innovation towards being satisfied from external rather than internal sources. As a result, a vicious circle develops in which there is no effective local demand for local scientific and technological inputs relevant to production. Therefore, local talent emigrates, or wastes its potential without adequate local backing, or retreats to the ivory tower, or may suffer from persecution and threats of dismissal in those cases where scientists (particularly social scientists) have used their fragile base of autonomy within the universities to protest against government policies. See Amilcar Herrera, 'Social Determinants of Science Policy in Latin America', in Charles Cooper (ed.), *Science, Technology and Development*, London, Frank Cass, 1973.
29. Maximo Halty Carrere, *Producción, Transferencia y Adaptación de Technología Industrial* [Production, Transfer and Adaptation of Industrial Technology], Washington, D. C., Department of Scientific Affairs, Organization of American States, 1971.
30. See for example 'Breaking up the Technology Package', Chap. VI of Junta de Acuerdo de Cartagena, *Technology Policy and Economic Development*, op. cit.
31. See especially Sábato and Botana, op. cit., J. Sábato, *Empresas y Fábricas de Tecnología* [Technology Enterprises and Factories], Organization of American States, 1972; Alberto Araoz, Jorge Sábato and Oscar Wortman, 'Compras de Tecnología del Sector Público: El Problema del Riesgo' [Technology Purchases in the Public Sector: The Problem of Risk], *Comercio Exterior* (Mexico City), Vol. 25, No. 2, 1975.
32. See Francisco Sagsti and Mauricio Guerrero, 'Lineamientos para Elaborar Políticas de Ciencia y Tecnología en Latinoamérica' [Lines for Elaborating Science and Technology Policies in Latin America]. *Comercio Exterior* (Mexico City), Vol. 25, No. 2, 1975.
33. For example the Andean Pact Technology Programs (PADTs) in such areas as tropical forest resources, metallurgy, petrochemicals. See also the recent publications of INTAL and UNDP on Latin-American multinational enterprises, and on technology enterprises as a mechanism for promoting 'horizontal' technical assistance.
34. See. J. Sábato, 'Bases para un Régimen de Technología' [Bases for a Technology

Policy Regime], *Comercio Exterior* (Mexico), Diciembre 1973; and J. Sábato, R. Carranza and G. Gargiulo, 'Ensayo de Régimen de Technología: La Fundición Ferrosa en Argentina' [An Essay on a Technology Policy Regime: The Iron Based Foundry Industry in Argentina], *Comercio Exterior* (Mexico), Vol. 26, No. 11, 1976.

35. For instance, the Brazilian Federal Government agency, Financiadora do Estudos e Projectos (FINEP) provides financial backing so as 'to strengthen the sector which supplies consulting services so as to promote the development on national technology'. Also local engineering and consulting industries are widely recognized in technology policies as candidates for special protection on the basis of the 'infant industry argument' and in view of their multiplier effects in building up local technology, permitting 'unpackaging' of imported technology, etc.
36. Cited from Jorge Katz, *Appropriate Technology for Developing Countries: Adaptation and Regulation* (Monograph No. 8), published by University of North Carolina School of Business Administration, Chapel Hill, May 1977. These results are in turn described in a much more detailed manner in Jorge Katz, *Importacion de Technologia, Aprendizaje e Industrialización Dependiente*, Mexico City, Fondo de Culture, 1976, and in Jorge Kotz and Eduardo Ablin, *Technología y Exportaciones Industriales: Un Análisis Microeconómico de la Experiencia Argentina Reciente*, Buenos Aires, Comisón Económica para América Latina, 1976 (BID/CEPAL/BA. 12).
37. See Jorge Katz and Ricardo Cibotti, 'The Framework for a Latin American Scientific and Technological Research Program', IDB-ECLA Technical Co-operation Agreement ATN/SF-1329-55, August 1975.
38. See John Roberts, 'Engineering Consulting, Industrialization and Development', in Cooper (ed.), op. cit.
39. Usually, the concept has been taken to refer to technology which is suitably designed for, or adapted to, a country's particular mix of production factors, or to its local raw materials, or to its local market characteristics, or to its local social characteristics or to some mixture of all four of these. By contrast technology which is 'inappropriate' implies a substantial waste of locally available resources, a reduction of employment opportunities per unit of investment, and the tendency for the domestic economy to produce products with excessive or luxury characteristics in relation to the income levels of the majority of the country's people.
40. See Ivan Illich, *Tools for Conviviality*, New York, Harper & Row, 1973.
41. Sven Lundquist, *The Shadow: Latin America Faces the 70s*, Harmondsworth, Penguin, 1972.
42. See the interesting polemical book by Oscar Varsavsky, *Estilos Technologicos*, Buenos Aires, Periferia, 1974.
43. See illustrations of this in Katz, *Appropriate Technology for Developing Countries*, op. cit.

Some historical considerations

A. Rahman

Chief, Centre for the Study of Science, Technology Development, Council of Scientific and Industrial Research, New Delhi, India

Some basic considerations

The discussion on the relationship between science and technology, a relationship which is undergoing rapid changes and social processes as we understand them through the current stage of development of the social sciences, may evoke different responses from scientists depending on the political systems of their country, their philosophy, and their cultural background. In this context, it may be worth while to emphasize a few points at the very beginning.

First, the diversity of the interpretations of the subject by Richta and by Hollander[1] is, in my opinion, neither accidental nor purely personal. Their interpretations reflect the way science and technology are viewed in socialist and capitalist societies. In socialist countries, science and technology are regarded and used as a part of the political process to attain a set of political and social goals. The literature on science and technology in capitalist countries reflects a tendency to consider them politically neutral. This tendency of the literature, particularly the management literature of the capitalist countries, is interesting, to say the least.[2]

Science as a method of acquiring knowledge and widening human horizons, and technology as an instrument of social change, have come to occupy positions of immense prestige in contemporary society. This prestige is being utilized in capitalist countries to justify, through the use of neutral words and scientific and technical terminology, highly political decisions which could not otherwise be justified or would not be acceptable to the people. For instance, 'speaking scientifically' or 'making a decision scientifically' is often meant to convey an impression, besides one of objectivity, of being without any political overtones.[3] And once expressions like these are used and material is presented by technical people, there is a greater possibility of the acceptance of highly controversial social and political decisions. This exploitation of science is rather unfortunate.

Just as the scientific method is used to justify irrationality, technology is also being used to justify production systems which would otherwise be indefensible. Under the pretext of the logical development of technology with its sophistication and its linkage with the size of the production unit, not only are monopolies justified, but the misuse of technology is also made to appear necessary.[4] In addition, concepts like unit cost of production, techniques like input-output analysis, and areas of research like time-and-motion studies are developed. In other words, instead of examining the social role and cost of technology or its dehumanizing effect, such studies not only justify it but further accentuate its disadvantageous character.

The second point on which I wish to focus attention is the general tendency in the socio-economic literature and in studies on science to divide the world into two categories : developed and developing countries. This is not, in my opinion, politically correct nor does it represent current social realities. It is, therefore, necessary to divide countries into three groups: capitalist, socialist, and former colonies or semi-colonies now loosely called the Third World countries.

The use of the word 'developing' is also highly misleading. The countries grouped in this category were highly developed before they became colonies or semi-colonies. Their colonization destroyed their technology and industry, and their economies were then developed to suit the interests—technological and industrial—of the colonial powers. In other words, they were developed, but only as satellites of the economies of the capitalist countries. Further, conceptually categorizing the countries as developed and developing evokes the image of the latter trying to follow the line of development of the former.[5] This again would be misleading, since the 'developing' countries are trying to free themselves from the stranglehold of the 'developed' countries not only politically and economically, but also scientifically and technologically. Their future development would differ, therefore, from their existing pattern, depending on the degree of freedom they are able to achieve and the self-reliance they are able to generate.

The overemphasis on the apolitical nature of science and technology, particularly of the latter, in the literature of studies on science of the capitalist countries is motivated by and intended to serve political objectives. This is briefly indicated by the sale of technology through the agency of transnational corporations in order to control the technological and industrial development of the Third World countries and to divert them from their social and political goals.

In view of the fact that it is relevant to the topic under discussion, it may be worth while to elucidate it further. The picture I am trying to paint can be briefly stated.[6]

Science and technology, particularly the latter offer an alternative to social and political problems faced by societies. Despite the prophecies of scientists, capitalist countries, by their emphasis on technology, have been able to abolish poverty to a large extent and usher in an era of prosperity. The only consideration given to a technology should be its technical merit and degree of sophistication. The latter is particularly overstressed in view of the fact that it is suggested that there can hardly be any difference in the technology developed under different social or political systems. The capitalist countries are the generators of most of the advanced and sophisticated technology. Their experience and expertise may be utilized for advice and import of technologies for development in the Third World countries. Further, this expertise is available through transnational corporations, which really belong to no country and are guided solely by technical considerations. Since they have the expertise in both generating technology and its sale, they should be utilized for purposes of technological development.

The implications of this conceptual approach for the Third World countries, who are struggling to free themselves from colonialism and neo-colonialism, should be obvious.

The third point to which I wish to direct attention at the very outset is the wide disparity in the (a) unfolding of the scientific and technological revolution in different parts of the world; (b) the impact of the uneven development of science and technology in different areas of the world; and (c) the possibilities which are opened up for different countries in the context of the present historical situation. Consequently, to think of a uniform, global impact and world solution would be unrealistic and the search for them unrewarding.

In view of the development of science and technology in Europe and North America and their implications for man and society, the problems thus created in European culture are exaggerated and depicted as world problems. For example, environmental pollution may be due to different reasons in different areas. In capitalist countries it may be due to the neglect of social problems, to the vested interests of different industries and their search for quick and high profits, or to the fashions set by the fancies of an élite class. In the countries of the Third World, it may be due to the lack of development, the use of science and technology, or sheer neglect of the problem. It could also be due to the desire of these countries to industrialize rapidly. In other words, the problem is not purely technical, as is often made out, but socio-technical. Further, the possible solutions will vary from country to country depending on their political set-up, social goals, scientific and technological potential, and capacity to utilize the latter. However, it may be mentioned here that to emphasize the national nature of the problem should not be taken to mean that there are no regional or global problems.

Science, technology, and development in the Third World

The Third World countries, in their fight for economic independence, soon realized the significance of science and technology as a powerful instrument for achieving self-reliance. This realization began slowly and haltingly, first as an imitation of the infrastructure of the capitalist countries, then as an instrument of import substitution or export promotion and, finally, as an instrument of social change, as an instrument in the fight against colonial and neo-colonial domination to achieve a set of social and political objectives.

The process of the realization of science and technology as an instrument of social change and the building up of the infrastructure and its gearing to social goals were retarded, however, by the advice rendered and the aid given by the former colonial powers to the former colonies. The advice rendered was in the area of science and technology institutions for education and research, science and technology policies and resource allocation, and technologies to be imported or indigenously developed. Such aid as was given was directly related to the advice given by the donor countries.[7]

Educational institutions were developed in the Third World countries following the models of the countries which were the former occupying powers, or of the United States which had emerged as a major capitalist power after the Second World War. Universities, institutes of technology and polytechnics were established with the help of equipment and guest faculties provided by the former colonial powers. It may be worth pointing out here that the universities, institutes of technology or polytechnics were established in Europe or the United States as a part of social needs and responses to them. Further, these had a long period of evolution, to chisel them to suit new and evolving needs. In the Third World countries, on the other hand, they were established as part of the advice of experts or as a result of the needs of the élites in these countries to be on a par with the institutions in the countries of the colonial powers. Consequently, they remained insulated from the needs of the society and social pressures. Those who came out of these institutions were more knowledgeable about the happenings of the capitalist countries and, therefore, more suited to solve their problems than those of their own countries. The net result was that, as the graduates resembled those who were trained in the institutions of the capitalist countries, there was large-scale emigration of trained personnel to meet the needs and demands of those countries.

What applies to educational institutions also applies to research institutions. They were also created in the image of overseas institutions and tended to tackle those problems on which people had worked during the period of their training abroad and those which were then currently fashionable in science and were relevant to the development of capitalist countries. Consequently, they also suffered from isolation and the fruitlessness of their efforts.

The result of the advice and the institutional support given as a follow-up to it, as well as its impact on tackling social problems, was marginal. Despite the technical assistance given to the countries, they were not able to move towards self-reliance. Their dependence on the donor countries continued, if not increased. It may also be worth pointing out here that the term 'technical aid' is incorrect, since in no respect was it aid. Though the word 'aid' was used, the transactions were really normal, commercial transactions which were highly disadvantageous to the Third World countries. The equipment passed on was either obsolete, or so sophisticated that the receiving countries could ill afford to maintain and run it and had to pay heavily for spare parts. As for the so-called experts, the less said the better.

The consequences of this process were: (a) the creation of a scientific and technical élite which was more at home abroad and shared the aspirations and values of the scientists of the capitalist countries; (b) the externalization of the value system; (c) the isolation of the new institutions from society and its needs; and (d) continued dependence.

Further, as a result of these developments, science and technology were separated from the political process. Science and technology made increasing demands on resources, but had little to offer by way of solutions to the pressing problems faced by the society, hence they could not derive the necessary support. The less support they received at home, the more they looked outwards, sought for, and often received some outside support. Thus isolation and dependence continued to grow.

The other effect was that the scientific outlook and attitude had no effect on the social outlook and political attitudes. As a result, the political leadership of the newly independent countries looked back to ancient and medieval values for guidance, leading to the spread of obscurantism and revivalist tendencies in these countries. Nehru in India was an exception among the Third World countries. He understood clearly that science and technology were instruments of social transformation, and must be used as such. This he tried to accomplish by making scientists socially conscious, by involving them in the decision-making process of the government, and by applying scientific conclusions as social decisions. His efforts, however, did not bear enough fruit, since the scientists of the country did not share his ideology.

However, as the struggle for economic independence in the Third World took shape, the questions, such as the relevance of the infrastructure of science and technology, the returns from the investment in this area, and the desirability or otherwise of different research programmes began to be asked. As a result, science began to be involved in political discussions and became an issue of some importance as a part of the political debate in the country.[8] In the course of this debate, for example, in India, the internal political struggle sharpened the debate on science and technology. Consequently, we find a definite shift in science and technology policy, where the concept of science and technology for rural development and for the benefit of the people came to the forefront and began to be reflected in research programmes as opposed to the earlier goals.[9] Another impact of far-reaching significance, resulting from the *rapprochement* of the scientific and technological process with the political process, was the internalization of the value system of science. The degree from overseas, the evaluation or recommendation from abroad began to have less value and prestige. The same attitude is now reflected in the choice of research programmes and the resources allocated to different projects.[10]

This trend and its further evolution will depend on the socio-political developments in the Third World countries and the manner in which the linkages of science and technology with the socio-political process are further developed. In this process, two factors are likely to play a major role.

First, to what extent will the scientists and technologists be able to change the outlook of the people from ancient and medieval attitudes to the outlook which is consistent with existing scientific knowledge. This would require them to take a stand on a number of issues, such as what the earlier savants of science—Galileo, Giordano Bruno, Thomas Huxley, etc.—did in Europe. In other words, they would have to make the scientific outlook and attitude a part of the public debate against superstitions, magic and miracle-mongering, a task from which the scientists and technologists of these countries have not only shied away but have actually done its opposite, becoming the promoters of miracle-mongering.

Secondly, the scientists and technologists have to make research programmes and investment in different areas of science a part of public debate. This can be done only by educating people on the implications of different developments. In this manner, science and technology will become part of the socio-political process of the Third World countries. In the absence of this, scientific and technological programmes and investment would continue to be decided as a matter of discussion between individual scientists and the administrator, with the consequences as indicated earlier.

Technologies, societies and the role of scientists

Having briefly indicated some of the features which are being ignored and need to be emphasized, it may be worth while to consider some of the basic features of the current scientific and technological revolution and its impact on certain aspects of current thinking.

One of the basic features of the current technological revolution lies in the use of raw materials, the degree of energy consumption, and the communication system. These features would become more evident if a comparison were made between the earlier agricultural revolution and the current scientific and technological revolution.

In the agricultural revolution, biological and agricultural raw materials were used for industry, and the source of energy was either animal or human, or natural renewable products. Non-renewable resources were used rather sparingly, primarily in the manufacture of weapons of war and a few luxury goods.[11] The production system was geared to meet local or regional needs. The communication systems developed were adequate to meet such needs as they were then. In contrast, the main thrust of the technological revolution has so far been the use of non-renewable resources and fossil fuels for energy requirements. The production system is being developed to meet world markets, hence there are well-developed communication systems for the transportation of goods, raw materials as well as finished products, and information.[12]

Like the earlier agricultural revolution, the present technological revolution appears to have run its course and is now faced with problems of its own creation. As long as the use of non-renewable raw materials and energy generation from fossil fuels was limited, the production system had no major problems. However, with the increase in sophistication, the size of production units, and the number of countries taking up industrial production based on these technologies, the possibilities of exhaustion of raw materials and energy sources in the near future appeared very distinct, and a major crisis was foreseen. As a result, three distinct responses were generated: first, attempts were made to limit growth so as to conserve resources and make them last longer, with the hope, in the meantime, that new technologies would be developed to meet the new situation;[13] second, serious questioning of the existing technologies was attempted from the point of view of their impact on man and society, and the future;[14] third, the search began for alternative technologies which could use renewable resources as raw materials and consume less energy.[15] Furthermore, the size of production units was also questioned.

These ideas are likely to have far-reaching consequences for the direction of the scientific and technological revolution. What direction it would take finally is difficult to foresee, but three factors would have a far-reaching influence on the outcome:

First, the course the Third World countries are able to adopt for their development; the degree to which they are able to free themselves from copying technologies and models developed in western Europe and North America and use existing scientific and technological knowledge to develop new technologies and production units to suit their requirements; and furthermore, the relationship they are able to build, in this context, with the socialist countries and the technologies and technological assistance these countries are able to offer for their development.

Second, the degree to which the Third World countries are able to exercise control over their natural resources, raw materials and fossil fuels. The technologies based on these sources and the economies based on their import are going to be profoundly affected, thus accelerating the search for alternatives as evidenced by the decisions of the Oil Producing Exporting Countries (OPEC).

Third, the effective role which scientists and technologists are able to play in the context of the present situation regarding: (a) defining social goals and gearing science and technology to attain these goals; (b) the development of new concepts and techniques to analyse and measure the development of science and technology, their impact on man and society at the present and in the future; (c) the development of machinery for implementation, monitoring, and evaluation, to give feedback to decision-makers with regard to the changes required in policies; (d) the development of new technologies based on the use of renewable resources including sources of energy; and (e) breaking the linkage of science and technology with the concepts developed by capitalism and its mode of production.

The last-mentioned point is particularly pertinent, though rather neglected at the moment.

At the time of the industrial revolution, science and technology provided an alternative outlook and tool to change society. Scientists and technologists became the *avant-garde* of a movement which attacked the feudal outlook and practices. Many had to pay for it with their lives, and fought for the emancipation of the human mind from the slavery of feudal society, and for the goals of liberty, equality, and fraternity.[16] This was nothing new in human history. The Ionians in ancient Greece and the Charvaks and Hastiks in ancient India had fought similar battles.

The way the scientific and technological revolution will evolve will depend, among the other points mentioned earlier, on the role scientists and technologists set for themselves. If they choose for themselves the role similar to that of their predecessors, the role of social revolutionaries, to fight against the capitalist outlook and its practices and for the emancipation of science and technology from the limitations of capitalism, then the possibilities are considerable; otherwise, as happened in earlier societies, science itself might be destroyed.

This fight, however, cannot be fought only in countries where science and technology are considerably developed, but also be fought in the countries of the Third World. So far, as stated earlier, the countries of the Third World have been subjected to ruthless exploitation. Scientists and technologists of the capitalist countries have been party to this exploitation; in fact, they served as an instrument for a greater and more ruthless exploitation and destruction of the natural resources. However, one redeeming feature of the situation has been the rebellion of some against this. Further, this rebellion has really been responsible for the rethinking which we now see in the capitalist countries. The future will depend on how this thinking becomes a part of the social movement in which the scientists and technologists of the Third World are partners.

The questions which we now face are: Can science and technology, instead of being a tool of exploitation, usher in rationality and equality? Can scientists seize this opportunity to bring about the necessary transformation? Can scientists become a community of men committed to certain goals and ideals?

Notes

1. Radovan Richta 'The Role of The Social Sciences'. (Originally entitled: 'The Scientific and Technological Revolution and the Social Sciences'—Ed.) A. N. J. Hollander, 'Science, Technology, Modernization and Social Change'. (Originally entitled, 'Notes on Some Relations between Science, Technology, Modernization, and Social Change.'—Ed.)
2. I am indebted to Leon Peres for drawing my attention to the problem of the use of neutral words.
3. See, for example, Clay Reeser, 'Making Decisions Scientifically', *Engineering Management Review* (IEEE), Vol. 4, No. 1 1976, p. 44–9.
4. John Langrish has quoted two production units to produce a chemical, each one capable of meeting the entire world's requirements! This also applies to some very high pollutant technologies.
5. George Basala's model actually gave a theoretical justification for it. cf. *Science*, Vol. 156, No. 3775, 1967, p. 611–22.
6. This picture is based on considerable literature published in English during the 1950s and 1960s.
7. A detailed study on this topic was carried out on Western Europe and North America in Third World countries at the Centre for the Study of Science, Technology, and Development, CSIR, New Delhi, in 1973. It is now being revised.
8. See, for example, the analysis of election manifestos of different parties in India for

the election of 1971. *Socialist India*, 1971. Also, 'New Congress and Science', *Citizen*, Vol. 1, No. 21, 1970, p. 29–33.

9. This is well reflected in the documents: 'An Approach to Science and Technology Plan', NCST, 1973; and the recent session of the Indian Science Congress Association on rural technology, January 1976.
10. This is reflected in the procedures of promotion, as, for example, in CSIR, and the criteria adopted by the Planning Commission on resource allocation.
11. A. Rahman, 'Goals of Applied and Basic Research under Different Social and Cultural Contexts', *Journal of Scientific and Industrial Research*, Vol. 34, 1975.
12. A. Rahman, 'Alternative Technology', *Journal of Scientific and Industrial Research*, Vol. 32, 1973, p. 97–100.
13. 'Limits to Growth' was an exercise in this direction. It evoked considerable controversy both with regard to its assumptions and its implications for the Third World countries.
14. For literature on technology assessment see, for example, Harry Rothman, *Technology Assessment*, New Delhi, Centre for the Study of Science, Technology and Development, CSIR, 1976.
15. For a detailed bibliography of the literature on alternative technology and reports on alternative technology experiments, see V. P. Kharbanda, 'Bibliography on Alternative Technology', *Current Literature on Science of Science*, Vol. 5, Nos. 1–2, 1976, p. 23–40.
16. For a detailed discussion, see A. Rahman, 'Some problems of Science in India: Social Role of Scientists', *Journal of Scientific and Industrial Research*, 35, Nos. 1–6, 1976; and 'Problems of Science and Technology Policy in Developing Countries', *Journal of Scientific and Industrial Research*, Vol. 34, 1975, p. 655–8.

Part Five

Social sciences and social change

Some methodological problems of comparative research on the mode of life

Jindřich Filipec

Institute of Philosophy and Sociology,
Czechoslovak Academy of Sciences

The social sciences have an irreplaceable function in that they clarify the meaning and aims of social activity in general and of scientific and technical activity in particular. This axiological or normative moment, for example, is clearly present in the category of the 'quality of life' or—as I call it—'the mode of life', that is, a certain mode of life suitable to mankind, whose realization is envisaged by us as one of the essential goals of our social system. Research must aim not only at understanding human needs but also at proposing and encouraging practical measures to be taken in order to ennoble them. In certain parts of the world today fundamental human needs are not satisfied; in others, people apply themselves not only to saturating inherited needs but also to cultivating them and raising them to a higher level.

Today, the ideas about the goals to be pursued are naturally very distinct. Some—for example, Daniel Bell, who in his time set forth a theory of the end of ideology—see the highest values of 'non-animal' life in religious transcendence; others, while they do not seek these values in the next world, nevertheless reserve them for the élites of this world whose choice operates one way or another. During a recent scientific conference in Moscow devoted to the problems of social indicators, Karl Deutsch proposed, as a desirable criterion, the development not of just one country but of all countries. Unquestionably, this idea is a step forward but it also leaves out the question of first and fundamental importance: is our objective the development of countries cut to the quick by class distinctions or is it a kind of development which tends progressively not only towards material well-being but also towards the multilateral realization of all individuals on the scale of the whole society and then on a world-wide scale?

I think that despite the profound differentiations of the contemporary world, the elaboration of this concept—this precisely from a perspective which wants the existing inequalities to be overcome—constitutes the essential orientation and the very basis of research in the social sciences. Hence the growing number of tasks which are posed (a) at the theoretical level, and (b) at the methodological level.

The theoretical task consists in defining both the values and the goals which establish a higher quality of life, and the forces and means of the social subject that will be capable of realizing this project. Consequently, the task of methodology is not only one of elaborating the procedures which allow an understanding of the quantitative aspects in particular, the empirical data, but also of following the tendencies of either the movement or the stagnation of this reality, as well as the tendencies of drawing near to or moving away from the desired concepts.

I shall illustrate this assertion with three models of research on the mode of life which show progress made in the direction of learning the importance to be found in the quality and dynamism of social facts.

During the last few years a whole series of taxonomic models of the mode of life, from the simplest to the most complicated, have been proposed in the literature, one of which has been elaborated by a group of research workers at the Institute of Philosophy and Sociology of the Czechoslovak Academy of Sciences. The structure of this model rests on four components, namely: human activities, social relations, social needs, and socio-psychic forms, proper to the members of a given social group at a given stage of its development. In the model human activities are considered to be the most dynamic, i.e. decisive component. On this basis, Table 1 was constructed: the analytic elements of the mode of life figure there as the point of departure for the choice and construction of indicators.[1]

This taxonomic diagram (Table 1) was used successfully to classify and select important statistical information and in an elementary way to characterize the mode of life of Czechoslovak society, i.e. it was used for analytical and descriptive ends.

But it would not suffice for tasks of the most demanding research, particularly for the analysis of the development of the mode of life and its planned formation. And so it must be replaced by a dynamic model capable of linking the dialectic of this social complex to the intricate structure.

One concrete step made in this direction is the diagram (Table 2) submitted by A. G. Zdravomyslov to the Conference of the International Commission of the Social Sciences and Unesco (Moscow, 1976).

The path followed to work out methodological questions is of fundamental importance for the perfecting of international comparative research projects. Experiments carried out in this field allow us to deduce certain principles for the approaches used in comparative research:

TABLE 1. Taxonomic conceptual diagram of the mode of life (intended to classify socio-statistical indicators)

	Formal components[1]			
Objective components	Human activities	Social relations	Social needs	Socio-psychic forms
1. Biological reproduction				
2. Health				
3. Work				
4. Participation in management				
5. Incomes				
6. Consumption and economics				
7. Housing		Analytical elements		
8. Mobility in space				
9. Education				
10. Culture				
11. Leisure				
12. Social communication				
13. Anti-social behaviour				

1. For example, what Andrews and Withey (1974) designate with the term 'perceived life quality' and other writers with 'the experienced' or '*erleben*', I shall consider as a subjective experience of real life constituting only one of the components of the human mode of life.

TABLE 2. Indicators of the social and cultural development of social systems

Indicators of social development (of the mode of life)			Indicators of cultural development (of social consciousness)		
Indicator	Alternative characteristic of the social system		Indicator	Alternative characteristic of the social system	
Social differentiation of class	Inequality	Equality	Nature of education	Religious Elitist	Lay Democratic
Goals pursued by production	Profit	Satisfaction of man's needs	Type of culture	Consumer	Creative
Nature of free time	Subordinated to production	Polyfunctional nature	Value of man	A means to attain other ends	Supreme value of the culture
Nature of human relations	Instrument	Autonomous value	Degree of openness and 'closedness' of the culture	Isolation nationalism chauvinism	Open (internationalism)

Each social phenomenon compared is inherent to the social system which acts first and foremost as a concrete socio-economic formation.

Certain analogous social phenomena brought about by the general movement of the scientific and technological revolution have different meanings and functions in different contexts depending on the systems; at the heart of different systems there appear, exist, and develop certain unique phenomena (for example, in socialist countries it is a question of the role of the Communist Party and of the working class in the programming and orientation of the processes of the society as a whole, and of the consequences stemming from the socialization of the means of production).

In different systems certain quantitative indicators have a different qualitative character with different meanings, and their dynamism is different in certain respects; certain phenomena in analogous forms have different contents.

An adequate comparison assumes the comparison of common phenomena not only in their static insertion in the system but also in their diachronic, i.e. historico-genetic movement; this means, *inter alia*, the application of comparative methods not only in space but also in time (from the retrospective, the present, and the prospective angle of each phenomenon); the historico-genetic examination requires (for example, as far as the tribe-horde relation is concerned) that the structural isomorphism and the essential differences stemming from different qualitative levels of the evolution of the totality of the material world be established correctly.

Each social phenomenon must be envisaged as a moment, factor, or component of the multidimensional system of relations (multidirectional and retroactive), and the hierarchization of components, not in the sense of purely functional and conditional dependence but on the determinist basis of the principle of causality conceived dialectically, here takes on a fundamental meaning of research and action.

Starting from these principles we can proceed to the shaping of models for comparing the mode of life, which are increasingly adequate for the needs of comparative research on the socialism-capitalism type.

On the other hand, I shall consider as entirely inadequate those comparative models which still are very common, which envisage the appearence of certain components of the mode of life at the heart of different systems practically as a mechanical reproduction in the sense of a more or less absolute identity. Figure 1 is an example.

This comparative model represents a mechanical transference of the comparison of phenomena, such as the pollution of the air, sea or earth, which, in so far as it turns out to be transitory, recalls, by its distribution, Euler's circles (or it still appears as a global phenomenon, above the systems). All the same, this comparative model is not suitable for

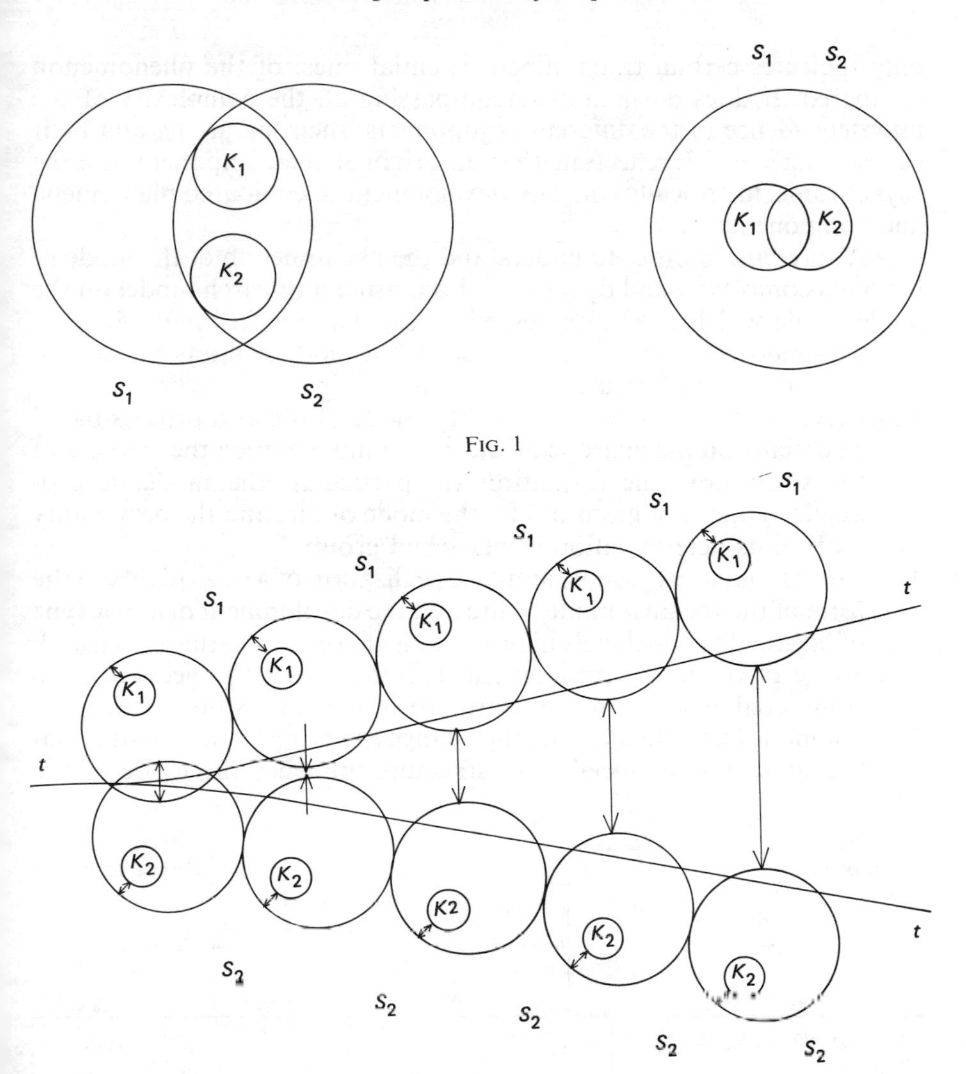

FIG. 1

FIG. 2

understanding these ecological phenomena (in respect to their causes and their social solutions), still less can it serve as an instrument for comparing increasingly divergent of life.

To express the historical movement of compared components, and to grasp their fundamental divergence as well as the appearance of specific, new original qualities of the phenomenon in the comparative diachronic dimension, assumes the development of a temporal model which would have more or less the form as shown in Figure 2.

It is appropriate to emphasize that even this temporal model, which takes into account the divergence of the trajectories of the mode of life, only indicates certain traits, albeit essential ones, of the phenomenon compared. It does not aim at encompassing all the complexity of the interdependence of the different components, their hierarchy, and their causal connection. It reflects neither the origin of new components nor the asynchronic, historically concrete development of compared phenomena and components.

At present, in order to understand the phenomenon of the mode of life in its complexity and dynamism, I am using a research model on the mode of life which can be expressed by the diagram in Figure 5.

This model (Fig. 3) allows us to study the mode of life as a historical and natural process because:

I conceive of the transformations in the mode of life as a process based essentially on the dialectic of the causal link between the changes of the socio-economic formation (in particular, the mode of production), the conditions of life, the mode of life, and the personality itself in its determination of class and group.

This process pursues a double aim: the realization of a new quality in the form of the socialist mode of life and the development of a new type of personality which acts in return as a factor stimulating social and, above all, economic development. This fundamental objective is also considered as the principal standard of the results obtained.

The new mode of production and the changes deriving from it in the social relations and the social class structure influence the mode of life

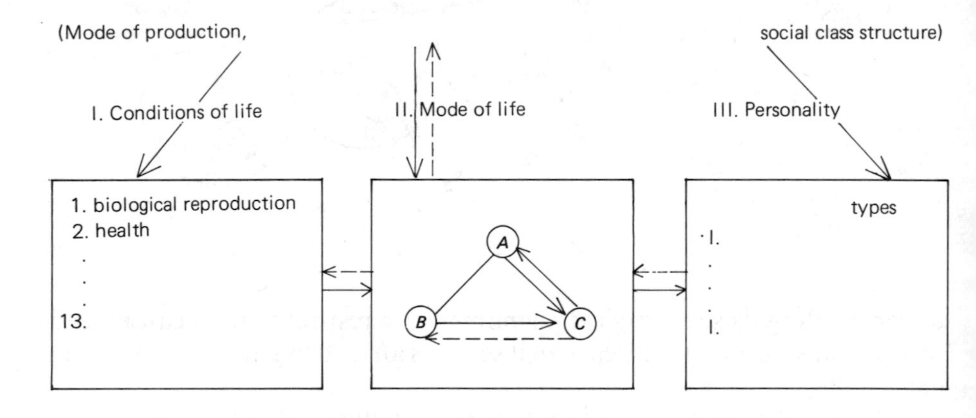

Fig. 3. Socio-economic formation. (The continuous lines indicate the fundamental causal link, the dotted lines, the retroaction which is in fact less a simple retroaction than a spiral dialectic movement. At the same time the whole of this process unfolds in the broader framework of the conditions of social reproduction and creation of man's life (natural, material, conditions of civilization and the whole complex of the cultural heritage). In: B. Filipcová and J. Filipec, 'The Socialist Mode of Life as the Object of Research and Management', Sociologický Časopis, 1977.)

directly as well as indirectly, i.e. they influence the change in the conditions of life.

From the point of view of the mode of life of the classes, groups and individuals, the conditions of life constitute an objective category in which the subject intervenes at the level of the whole society, having penetrated its objective determinations in an increasingly profound way. They express the degree and mode of satisfaction and development of the needs, which are indispensable for the subject's reproduction and development. In socialist society, they represent, above all, the development of new social relations (of production, in particular) which, with the progressive emphasis placed on the results of the scientific and technological revolution, leads to the *rapprochement* of the conditions of the mode of life at the heart of different socialist countries and in the framework of the world socialist system as a whole.

The mode of life, whose activities (among which work occupies a predominant position) are the determining component, is an objective category. The subject (determined by class and group) (a) appropriates the transformed conditions of life to himself; and (b) participates, in turn, in their transformation to a different level of activity.

The personality is the last and at the same time final link in this causal-dialectic chain outlined above. Changes in the mode of life gradually bring about and fix new needs and new habits of behaviour and thought which are stabilized in the form of new qualitites of the personality. In this sense, particular qualities and traits become of primary importance as they are transformed by retroaction into stimulants to the development of the social system, particularly the economic sphere as principal instrument of transformation of the mode of life.

To put this whole process into operation means searching for indicators not only for its different links but also for their mutual and contradictory relations and the relations at the heart of these links.

If we consider, for example, the relation between the conditions of life (I) and mode of life (II), we must say that no automatic mechanism intervenes there, which would mean that a positive change in domain I automatically also brings about desired changes in domain II and eventually in domain III. If we bring about change A in domain I, we will thereby open up a field of possible changes in domain II (from A' to A'') whose limits are determined by extreme combinations of the best and worst consequences. We must not only express this field hypothetically, but we must also look for relevant indicators which will allow us to measure the results attained and to optimize the whole process with the help of effective social mechanisms.

We can make some further progress toward expressing the inner dynamism of the mode of life by indicating the mutual relations among the following three fundamental circles of components: *A*: strategic components, namely work, participation in the management of social processes, education in the sense of assimilating scientific knowledge and cultural values; *B*: the components which represent primary biosocial needs and the activities which depend on them; *C*: the components which represent needs and the highest, socially created activities.

The model of optimal behaviour (from the point of view of the objective pursued) is characterized by two criteria: (a) the transfer of the centre of gravity of man's activities and needs from circle *B* to circle *C* (including the progressively cultivated activities and needs in circle B); (b) the satisfaction of individual needs through the expedient of active and creative participation in the activities in circle A.

For the moment, I am using here as often as not indicators deduced from research on budgets of time and income. It is true that these last items furnish important information but of an elementary nature which cannot take into account what is essential, i.e. the level of value, content, function, and historical meaning (intention) of different activities.

The purpose of this model is also to distinguish the tendencies of development of the activities and needs in the perspective of their relation to the desired values of the mode of life. This must facilitate the study of the axiological aspects in time as well as the definition of the criteria of appreciation of the movement evolving between the ascension, stagnation and regression of the phenomena examined.

We can logically assume, in terms of the model of the mode of life thus conceived, the creation of several fundamental personality types and try to establish which qualities they stabilize under the circumstances: the centre of gravity of life in circle *B* with minimum participation (strictly necessary) in circle *A*; centre of gravity of life in circle *C* with minimum participation in circle *A*; centre of gravity of life in circle *B* with maximum participation in circle *A*; centre of gravity of life in circle *C* with maximum participation in circle *A*.

If it is true that, in research on the mode of life, we must pass from the taxonomic model to the dynamic model, it is nevertheless true that we must consider our attempts outlined above all working attempts, subject to discussion under more than one aspect.

This dynamic research model on the mode of life can be no more than one new step in the process of cognition but one which nevertheless aims at more effective communication and co-operation in the elaboration of comparative international research projects. But more than a research instrument we need the means to reconstruct society, and thereby an instrument for the desired transformation of man and his transformation of himself. We need to facilitate the solution of the hypothetical problem of the alternative, namely the humanist assimilation of the conquests of the scientific and technological revolution, with the concrete aim of assuring not only the well-being but also a harmonious flowering of all members of society.

Note

1. M. Illner, *et al.*, *The Indicators of the Mode of Life in Czechoslovak Statistics*, Prague, Institute of Philosophy and Sociology of the Czechoslovak Academy of Sciences, 1975.

Automation, technological, social and economic change: the case of the Federal Republic of Germany

Günter Friedrichs

Federal Republic of Germany

What is technological change?

The computer is a very important tool of automation. However, automation means much more than computerization. And technological change means an even wider range of different and new technologies. It is impossible to mention all of them.[1]

In agriculture, we have specialization, mechanization up to automation of production, the introduction of new products and the improvement of fertilization. In industry, there are many processes being replaced, and in many cases materials are being substituted. Especially, plastics and non-ferrous metals are replacing conventional materials, while oil and atomic energy are substituting for coal.

The main objective of technological change is to realize savings. The aim is to have more, or at least the same output with less input. New technologies are not introduced from one day to the other, but in a permanent process, affecting nearly all sections of the economy.

SAVING OF MANPOWER

In most cases, the introduction of improved technology is done to increase productivity and to save manpower. This may be achieved simply by intensifying human work, or, as is in the case of automation and computerization, substituting capital for labour. Both ways are labour-saving and therefore of importance.

In 1950, the manufacturing industry of the Federal Republic of Germany required an average of 191 working hours (blue- and white-collar workers) to produce goods at a net value of 1,000 DM (in prices of 1962), while in 1975 only 50 hours were required to produce the same value at constant prices. Within twenty-five years, the manpower requirements to produce 1,000 DM decreased to a little bit more than one-quarter.

In 1972 the labour force of the manufacturing industry of the Federal Republic of Germany had about the same size as in 1962 (Table 1). However, the labour input was about 7.3 per cent less, due to the reduction of working time. But the production increased about 64 per cent.

In comparison with 1962, in 1972 out of forty-three manufacturing industries twenty-four had declining, seven stagnating and only twelve expanding employment conditions. Due to the recession, it is not possible to make adequate comparisons with present figures. However one effect can be shown: in June 1976 industrial production was about 10 per cent higher than in 1972, but industrial employment had dropped by nearly 1 million.

SAVING OF CAPITAL AND MATERIALS

The capital per output ratio of the manufacturing industry in the Federal Republic of Germany is increasing (1960–72 = + 14 per cent). But in some branches it is decreasing. The most striking examples are oil processing (– 47.5 per cent), chemistry (– 44.6 per cent) and petroleum manufacturing (– 10.4 per cent). In highly automated industries, it might be more promising to save capital than labour, since there are only a few workers left.

Much more important is the saving of materials. Many kinds of automation tend to be extremely material saving. However, there are no statistics available. But the saving of material, next to labour, is among the most important motivations for the introduction of improved technologies. The experience of the oil crisis will increase these tendencies.

CHANGES IN JOB CONTENT

Changing technology also changes the content of jobs. Certain jobs disappear totally, others are newly created. However, new jobs are relatively rare and offer chances only to a limited number of people. This is true at least as far as industry is concerned. The situation is a little different in the service industries. Many of the new jobs in services, like teaching, scientific work, advising, or medical care are restricted, since the educational requirements are very high. The same holds true for today's

TABLE 1. Decreasing employment due to technical and structural change: decreasing and increasing employment in forty-three branches of industry in the Federal Republic of Germany, 1962–72, including West Berlin

Branch of industry	Persons employed				Net-production in 1962 prices (%)	Working time per year and per person (%)	Labour productivity per man/hour (%)
	1962	1972	Increasing or decreasing: Absolute	%			
DECREASING EMPLOYMENT							
Iron-ore mining	15,650	3,367	– 12,283	– 78.3	– 60.2	– 13.1	106.2
Leather-producing	30,595	11,603	– 18,992	– 62.0	– 15.4	– 2.7	123.0
Coal mining	486,723	251,969	– 234,754	– 48.2	– 25.9	– 14.1	65.6
Potash and rock-salt mining	22,665	12,799	– 9,866	– 43.5	40.6	– 6.5	166.5
Petroleum extraction	11,704	6,658	– 5,046	– 43.1	180.6	– 0.1	398.9
Tobacco manufacturing	48,520	28,507	– 20,013	– 41.2	48.0	– 7.1	171.4
Other mining	12,732	7,800	– 4,932	– 38.7	7.8	– 6.4	88.3
Milling	15,047	10,553	– 4,494	– 29.8	– 9.1	– 4.5	35.5
Cellulose and paper manufacturing	82,812	65,762	– 17,050	– 27.3	60.8	– 7.2	118.3
Sugar	15,930	11,808	– 4,122	– 25.9	44.3	– 8.9	113.8
Shoes	102,963	77,124	– 25,839	– 25.0	– 6.5	– 7.6	34.9
Iron, steel and metal casting	145,617	111,265	– 34,352	– 23.5	– 0.4	– 7.9	41.5
Textiles	589,590	458,108	– 131,482	– 22.3	42.5	– 5.6	94.4
Saw milling and woodwork	86,655	67,723	– 18,932	– 21.0	50.5	– 5.0	103.0
Shipbuilding	92,973	75,324	– 17,649	– 18.9	33.4	– 4.7	72.9
Fine ceramics	89,337	75,719	– 13,618	– 15.2	20.5	– 8.4	55.4
Iron-producing	362,515	311,893	– 50,622	– 13.9	40.0	– 5.4	72.1
Oil milling and margarine	16,918	14,895	– 2,023	– 11.9	28.4	– 4.6	52.8
Leather manufacturing	40,925	37,226	– 3,699	– 9.0	16.8	– 10.4	43.5
Musical instruments and toys	58,606	56,439	– 2,167	– 3.7	39.6	– 9.1	59.6
Stone and earth	263,510	241,457	– 22,053	– 8.3	51.5	– 2.4	69.4
Steel construction	227,712	210,995	– 16,717	– 7.3	22.7	– 4.2	38.4
Drawing and cold rolling mills	72,205	68,700	– 3,505	– 4.8	– 66.7	– 1.2	77.3
Clothing	384,159	372,152	– 12,007	– 3.1	37.7	– 9.5	57.3
SUB-TOTAL	3,276,063	2,589,846	– 686,217	– 20.9	34.7	– 6.5	

			STAGNATING EMPLOYMENT				
Precision engineering and optical	154,003	153,930	– 73	–0.0	38.0	–8.5	50.9
Steel moulding	140,257	142,083	1,826	1.3	25.8	–4.9	30.7
Glass	92,743	93,957	1,214	1.3	81.2	–7.8	93.9
Breweries and malting	87,963	89,282	1,319	1.4	46.5	–6.4	54.3
Metal goods	406,659	412,483	5,824	1.4	73.6	–3.5	77.5
Non-ferrous metal recasting	84,581	87,143	2,562	3.0	68.5	–4.8	71.8
Other food and drink trades	339,613	351,385	11,772	3.4	52.5	–4.8	54.9
SUB-TOTAL	1,305,819	1,330,263	24,444	1.7	75.4	–60.1	
			EXPANDING EMPLOYMENT				
Woodworking	221,767	237,001	15,234	6.8	96.3	– 5.0	93.4
Mechanical engineering	1,056,665	1,154,323	97,658	9.2	31.4	– 6.8	29.2
Printing and duplicating	201,335	220,482	19,147	9.5	62.6	– 8.8	63.0
Paper and cardboard manufacturing	119,997	132,519	12,522	10.4	70.7	– 7.6	67.4
Non-ferrous metal casting	27,236	30,110	2,874	10.5	39.5	– 6.3	34.7
Electrical engineering	919,607	1,058,415	138,808	15.0	104.8	– 8.0	93.5
Chemical	502,245	584,164	81,919	16.3	163.3	– 8.0	146.1
Mineral-oil processing	31,513	37,136	5,623	17.8	127.0	– 6.7	106.8
Rubber and asbestos processing	112,552	135,657	23,105	20.5	68.1	– 8.5	48.0
Vehicle construction	437,737	609,201	171,464	39.1	82.5	– 9.1	44.4
Aircraft engineering	25,967	40,143	10,176	54.6	100.0	–10.9	45.9
Synthetics manufacturing	100,731	180,979	80,248	79.6	270.9	– 5.2	117.7
SUB-TOTAL	3,757,352	4,420,130	662,778	17.6	95.1	– 7.6	
TOTAL[1]	8,339,234	8,340,239	1,005	0.0	64.4	– 7.3	76.5

1. Without building construction trades and public energy supply.

Source: Calculated by Automation Department of the Metal Workers' Union from: Deutsches Institut für Wirtschaftsforschung, *Produktionsvolumen und -potential, Produktionsfaktoren der Industrie im Gebiet der Bundesrepublik Deutstchland einschl. Saarland und West-Berlin,* Statistische Kennziffern, 14. Folge, 1961–72, Berlin, Dez. 1973 (Rolf Krengel, Egon Baumgart, Arthur Boned, Rainer Pischner und Käthe Droege).

most fashionable job, the programmer. In the beginning of the computer age, it was the most attractive job for up-grading. Today, it is changing more and more to a routine job because the supply of prefabricated software reserves really interesting work for specialists with higher education, like system analysts or organizers.

In manufacturing, we find the tendency to make some jobs easier, in terms of physical conditions as well as in terms of training. This is even true for many activities in regard to maintenance. There we have a development that tends to replace repair work by exchanging parts, as has been customary in the case of the automobile. On the other hand, we have the development of increasing psychological stress. Obviously, those who just push buttons or look at certain measurement tables, experience a psychological disorder unknown in the past. Most important: technological change did not overcome Taylorism and its negative effects on human work. It even created new working places that are not acceptable, like keypunching and others.

CHANGES IN STRUCTURES

One of the main results of technological change is the permanent change of structures, as shown in following examples.

Between 1950 and 1975 we had considerable changes within our working population. The percentage of agriculture dropped from 24.6 to 7.2 while the percentage of manufacturing increased in the same period from 42.6 to 45.9 and the percentage of service industries increased from 32.7 to 46.9. For some years manufacturing has been losing while the service occupations have been winning.

There is also a change in the structure of the work force. In manufacturing, the non-manuals increased considerably in spite of increasing computerization. The same is true with female workers. The new technology made some jobs easier than they were in the past. Therefore, in certain industries men were replaced by women.

Union activities

STIMULATING RESEARCH

In the Federal Republic of Germany it was a labour union that made technological change into a national issue, because it was not prepared to accept the negative effects of labour-saving devices. Already in 1963, the Industrial Union of Metal Workers organized its first international automation conference. This was followed in 1965 by a second, and in 1968 by a third international meeting.[2] In all three cases, the best scientists and other experts on the social and economic effects of technological change from industrialized countries and international organizations were invited. The aim of these conferences was not only to learn from international experience but also to create interest in the public as well as in social sciences.

Consequently, the unions were able to initiate big research activities. The Federal Employment Agency (Bundesanstalt für Arbeit) founded a special institute for research on labour market and vocational developments. The National Productivity Centre (Rationalisierungskuratorium der Deutschen Wirtschaft) started a whole series of empirical research projects on the social and economic effects of changing technology. One of them belongs to the biggest projects of the social sciences in the Federal Republic. Investigators comprised five institutes of different disciplines. They covered sixteen branches of the economy and forty different plants.[3]

COLLECTIVE BARGAINING AND LEGISLATION

By stimulating research in the social sciences, the unions tried to learn about the risks and the chances of technological change. Considered as risks were unemployment, downgrading, changing of working places, and retraining. Considered as chances were higher income, less work, more leisure time, growth of wealth and a better quality of life for the workers. And it was the strategy of the unions to realize the chances and to prevent the risks of modern technology.

The unions in the Federal Republic consequently used collective bargaining to increase pay, and to obtain a 40-hour week and four to five weeks of vacation. Partly by collective bargaining, partly by legislation, a system of social measures was created that considerably improved protection against the negative impacts of technological change.

THE QUALITY-OF-LIFE ISSUE

As the unions stepped further into the wide area of technological change, they had to recognize that they were dealing pretty soon with all aspects of economic and social change. They finally learned that modern technologies may create not only difficulties at the working place but also in many aspects of human life. They started to question the concept of 'qualitative growth' very early. The Metal Worker's Union already decided in the autumn of 1970 to have another international conference on the subject: 'Quality of Life: The Challenge of the Future'. This conference was opened by the president of the Federal Republic of Germany in April 1972. It dealt with subjects like education, transport, environment, health, regional development, qualitative growth and democratization.[4]

The Commission on Economic and Social Change

COMPOSITION AND TASK

Due to union pressure, in 1970 the Federal Government appointed a Commission on Economic and Social Change (Kommission für wirtschaftlichen und sozialen Wandel). Seven members represent different disciplines of mostly social sciences, ten members were nominated, five by employers' associations and five by the unions. The commission finished its job in November 1976 by presenting a report with findings and recommendations for the next ten years. The commission has its own budget and is completely independent. The government never tried to interfere.

THE RESEARCH PROGRAMME

The commission spent about 8 million DM for a programme of research in the social sciences. Out of 189 projects, 140 research reports will be published. The publication of research reports is done without any comment by the commission. The commission presents its views only one time in its own final report.

The research programme was prepared by systems analysis. A relevance tree was used to identify those areas where research was needed. In long and intensive discussion the programme was developed and described. Each time when a specific research area was finished, the projects were advertised in a weekly, claiming to have many academically trained readers. In this way, everybody was invited to apply if he was able to prove his qualification. Some ministers of the federal government learned by reading this magazine that the commission had decided to investigate their activities.

By inviting all qualified persons or institutes, the commission in nearly all cases had a choice between several different applicants. Very often, it was not the established professors and scientists who got the jobs, but very capable younger persons. A few projects went to Switzerland, Austria, the United Kingdom and the United States.

THE FINAL REPORT

The preparation of the commission's own report proved to be difficult for several reasons. One difficulty was the composition of its membership. The three represented basic groups were very selfish, each suspecting the actions of the others. While the scientists very often claimed that the representatives of the unions and of the employers would behave extremely selfishly, they themselves never hesitated to ask for and to accept any type of majority, as long as it favoured their own interests. Under those conditions a unanimous report was impossible. All groups introduced some important minority votes.

A much bigger difficulty was created by the dimensions of the task. The commission had to deal with too many different things. For this reason the final report is not very original in the terms of specialists. However, the report concentrates on the interdependencies between different subjects. A typical example is regional development. Involved here are factors like industrial branch policies, housing, transportation, education, health, environment, working conditions, distribution and so on. Those who want to shape regional development need to consider all factors and their interdependencies. If they change only one factor, all the others might be affected also. However, the main problem in regional development is that in charge of each single factor is not the same but in most cases a different administration. We cannot improve planning without considering those interdependencies. Another and most important difficulty proved to be the information lag. In some areas, there are no figures available at all, in others too many, and in others again not at the time when needed. The methods of forecasting are still extremely poor. The lag of information and of methods to process information adequately is still limiting the possibilities of successful planning.

Another difficulty was the time factor. The commission started at a time of full employment, low inflation rates, no major budget problems and a general feeling in the public that there is not only a need for reforms but that the country can also afford the needed reforms. In the period when the main part of the report was written, the scene had changed completely: deepest recession since the war, with a million unemployed, big structural changes, high inflation rate, major budget difficulties, high prices for energy and other natural resources.

The report passed by the commission at the end of October 1976 contained chapters on the following: shaping of social change; policies for growth and structure; accounting of distribution; policies for long-range stabilization; policies for research and technology; policies for the shaping of regions; policies for the environment; policies for trade and competition; policies for consumers; shaping of working conditions; social security; education and employment; increasing the efficiency of the public administration, increased need for additional information, further development of the participation of citizens.

Finally, I want to take up only one subject from the report that might be of specific interest. According to the tenor of the text, the Federal Republic of Germany needs to make strenuous efforts to reach full employment again. This will be true in spite of the fact that the labour savings due to introducing modern technology (due to the increase of productivity) will slow down a bit in the future, since the service sectors gain growing weight within the economy. However, the country does not have to deal only with increasing productivity. It needs to employ the million at present unemployed plus another million, who will enter the labour market within the next ten years. Another (up to) 700,000 presently existing working places in industry are expected to be lost in the same period, because certain branches with labour-intensive mass production will move at least some of their factories into low-wage developing countries.

The commission did not ask for protective measures to prevent such an exodus. However, it did demand a policy of economic expansion through the creation of the new technologies, products and structures that are needed in the coming decades.

The idea is to make much more intensive use of the available highly qualified work force. Since the Federal Republic of Germany does not have natural resources, its only comparative advantage can be the labour force. Therefore, the commission asked for stimulated research and development activities, including transfer mechanisms that make sure that R&D results will be followed by innovations. Such policy would not only strengthen the economic potentials, but it would also help to realize the change from quantitative to 'qualitative growth'. The really important products and technologies of the future must be: (a) acceptable for the environment; (b) energy saving; (c) capital saving; (d) resource saving; (e) raising the quality of working places; (f) increasing the quality of efficiency of services and administrations; (g) friendly for users; and (h) meeting the real needs of developing countries.

The quality of working conditions

The demand for radical improvements of working conditions came up in 1972 in connection with the quality-of-life issue. Unions and workers no longer accept jobs that are dangerous or that require extreme physical or psychological stress. They are no longer prepared to adjust themselves to technology or to the organization or work. Both of these have to be adjusted to the needs of human beings. Modern methods of technology and of the organization of work permit alternatives that are acceptable to workers.

COLLECTIVE BARGAINING AND STRIKES

Already in 1968 a special collective agreement against the negative effects of labour-saving devices (*Rationalisierungsschutzabkommen*) was concluded by the Metal Workers' Union. It introduced special protection measures against downgrading, transfers within or between plants, regrouping, retraining and dismissals especially for workers older than 40 and with a company service of ten years. According to this agreement, workers between 55 and 60 years of age and having at least ten years of company service could not be dismissed at all in connection with the introduction of labour-saving devices. This type of agreement was taken over by many other unions of the country operating outside of the metal trades.

In 1973, in one larger region of the metal industries, the union won a basically very important agreement after a strike and a lockout of about half a million metal workers. Under the terms, workers older than 54 years with only one year of company service could not have their income reduced till the end of their working life and will have all advantages of wage increases in the future. Workers older than 53 and with at least three years of company service cannot be dismissed at all. At least two other basic breakthroughs should be mentioned. New technical equipment is only accepted if it permits a minimum working cycle per worker of 1.5 minutes. Another important advantage was the regulation of paid breaks. All workers on piece-rate work are entitled to a minimum of eight minutes break per hour.

The German Federation of Labour (DGB) organized a major conference in 1974 on the subject of improvement of working conditions.[5]

THE GOVERNMENT-RUN PROGRAMME

In 1974 the Federal Government introduced a special programme for the improvement of working conditions (*Programm Humanisierung der Arbeit*). This programme will continue for several years. It provides financial aids for research projects in those areas where working places and working conditions are still dangerous or not acceptable (pollution, heat, noise, extreme physical or psychological strength and others).

One part of this programme is of specific interest. Plants who wish to experiment with new types of work organization, especially with job enrichment, job enlargement, replacing assembly lines by group work (semi-autonomous or autonomous) can be refunded up to 50 per cent of their expenditures if they accept two conditions: (a) they have to guarantee co-operation with a workers' council and with the responsible union; (b) they have to accept free access of scientists (economists, technicians, different types of social scientists) for checking the programme permanently. The main objective of the so-called accompanying research is to insure the transfer of useful results to other plants.

In August 1976, the federal government reported a total of 140 different research projects within the programme. The projects for restructuring human work are not so many, because they are extremely expensive. However, the electrical engineering industry alone is operating fifteen different projects, directly affecting 1,800 workers and indirectly (in later stages) more than 5,000 workers of the participating companies. It is certainly expected that the transfer of results to other plants and industries will affect a much bigger number of workers. The government expenses within electrical engineering are about 24 million DM.

Notes

1. This article does not deal with theories, but with practical experience.
2. *Automation in Deutschland und in den USA* (1963); *Automation, Risiko und Chance*, 2 Vols. (1965); *Computer und Angestellte—International Edition*, 2 Vols. (1971); all published at Frankfurt am Main and edited by Günter Friedrichs.
3. Rationalisierungskuratorium der Deutschen Wirtschaft (RKW), Vol. 1. *Sieben Berichte* [Abstracts of Results]; Vol. 2, L. Uhlmann, G. Huber (Ifo-Institut für Wirtschaftsforschung); *Technischer und strukureller Wandel in der wachsenden Wirtschaft*; Vol. 3, K. H. Oppenländer *et al.*, (Ifo-Institut für Wirtschaftsforschung): *Wirtschafliche Auswirkungen des technischen Wandels in der Industrie*; Vol. 4, W. Fricke, H. Lindner, A. Mohr, G. Stümpfig, P. Thelen, and K. H. Weimer (Forschungsinstitute der Friedrich-Ebert-Stiftung), *Auf dem Weg zur Dienstleistungsindustrie*; Vol. 5, G. A. Koch and W. Hackenberg (Forschungsinstitut für Rationalisierung an der RWTH Aachen), *Technisch-organisatorische Umstellungen in der industriellen Produktion*; Vol. 6, G. Koch *et al.*, (Forschungsinstitute für Rationalisierung an der RWTH Aachen), *Vernderung der Produktions- und Instandhaltungstätigkeiten in der industriellen Produktion*; Vol. 7. W. Rohmert, J. Rutenfranz and E. Ulich (Insitut für Arbeitswissenschaft an der TH Darmstadt), *Das Anlernen sonsomotorischer Fertigkeiten*; Vol. 8, H. Kern and M. Schumann (Soziologisches Seminar der Universität Göttingen), *Industriearbeit and Arbeiterbewusstsein*; Vol. 9, H. Kern and M. Schumann (Soziologisches Seminar der Universität Göttingen), *Der soziale Prozess bei technischen Umstellungen*; all published at Frankfurt am Main at the RKW; forty-two basic reports, dealing with sixteen branch and about forty plant investigations.
4. Series 'Aufgabe Zukunft—Qualität des Lebens': Vol. 1, *Qualität des Lebens*; Vol. 2, *Bildung*; Vol. 3, *Verkehr*; Vol. 4, *Umwelt*; Vol. 5, *Gesundheit*; Vol. 6, *Regionalentwicklung*; Vol. 7, *Qualitatives Wachstum*; Vol. 8, *Demokratisierung*; Vol. 9, *Zukunft der Gewerkschaften*; Vol. 10, *Register*, Abstracts-Résumés (217 p. English and French Abstracts); all published at Frankfurt am Main (1973–74) and edited by Günter Friedrichs.
5. H. O. Vetter (ed.), *Humanisierung der Arbeit als gesellschaftspolitische und gewerkschaftliche Aufgabe*, Frankfurt am Main, 1974.

Integral management of the science-production cycle

L. M. Gatovski

Academy of Sciences of the USSR

The process of the organic union of science and production, that is, the process whereby science is increasingly transformed into a direct, productive force, is one of the main manifestations of the scientific and technological revolution. In the Soviet Union, this objective is expressed by the ever-broadening integral management of all the phases of the 'science-production' cycle taken as a whole. The present article tackles certain methodological problems pertaining to this management.

In the Soviet Union now, the development of the planning system—stimulation as well as the organizational management structures—is increasingly determined by the following characteristic regularities arising from the scientific and technological revolution. Scientific research and technical studies (at the level of the elaboration of projects, technology, and experimentation) are part and parcel of unified 'science-production' cycles and are organically linked to production. Science is transformed into a reference point for the development of production and becomes an increasingly indispensable factor in actual production (active participation in the management of technical processes, their verification and inspection, modernization, realization of programmes, etc.). At the same time, technical study and research necessarily imply the direct participation of specialists working in production. In the final analysis, all this stems from the modifications brought to bear on the very character of production; from the enrichment of the latter by the ever-evolving scientific content; and from the constantly rising scientific and technical level of the processes of production as well as the 'scientific capacity' of the latter (wherein investments assigned to research and study are directly linked to its realization).

Under present conditions, scientific activity, generally speaking, cannot be reduced solely to the creation of knowledge. It is a characteristic of the scientific and technological revolution that science assumes new functions which are always developing, that is, science participates directly in the realization of the new knowledge it has created. We are witnessing, then, a process of interdependence: science is applied to production which, in its turn, is an integral part of the 'science-production' cycle, including in the form of the rapidly evolving science industry. Numerous modern technical articles have been used first in science. The priority given the development of science is expressed by the fact that it offers perspectives on the progress of technology, production, enterprises and culture, as well as by the fact that its proportion of social expenditures seems to increase continually as the investments assigned to these ends are covered, thanks to the application of the attainments of science. In the Soviet Union, the credits granted for scientific purposes substantially exceed the growth rate of the national income.

Hence in the Soviet Union the appearance of modern integral management of the 'science-production' cycle results from the fact that, in the conditions of the scientific and technological revolution, the processes of preparation, functioning, and development of modern production involve the direct participation of science; the role which has devolved upon science stems from the progress of basic research. This is a necessary condition for the considerable optimization of all the levels and spheres of technology management of production as well as the management of economic ratios and social processes. We are witnessing the unified and planned management of different processes such as the accelerated progress of the sciences and technology, the utilization of the latter in order to increase the profitability of production and to resolve social problems on this basis.

In this way, we have recourse to a comprehensive approach in the combination of scientific research, technical studies, production of new techniques, and their application to production and services. This approach is expressed by the continuity of the cycle as a whole and by the harmonization, succession, and simultaneity of the functioning of its phases and their interaction. The saving of time which results, and consequently the reduction of costs per unit of useful end-product and the growth of the profitability of technologies arise from the combination of three essential indicators: reduction and suppression of the time intervals between phases; shortening the time per given operation in each phase; partial simultaneity, parallel work of phases with a view to realizing a certain technical operation. Each preceding phase plays a determining role in terms of the functioning of the succeeding phase. The increasingly close economic links between phases are part of a complex and, between the phases and the national economy, they are realized in production as well as in the action exerted by scientific and technical progress on final, non-productive consumption.

The integral management of 'science-production' cycles is realized on three levels.

As a whole, the national economy level allows, on the one hand, the orientation of science and technology towards raising the technical and economic level of production, and, on the other hand, the orientation of the national economy towards the immediate realization of the attainments of science and technology. Here we have the organic union of science and technology planning with all the sections planning the economy and culture. The State level allows the determination of crucial problems confronting the national economy; the line to be followed in science and technology; the bases which allow the development of enterprises while bearing in mind their harmonization; the important inter-sector programmes, and others.

The level of different sectors of the economy makes it possible, following the example of electronic technology and other branches of industry, to pass to the integral planning and financing of the scientific, technical, and production phases, taken as a whole, at the sector level. At this level the concrete application of the science and technology policy elaborated for the sector as well as for its subsectors and indicators is provided for; in addition, here we proceed to the elaboration of programmes and set the fundamental tasks for the phases of production.

The groups of production (firms) effect the transition to the unified management of research institutes, experimental, technological, and project operations, processes linked to the introduction of new techniques, the utilization of the latter in production, and their distribution. At the level of groups and enterprises, there exists a whole profitability system of investments and self-financing (the latter is already applied to certain sectors); the way to the immediate application of technological progress, that is, the manufacture and utilization of new technologies, is provided for.

Such is the unified 'chain' of management of the 'science-production' cycle where the principal role (as in the management of the whole economy) amounts to this aspect of the national economy which is inseparable from initiative, operational autonomy, responsibility, and interest witnessed by the lower links in the chain.

This type of management embraces the complex system of multilateral economic relations which are linked to: long-, mean- and short-term planning at different levels; the realization of indicators and norms fixed by the plan; expenses and economic results; price formation; the dynamism of prime costs and profits; the effectiveness of new techniques and investments; the problem of resources (materials, financing, and manpower); the system of material encouragement; the regulation and remuneration of work; the application of production standardization to all levels of management and the classification of manufactured products by categories according to quality; the concentration and specialization of production; the unification and scientific organization of production, etc.

The integral management of 'science-production' cycles is called upon to push to a greater extent the enterprises and their personnel, the groups, and sector organisms to proceed to a rapid and highly efficient technical renovation of production, which is made possible by planning; appreciation of the activity involved in production; material encouragement; verification and inspection; and interwoven economic and organizational factors.

For this purpose particular importance is attached to the final results as reference points and to the orientation of plans and incentives towards indicators and norms of a better quality and profitability as well as of a higher technical and economic level. In addition, quantitative indicators must be based on qualitative indicators. Only such an orientation can allow the fusion of science and production given the fact that the achievements of science and technology which find their expression in quality and profitability indicators serve as norm and reference point for the activity of production and the economy.

The organic union at all levels of planning the development of science and technology as well as the organic union of production and capital investments (bearing in mind all the particularities of these spheres) has primary importance for the integral management of the 'science-production' cycle.

The plans for the development of science and technology serve as the point of departure for all the other sections of economic planning which are drawn up later.

This involves the introduction of the system based on a particular management of profitability indicators and norms in the whole chain, which are fixed by the plan, i.e. research-study-development (anticipated results); planning, fixed norms; inspection of results obtained; incentives for having obtained the best profitability. The system of indicators and norms is expressed first in the creation of prices, then in the standardization and classification of production in terms of categories based on quality.

The conception of new techniques and the evaluation of their profitability are considered as the 'finished products' of the scientific and technical phases in the form of projects and industrial and experimental samples. The utilization of new techniques and their profitability is the 'finished product' of the chain of production. Only attaining a unity of the object of the functioning of all the phases (these are the new techniques of this or that profitability which were first conceived, then manufactured and finally utilized, and which are profitability's object) allows integral management of scientific and technological progress—starting with the choice of the theme destined for study and the research institute, and ending with the material achievement of the conclusions obtained at the end of research—in mass production and in mass utilization of the most effective new techniques.

The co-ordination of the development plans of science and technology with those of production and the utilization of new techniques is made possible by the following conditions:

The plans for scientific and technological development must contain final, adequately verified technical and economic results (including general economic results) of the studies and research, taking into account the possibilities offered by their utilization in production on the basis of balancing the accounts.

Production plans must be based on the end results (technico-economic and general economic) of the functioning of the scientific and technical phases, bearing in mind an eventual restructuring, capital investments, capacities, manpower resources, finances, and material and technical possibilities. Thus, the plans of scientific and technical development and those of production have the same indicators which unite them and which have been foreseen at the scientific and technical stage preceding production and are realized after production.

The technical and economic indicators must appear in concrete form in the appropriate norms of the management of the outlay of resources per unit of useful end-product (e.g. per unit of production, taking quality into account).

The management of resources must be reflected in the final plans which are studied and established, bearing in mind the control already made possible by scientific and technical progress.

The organic union of science and technology development plans and those for the application of scientific and technological achievements to the national economy are based on the system of indicators and norms characterizing: the quality and effectiveness of new techniques; the volume of their production and utilization; the time allotted for the attainment of the final results of the progressive functioning of all the phases of the 'science-production' cycle; the stages related to the application of measures fixed by the plan; and the necessary resources, including the funds for incentives and their concrete utilization.

We are dealing with a unified system of indicators of technical progress and its effectiveness, a system which rests on the appropriate norms and represents the combination of: (a) general indicators of economic output which generalize the final results; (b) social indicators and norms (improvement of working conditions, purification of the surrounding environment, etc.) which must be included in the predicted positive result which allows one to evaluate the management of expenses;

(c) technical and economic indicators and norms characterizing certain aspects of the efficiency achieved because of new techniques, namely: prime cost; work productivity; amount of work involved in the manufacture of a given article and the recovery of manpower; management of materials; fixed funds; specific investments; the technical-economic levels of production and manufactured goods; the objectives to be resolved regarding time spent in replacing an outdated mode of production with a new one; the degree of utilization (given the fixed range of products) and the specific importance of new techniques (new means and objects of work and technology, new methods of organizing production) and their effectiveness; the tasks related to the mechanization of manual labour, automation, etc.; and (d) perfected technical parameters. Most of the technico-economic indicators and norms, as well as the technical indicators and norms, differ according to the sectors and types of production.

The system of indicators is determined by the efficiency criteria of social production, the 'science-production cycles'. Given criteria express, in their turn, the global character of the management of the economy including the 'science-production' cycle; such a comprehensive approach rests above all on the following fundamental principle: everything which responds to the interest of the entire people, in other words, to that of the national economy, must be advantageous for the enterprises (group-firms), for their personnel, and for certain workers. The greater their contribution to the efficiency of the national economy, the greater the economic appreciation and encouragement of the results of the work furnished.

The principle of the harmonization of interests with that of the national economy is the point of departure, and serves as a decisive element for making use of the criteria of efficiency and the system of economic indicators, norms, and incentives. Consequently, the interests of the national economy are presented in terms of the efficiency of the national economy in so far as they constitute the frame of reference for the economic functioning of the enterprises and play a fundamental role in their interests.

Given the available resources, the interests of the national economy consist in the maximum satisfaction of the material and cultural needs of the entire population, needs which are always growing. This explains the tendency to intensify the economy on the basis of accelerated scientific and technical production, the progressive realization of the scientific and technological revolution; this allows the solution of the important problem, namely the reduction of expenses per unit of useful end product (production and services, taking into account their quality), and the creation, thereby, of more important resources which favour the growth, on a national scale, of the socially useful end-product.

As a result of this, two aspects are expressed in the efficiency of the national economy regarding scientific and technical progress (and regarding all social production).

First, there are the results deriving from scientific and technical progress, i.e. the growth of resources and the possibilities of a more complete satisfaction of society's needs (including new needs); it is a question of increasing production and services, perfecting their structure, and raising their quality (bearing in mind concrete demand) as well as improving working conditions and purifying the surrounding environment.

Second, there is the management of expenses on the scale of the national economy (starting from the useful end product) when the economy obtained from each unit of new techniques is multiplied by the mass of such techniques.

These two aspects of the efficiency of the national economy are determined and co-ordinated in the integral management of the 'science-production' cycle, bearing in mind the progressive, iterative checking of their interaction. The levels and dynamism (in absolute sizes and specific weights) of the structural indicators of the volume of production are established together with those of the utilization of new techniques and the growth of production and services stemming from the use of new techniques, all the while taking into account the management of expenses.

The first aspect of efficiency is realized by means of classifying needs always bearing in mind the availability and outlay of resources; in order to do this, urgent and less urgent needs are distinguished, using the appropriate scales. This aspect of efficiency can be considered the goal, the result useful to the second aspect of efficiency, namely, management on the scale of the national economy, which makes possible the realization of the first aspect.

Let us examine in more detail the second aspect of management resulting from the use of new techniques.

Economic output is expressed above all in terms of the growth of revenue obtained by the society thanks to the use of new techniques. The growth of revenue is based on the increase of the active stock coming from the results of the utilization of these techniques less the investments granted for their creation and use. On a national scale, this is expressed in the growth of the volume of national revenue obtained from the management of the use of new techniques.

Global economic output obtained from such management is evaluated by taking into account the final result (in the area of the utilization of these techniques). It synthesizes the elements composing it—the management of certain types of resources of which we have already spoken—and is used for the choice of the most efficient variant of the techniques for which production is forecast. This indicator of economic output resulting from the use of new techniques is taken into account by the planning and statistical bodies at all levels, by the enterprises, groups (firms) as well as by the sector organizations, scientific research institutes, study units, and experimental production. Thus, such a global economic output is made use of at all levels of management, from management on a national scale, to the enterprises.

This output characterizes the entire management and economy of all the resources of production (manpower, raw and other materials, fixed funds of production, capital investments) obtained thanks to the use of new techniques starting from the minimum expenses of reference (the sum of the prime cost and capital investments established for a term of one year according to the formula for expenses is presented as follows:

$$e = P + E_n I \qquad (1)$$

where:

e corresponds to annual expenses per unit of production (in roubles).
P is the prime cost of one unit of production (in roubles);
I corresponds to specific investments allocated to production funds (in roubles);
E_n is the normative capital investments efficiency coefficient (12 per cent at present).

The sum of current expenses plus installation costs forecast for one year is evaluated according to the chosen variant of new techniques (basic techniques) which serve as the basis for comparison. The difference of these total expenditures between new techniques and basic techniques serves as the basis for the choice of the variant of new techniques with a view to including it in the plan of production.

The annual economic ouput obtained thanks to the utilization of new techniques and representing the difference between total and current expenses and the installation expenses of new and basic techniques (the latter serving to establish the comparison) is presented in the following formula:

$$R = (e_1 - e_2) \cdot V_2 \qquad (2)$$

where:

R is the annual economic output (in roubles);
e_1 and e_2 are the total expenses per unit of basic techniques and new techniques (in roubles) which are evaluated following formula 1;
V_2 is the annual volume of production of new techniques (in natural units).

When the output is evaluated in terms of the variable, one must take into account the fact that capital investments are allocated in different time periods, whereas current expenses and the results of production are modified according to the years of exploitation. In order to take into account the time factor, expenses allowed for the creation and utilization of techniques and the results obtained thanks to their utilization must be reduced to one and the same moment (beginning of the period of calculation) following the formula below:

$$a_t = (I + E_n)t \quad (3)$$

a_t is the reference coefficient;
E_n is the normative efficiency coefficient;
t is the number of years separating the expenses and results of the given year and the beginning of the reference year.
The expenses allowed and the results obtained after the reference year are multiplied by the coefficient (a_t), while after the beginning of the year of calculation they are divided by this coefficient.

When one proceeds to the choice of a variant of new techniques with a view to producing them, one has recourse to the indicators of thc best techniques (which have already been put into use or whose utilization is forecast) which serve as the basis for comparison.

The similarity of the variations of new techniques and basic techniques, whose comparison serves to establish economic output must be assured with the help of adequate calculations, that is, it must be assured in accordance with the time factor (one counts from the moment when the chosen variant of the new technique is made and put to use), the volume of products manufactured or work furnished, the material structure and quality of production as well as the social consequences of production and the exploitation of techniques (working conditions, action exerted on the surrounding environment).

Economic output arising from the use of new techniques is evaluated according to each area of their use; then, one makes the sum of the output according to the areas.

Consideration of just the interests of the national economy is insufficient if one does not also take into consideration the economic interests of the economic enterprises and groups, given the fact that thc economic planning system in the Soviet Union involves self-financing.

Consequently, the economic output obtained from the utilization of new techniques, which expresses the interests of the society as a whole, must be expressed in the indicators characterizing the economic advantage obtained thanks to the production and utilization of such techniques by the enterprises, which either produce or use them and which thereby obtain the so-called effect of self-financing calculated separately for the producers of techniques and for their consumers. The economic mechanism has as its objective the increase of the importance of the enterprises' economic advantage, because of the management of the more considerable expenditures stemming from technical progress as well as because of the principle of the reduction of expenditures and of the price per unit of the value of use if its application is increasingly important. In the final analysis, such management and output obtained through new techniques are expressed, at the level of enterprises and groups (firms), in the form of the growth of profit and net profit (profit less deductions).

Thus, economic output deriving from new techniques is directly related to the entire economic practice and is realized in it. The planning indicators for greater efficiency in technical progress and control of the output really obtained are expressed in the indicators of the results of economic activity.

This means that the following outputs are found reunited in one system of comparable indicators and norms: the expected output (that which is anticipated at the pre-industrial stage), the planned output, and the real output (at the industrial stage). In order to establish the planned and real outputs, the new techniques must be compared (according to the expenses and results) with the existing techniques which must be replaced. The management of expenses arising from new techniques is compared with the expenses which would be granted if the old techniques were to continue to be produced. The management is calculated for the entire period equal to the time periods foreseen for the manufacture and exploitation of the new techniques.

The sum of annual outputs is made according to the periods forecast for the plan (five-year and others).

The output resulting from new technical products is calculated at the level of the complexes of production, groups, technical divisions, enterprises, groups, sectors and at the national economy level as a whole.

Hence, the criteria which determine the economic efficiency of new techniques have considerable practical importance for the harmonious management of the 'science-production' cycles. First, as I have already said, these criteria constitute the basis for technico-economic motivation on which is founded the choice of an optimal variant for the creation and use of this technique. Second, its effect so determined is reflected as an important indicator in the technico-economic norms of resource consumption (expenses, time factor, results); in the plans of the enterprises, groups, ministries, and finally in the plans of the national economy as a whole. Third, starting with these criteria, the output actually obtained because of the new technique is evaluated and an appraisal of the economic activity is also given. Fourth, the sums allocated for material incentives for the enterprises, their personnel and certain workers are defined in terms of this output. Fifth, the output produced by the new techniques is taken into consideration to justify its price economically, the sums deriving from the output being distributed between the enterprise-producer of the technique and its user by the expedient of price.

The various multiform incentives to scientific and technical progress are realized above all by means of bonuses regularly given to those who perform in accordance with the above-mentioned system of indicators and norms, which express the economic output obtained because of the achievements of science and technology; the incentive is also expressed in the recourse to quite severe economic sanctions when the indicators and norms are not respected. For this purpose, the system of economic levers is used to accelerate technical progress and this comprises funds for incentives at the national economic level as well as special incentive funds.

A particularly important role comes up in the following problem, namely, how to make the manufacture of machines profitable for the enterprise (group), which can have a very great effect on the national economy as soon as their production has been launched, done so that the enterprise does not suffer inevitable, though temporary, losses following the initial stage of production.

The essential thing is to guarantee the enterprise the technical conditions to reduce the troublesome consequences arising from the introduction of new techniques to a minimum and overcome them. Thus, it is necessary to compensate for the socially necessary expenses arising from getting production underway in conformity with established norms. Provided that the enterprise is working well, such a compensation must guarantee not only work without deficit in this period and normal profits, but also the supplementary profit as a special incentive for the enterprises producing the new technique of high quality. The essential source of the supplementary profit must be the increase of wholesale prices of products of higher quality.

The conditions necessary for increasing the role of prices in the acceleration of scientific and technical progress are the following: consistent application in price formation of the principle of price reduction per unit of useful end-product obtained by using the new technique; the dynamism of the efficiency of the new technique must substantially surpass the dynamism of its price, which would make its use more advantageous and increase the importance of the economic incentives used to decrease production expenses; the timely and progressive lowering of the prices of old machines for their producer; the strict respect for this state of things where the projects of the new technique are not valid where its maximal co-ordinated price is lacking; the impossibility of arbitrarily lowering prices (then, the projects of the enterprises) of efficient new machines and of refusing the producer, by means of the expedient of prices, the 30–50 per cent of the output obtained because of this technique; the transition to the system of graduated price reduction.

This system was introduced gradually into the Soviet Union, first to stimulate the production of the most efficient new technique by raising the wholesale price of products of superior quality. As such an increase is temporary (about three years), it is in the interest of the producer of the new technique, using all possible means, to accelerate its manufacture in order to be able to produce the maximum number of products of superior quality, thereby justifying the price increase. Furthermore, this increase encourages the producer to perfect the model so that he claims higher quality for the longest possible time in order to be able to maintain the right to the price increase for the longest possible time. Second, when the right to the price increase is cancelled, i.e. when the price is lowered for the first time (the new technique being considered as the first category of quality), the producer is economically obliged to obtain a sufficient profit within the framework of lowered prices by expanding production, that is, by accelerating the mass distribution of the new technique. The producer also has an interest in producing the new technique to the maximum extent before an exceptional lowering of the wholesale price, called a 'forfeit', is applied, following the technique's ageing. Third, lowering the price urges the producer to abandon the manufacture of the old technique.

In order to increase the role of profit and net profit as an effective means of accelerating technical progress and raising its efficiency, one has recourse to the factor analysis of profit and net profit by evaluating the work of the enterprises (groups) as well as the formation of incentive funds. Factors such as efficiency, decreasing expenses following the utilization of new techniques (which must be encouraged, naturally) are separated from factors such as exaggerated price increases, changing the range of products, and using costly raw materials in the interests of the national economy.

The role of the first plan amounts to the transition to incentives for enterprises and their personnel based on the definitive results of their work and the time allotted for their attainment. The quantitative results must be taken into consideration and stimulated only in relation to qualitative factors (efficiency, quality of production, management of expenses).

An important economic factor contributing to the organic fusion of science and production is that of incentives for research-and-development groups based on the output obtained in the production which uses the final results of their work. This factor is linked to the development of ever-closer ties based on economic agreements between research-development groups, technological groups, experimental and mass production, accords which take the place of their economic and juridical particularities within the global 'science-production' cycle. The financial autonomy of applied research institutes, the development of the system of economic accords with clients are as much the necessary conditions to create close ties between research, design, and production groups. It is important that the balanced management system does not weaken research but, on the contrary, contribute to its development by using special funds to this end. The particular role in the integral management of the 'science-production' cycle belongs to the establishment of a unique system ('integral') where work would be carried out by all the groups of the cycle (at the inter-sector level and within the sector and group) based on the manufacturing orders which specify the conception, experimental introduction, and diffusion of the new technique with consumer participation, and which would assure the unity of material, technical and financial supply, and would create unique funds for the development of science and technology.

Manufacturing orders are thematic charts (technico-economic) which show the sequence of work to be done by each of the producers in conformity with a work schedule, from research to the mass utilization of the new technique under the direction of the research institute. The orders unite all the groups of the 'science-technology-production' cycle with a task for each group that is at once common and distinct because of each group's particularities.

The manufacturing orders must include all the work following the theme, plan and dimensions of their realizations; they must forecast the final results; the economic output guaranteed by the utilization of the end product in the national economy; and the time and expenses necessitated by working in stages; they must have the juridical force of economic accords.

The management of the 'research-production' cycle demands better co-ordination of the activity of the research-development and production groups within the organizational structures and their concentration. In the Soviet Union, we are in the midst of this process of reinforcing the organic ties between research and production at the level of industrial branches and groups with the creation, if necessary, of hitherto non-existent groups or the reinforcement of existing ones. Large groups are attached to organizations specializing in research-development, study, technology, such as the experimental factories. This is accomplished first of all in the branches which play a determining role in the acceleration of scientific and technical progress.

The processes for the structural co-ordination of the research-production cycle develops particularly quickly at the level of the groups. Giant complexes are created which are composed not only of units of production but also of research development, pilot research departments, and factories for experimental production of the new technique; other groups make every effort to develop research groups.

On the base of the principal research institutes and departments there develops a network of research-production groups composed of research-development groups and experimental production groups, which assure initial mass production and the introduction of the new technique. The direction of the 'research-production' group by an important research institute must substantially stimulate the work of conception and production of the fundamentally new, efficient technique. In the final analysis each division and subdivision of an economic branch must have large research-production groups.

At the same time, the network of research institutes develops with all the means necessary for study, technological research, and experimental production.

Gradually, research centres are developed for each technical division with the sectors and subsectors. The role of such centres is assumed by the most developed research-production groups and the principal research institutes. To a certain extent these centres must be responsible for: (a) raising the technico-economic level of production within a given division; (b) shaping science and technology policy and the long-term forecasts for the development of science and technology; (c) working out concrete, long-term research-development projects; (d) using inventions of great importance in the national economy; and (e) shaping technical and economic demands with regard to new products and technological procedures which can assure great efficiency for the national economy.

The centres must carry out research which will make possible the creation and launching of new types of products; the introduction of technical material and procedures; fundamentally new types of equipment; and the accumulation of scientific experience capable of being used in the future. The creation of research centres must contribute to the concentration of material resources, the most important axes of scientific and technical progress.

Of considerable importance within the groups is the concentration of research-development and divisions concerned with experimental production at the heart of the research centres, which co-ordinate the efforts, radically perfect technique and technology, and contribute to their immediate introduction and utilization. This presupposes the creation at the heart of the groups of common services for integral planning and management of scientific and technical progress, services which must guarantee the co-ordinated activity of all the technical departments, the co-ordinated work of all the technologically advanced builders and engineers who must situate the perspectives of technical progress in production and raise the responsibility of the producers of the new technique to attain greater production efficiency.

The growing importance of basic research goes hand in hand with the reinforcement of the practical ties between science and production, economic development, and social problems: such is the law of the scientific and technological revolution.

In order for science to be able to play a pioneering role in production, a determinant role must be reserved for basic research, which assumes the guiding function within science. In this connection, there is another law of the scientific and technological revolution—the growing cohesion between basic and applied research. The guiding role of the former is the first and one of the most important conditions for the accelerations of scientific and technical progress and the growth of its effectiveness in social production. The success of the applied sciences depends on their close ties with production but also on their continual efforts as much as possible to use the results of basic research which can be utilized in the field of applied research. This allows applied research institutes to promote the latest achievements of science in industry and to play the guiding role in the concrete realization of technical progress.

At the same time, one of the logical consequences of the scientific and technological revolution is the intensification of applied research on the part of basic research institutes, i.e. their direct intervention in the shaping and introduction of the new technique (if it is a question of important and basically new works).

Such a practice in no way restricts fundamental theoretical and basic research, which remains their primary task. At the same time, applied research and study in such institutes are very often a necessary and very important factor in obtaining the important basic results. The close ties between basic research institutes and applied research institutes have always made use of this factor. The development of basic research (that which can be realized in the immediate future in technical projects) carried out by principal research institutes belonging to the sectors of production, is called upon to play an important role in the acceleration of technical progress.

The co-ordination of basic research, the guiding, driving force, with applied research, the technical projects, and with the final material realization of science and its utilization in the national economy, has primary importance for the entire system of the integral management of the 'science-production' cycle, for its acceleration and for the consistent realization of the scientific and technological revolution. This demands the creation of well 'regulated' mechanisms which, on the basis of long-term plans, would continually assure the systematic realization of the results of basic research by applied research, research departments, technological departments and experimental production, and the sectors of the national economy; the obligatory plans for the accumulation and opportune utilization of research results are imposed for the creation of new techniques at the inter-sector and sector level.

Today, considerable importance attaches to material and financial provisioning, to the organization of work and the formation of the cadres necessary for the industrialization of the results of basic research; to the condition that this work is well organized and that the forms of fruitful practical co-operation between the basic research institutes and the applied research institutes and production is perfected; and to the opportune implementation of the results of basic research by the institutes belonging to a branch of industry and by the production units with the assistance of the basic research institutes.

It should be noted that the institutes of the Academy of Sciences of the USSR and the university research centres maintain increasingly close relations with the sector institutes, the large pilot research departments, experimental production, and directly with production.

In the Soviet Union today particular importance is attached to (a) the elaboration, based on fundamental research, of a long-term strategy of scientific and technological progress and of forecasting the future of production; and (b) the definition of the socio-economic premises and the results of the 'horizontal' and 'vertical' diffusion of the scientific and technological revolution. This requires consideration of the probable dynamism of: needs and resources; expenses; efficiency factors; demographic growth; increasing material well-being and raising the cultural level; education; formation and recycling of qualified manpower; changing the character of work and the level and mode of life; the problems of ecology and ergonomy; and the dynamism of the forms and methods of management, etc.

The global solution of scientific and technological problems, on the one hand, and socio-economic ones, on the other, confirm in an eloquent way the growing unity of the natural, technical, and social sciences and the tendency towards the integration of their research. This co-operation between the sciences (and at the same time, their specialization) is one of the conditions *sine qua non* of the integral management of the 'science-production' cycle.

The scientific and technological revolution is not only the highest form but also the most important trait marking contemporary scientific and technological progress in general, its character and its tendencies; it serves as the basis for scientific and technological development strategy and for putting its accomplishments into quick and mass practice. But it would be wrong to consider the scientific and technological revolution as well as other traditional forms of scientific and technical progress in isolation. This revolution results from a set of qualitative changes taking place in science, technology and production. It guides scientific progress but at the same time it depends on it, gradually extending its influence over all the sectors and types of production.

There are a number of intermediary stages in the transition from traditional technology to the technology brought to the world by the scientific and technological revolution. It is as a consequence of the development of such intermediary stages that the premises for the generalization of the scientific and technological revolution are created.

This is shown in the creation of a number of key industries coexistent and co-ordinated with traditional industries. In order to apply the achievements of the scientific and technological revolution, it is necessary to create favourable production conditions for it by radically perfecting traditional industries. Otherwise, the integrity of the technical re-equipment of the industry is impossible, which will check the realization of the latent economic advantages provided by new techniques and technology.

It is for this reason that the application of the accomplishments of the scientific and technological revolution in this or that branch of the national economy, on the one hand, and the substantial lifting of the scientific and technical level of the traditional industries, on the other, and their most effective integral combination in a given period play a primary role in shaping science and technological policy in the Soviet Union.

Among the laws of the scientific and technological revolution we can also cite the unceasing and accelerated rise of its importance for global scientific and technical progress (which also assumes the development of the traditional sectors). This appears in the priority given the development of new industries and types of productions created directly by the scientific and technological revolution; in its penetration into traditional industries and their 'internal' restructuring; in the changing character and accelerated rhythms of scientific and technological progress in general.

The science and technology policy of the Soviet Union aims at making this law more effective and, on this basis, spreading technology far and wide, which will surpass the level of the best international standards (bearing in mind the expenses, results, and time factor).

Today, the foremost aspect of long-term planning and forecasting is the shaping and perfecting of an integral programme of scientific and technological progress and its socio-economic consequences by scientists and experts. The direction of this work is assumed by the Academy of Sciences of the Soviet Union and the State Committee for Science and Technology. This programme would be systematically brought up to date in each five-year period so that the global document integrating the data of science, technology, the economy, and social progress is constantly active.

The establishment of the stages of the realization of the programme is very important. This must co-ordinate the strategic line with the immediate tasks which is important for defining the most rational proportions between the sums allocated for future problems and those facing us in the present.

The integral programme, like the Soviet Union's science and technology policy in general, emphasizes the 'horizontal' optimization of industries, that is, the choice of the most efficient proportions among those conforming to the interests of the national economy.

An important role belongs to planning of the 'programme-objective' type based on definitive results, by wide use of integral programmes of long-term scientific and technological progress set up at the intersector and sector level and aimed at the solution of social problems. As opposed to previous co-ordination plans, the programmes indicate concrete paths to their practical realization in production and the non-productive sphere. The programmes for scientific and technological progress must occupy the foremost position in the entire planning system, which demands that several organizational problems which are tied to the management, elaboration, and realization of the programmes, to the responsibility of the performers, and the distribution of resources be solved.

Today, the Soviet Union is applying itself to hastening the transition from plannning products and isolated technical measures to integral operational planning, allowing different combinations which assume broad technical and economic ties and which embrace different types of technology and entire areas of scientific progress. This stresses the intersector character of production planning, and the utilization of new technology, and hastens the process of forming large technological systems. The scientific and technical achievements are used to form territorial complexes of production.

We are hastening the transition from the creation and industrialization of isolated machines and technological processes to the shaping, manufacture, and massive introduction of machine systems, equipment, and apparatus of the highest efficiency.

The increasingly global nature of the 'horizontal' management of the 'science-production' cycle contributes to the perfecting of a unique science and technology policy, making it more justifiable and more operational.

The elevation of the role of social organizations in the integral management of the 'science-production' cycle is also characteristic of the management system in the Soviet Union. The technical societies and the Society of Inventors and Planners of the Soviet Union include millions of people. The network of social organization such as the departments of pilot and technological research, and patents; as well as different laboratories, exhibitions, inspection posts, groups for realizing inventions and rationalizing projects, etc., is extensive. Mass socialist competition is oriented more and more towards improving the indicators of scientific progress and raising production efficiency.

The shaping and planned achievement of integral management of the 'science-production' cycles, oriented towards the solution of social problems, pose new and higher demands for the social sciences and offer them new perspectives.

Science, technology, modernization and social change

A. N. J. Hollander†

University of Amsterdam, The Netherlands

Contemporary Western society is permeated by rationality as no other society has ever been. It manifests itself in many spheres of life but it is most spectacularly, most visibly evident in its technical equipment. The technics of man are always tied to his three capacities. Man wants to know and understand—he investigates. Man can also create fantasies, dream up solutions for satisfying certain needs—he invents. He can also realize his inventions, he can make things—he is *Homo faber*. Only man possesses these three capacities in one being. It has brought him a long way and he may go still further. But even if he does not, he has already made a world in which major influences are at work which will make it, at the end of the present century, quite different from the one we are living in now.

Modern urban industrial societies tend to depend increasingly on human ingenuity, particularly scientific explanations and predictions. Magic and religious beliefs persist, but are rarely used in determining practical activity. The modern, scientific perception of cause and effect, the precise sense and measurement of time and space, the systematic relation of past, present and future, the administration of large organizations and the maintenance of trading, banking, even diplomatic relations all depend on knowledge, stored in books, files, computers, contracts. Modern man has to learn this written culture, must know the ways in which it is expressed, the symbols with which to handle it, and nowadays he has also to be skilled in the electronic means by which he can store, recall and manipulate immense amounts of accumulated information. In modern industrial societies therefore, primary and secondary production have decreased in importance, as any statistical grouping of the gainfully employed in a modern nation, compared with succeeding decades, convincingly demonstrates. In such societies, marketing, financing, insuring, advertising, transporting and the production and distribution of new knowledge provide the fastest growth of new opportunities.

The 'knowledge industries', as they have been aptly named, that is those branches of endeavour which produce and distribute data, information, ideas rather than goods and services, account for a constantly increasing share of the total national product of modern nations. We need more and more workers in the field, and countless engineers, researchers, programmers, teachers, accountants are being trained each year in order to swell the ranks of the knowledge-workers for whom the demand seems to be almost unlimited.

Enlargement of the scale of enterprise and dwindling distances are bringing about major changes in the world economy. These are spearheaded by the multinational corporations. The contrast between the rich parts of the world and the poor ones is becoming more and more obvious and for all we know, the gap is widening. New technologies are giving rise to new industries, accentuating the division of our world into advanced and underdeveloped countries. Another factor reinforcing the trend is the ever larger reliance on the laboratory for providing the raw materials we need. Plastics have displaced metals and wood, the test tube is becoming more important than the mining of minerals. A new technology, based on our ability to design composite materials, tailored to suit specific needs, is already emerging. The old resources have not lost all of their former importance—far from it. Oil is still crucially needed, as the Western world has in recent years been forced to realize. Still, there is an undeniable shift: knowledge, acquired by organized training, is becoming a central economic resource.

In toto, this shift means an extremely important divergence from long-established patterns. It means that in our social, political, educational outlook we have to bear in mind that the development of the information industry is likely to go on for quite some time to come and that we shall have to adjust social institutions, social processes to the trend. It has enlarged the total productive capacity of advanced nations to such an extent that its masses have been able to participate in some measure in satisfying less than elementary needs, to acquire a decent life style and to be less preoccupied with that menacing ghost of a former century: the war between the classes. It may well have been replaced by the growing gap between the developed and developing parts of the world. In an objective sense, the difference is not new, it has existed for centuries. But the consciousness of it is new, now that everybody knows how everybody else lives. This, again, has been brought about by forces active in the Western world, not least by its achievements in the communications industry.

Here we meet one of the most obvious consequences of the triumph of the scientific, technologically oriented mind: enlargement of scale. A communication network encompassing the whole world has replaced the town crier and the local pamphleteer. The size of government buildings now would seem unbelievable to those who lived and worked a century ago. Since the early years of this century most of the social structures in modern societies have become large and complex. Their size and power bring many advantages. The co-operative endeavour of a large number of people make great projects successful. The pooling of resources, human and otherwise, has speeded up research and has turned new insights into technological ventures which otherwise might have been impossible to achieve. These, in turn, had an impact on society by enlarging the scale of existing social institutions. The pooling of means has also made it possible to divert energy and work capability from immediate activity to planning for the future, thereby increasing efficiency in the effort to reach significant goals. Enlargement of scale has also built up concentrations of power which have made the patterns of decision-making, the control of individual opportunities, the division of wealth and the distribution of political influence different from former times, and not to everyone's satisfaction.

Enlargement of scale has also been spectacular in the size of human settlements. Everywhere in the modern world the village and the small town is either stagnant or is disappearing, ever larger concentrations of people becoming the rule. It is probable that a sizable share of modern humanity will soon be packed in conurbations—or, as our American friends say, in metropolitan areas—similar to the great American city complex on the American east coast, beginning in Maine and extending to Norfolk, Virginia. It contains the major financial, educational and industrial centres of the world, it is vital to the communication network that binds the world together—it is an astonishing creation, repeated by the Tokyo region, by various areas in Europe, by similar stretches elsewhere in the United States. Our new rational, technical civilization has made possible and is itself influenced by a revolution in environment which was unthinkable a century ago.

In these enormous conglomerations, organizations have proliferated in all spheres of life. They must enable people to solve the problems of urban, industrial living. Industrialization is, of course, more than the spread of factory production. Although in its strictest sense it refers to the creation and integration of roles organized around manufacturing, it is in a wider sense also the expansion of all productive enterprise and of the mental attitudes that go with it, as the integrating factors in social life. It aims at rational decision-making and management, which often has to be bureaucratic, at optimum organizational size and efficient utilization of resources as common concerns. Modernization not only alters the scale, technology and tempo of life and work, but also changes the expectations that rule behaviour itself. Industrialization and urbanization involve men in relations in which it matters less who you are than what you do. Social interaction becomes largely temporary communication with strangers with whom one exchanges information but does not identify as real persons. For rural migrants to giant cities, it is imperative that they impose upon themselves this new urban, industrial discipline. They will soon learn that in their new environment the seasons mean less than they did in the country, that urban life depends on the clock, the factory hooter, the time-table, the traffic light. They may not take kindly to the precision and urgency of urban life, demanding a new personal and collective discipline, but they have now entered modernity and they will have to pay the price for it.

Studies of the scientific and technological revolution and what it has done in promoting 'modernity', tend to concentrate attention on urban life and pursuits. There are understandable reasons for this. Still, it may be well to point out that modernization has sometimes come about through the important increase of agricultural productivity and the rational exploitation of primary resources, as in New Zealand, Australia, the rural regions of the Netherlands, and in so many other parts of the world. The technical complexity of modern industry and the need of expert know-how sometimes oblige latecomers to rely on improvements in agriculture in order to provide necessary resources. Agriculture and mining will then leave old-established ways and will tend to become industrial in technique and organization, capital-intensive, mechanized and 'scientific', a fundamental departure from the agriculture which used to be the way of life of the great majority of mankind, having subsistence-living as its main objective. Regions that for some reason cannot make the switch to modernity may not be able to face the competition of the modernized area and disappear from market production. In many countries the process has caused one of our greatest economic and social changes.

In most cases however, it has been the industrial-urban sector, large and small, that have moved fastest from traditional to modern ways of living. As traditional societies modernize, old habits, old patterns of authority, old relationships and old values are challenged, disrupted and replaced. The transition has taken place with surprising speed. The technological society of high mass consumption has little in common with the traditional society from which it developed, yet the change has taken only a minute part of the total span of human history. In time the values of the urban, industrial way of life were accepted as normal by a majority and pervaded the village as well as the town. In a singular way, village folk, whose parents in their days of illiteracy relied almost exclusively on things heard, spoken and seen, i.e. on audio-visual communication of sorts, went through a phase of written culture which is now being replaced by one based on electronics, audio-visual again, although on a different level of sophistication from the old one. Where once reading and writing were out of their reach, these skills are becoming partly superfluous in an era of transistor radios and television.

The world-picture of modernization varies considerably. No common pattern is discernible throughout the world. The speed of change has varied between cultures. In the decades during which Japan underwent revolutionary change, Java hardly changed at all. Neither are mature industrial societies homogeneous. Each has its own distinct pattern of values, norms, institutions, and life-styles. Many studies have been made of the factors facilitating and hampering modernization of a given society. Still, we do not really know what type of social organization is conducive to social and cultural change and what type of organization resists alternation or produces only a low rate of change. We know that in traditional cultures methods of production are always part of an institutional framework within which innovation is frequently difficult.

Sociologists have long been preoccupied with those features of traditional culture and belief which affect the direction of change and the receptivity of a society to innovation. The picture is extremely complex. We have learned to recognize that traditional factors, which under some circumstances seem to create immobilities in social processes, that abort or minimize innovation, at other times or elsewhere can open the door to entirely unexpected novelties. The female physician made her entrance in Europe as part of a general modernization of professional life and emancipation of women. In India, I have been told, she rapidly came to the fore because of the lasting strength of an age-old taboo and the absence of rapid emancipation of women.

Commentators on the immovability of so many elements of traditional culture, would also do well to consider irrational resiliences in Western culture that oppose adaptation to changed circumstances. Even in the most modern social organizations little thought is usually given to the provision of avenues for social change. Everybody knows that things do not remain the same in a dynamic society, yet as a rule no one thinks of designing paths to influence an organizational structure towards change. Our rationality is not absolute, one sometimes wonders whether it is not still very spotty. In this connection it is, however, only fair to realize that by their very nature, large organizations cannot be versatile, and that they have limited flexibility. They are effective through their mass rather than through their agility and adaptability.

We know that history abounds with instances of spectacular and swift transformations. They are sometimes accredited to special social groups. We know that historically the social position or religious beliefs of some groups seem to have motivated them to take advantage of new opportunities. Calvinists, Jews, Parsis, socially peripheral groups such as the merchants of Tokugawa Japan, nationalists in colonial territories, have been innovators in otherwise traditional societies. We also know that the marginal man is by no means absolutely destined to become an agent of change. He may very well remain an underdog and as such demonstrate the extreme conservatism of very poor people who rightly fear that any change may make a life, which already is precarious, fatally impossible. We also know that innovators are often deviants. They operate in the face of opposition from agencies of social control. This does not mean that a deviant will be an innovator. In most instances he will just remain a deviant, at best a person who is only accepted with a social discount, at worst a criminal. It is impossible to make any predictions in these matters and even generalizations after the fact are hazardous. It has been remarked by a historian that in practice, historians do not assume that events are inevitable before they have taken place. It is well to bear this in mind when we are hoping for parallel developments. The factors influencing the probabilities of entrepreneurial activities must be determined anew for each subculture, every nation and decade. A low probability for one nation at one time is no cause for despair for other countries, nor for the same nation at another time. We can only say that innovating groups pressing for modernization have arisen from very different social conditions and have had a variety of objectives. Their success depends on the existence of a social order that not only tolerates their disruptive efforts, but enables them to strengthen their own position. A fairly common motivation of nationalists in developing countries is their view of industrialization as the solution to a number of problems raised during the colonial period. As they see it, industrialization promises to increase wealth, social mobility and opportunity. Education can then expand, national status improve, national dependence decrease.

Even when the efforts are even partly successful, a process of social and political mobilization occurs through a combination of agitation from subordinate groups and action induced by those in power. Many other processes will be set in motion. Partly, these will only repeat what happened earlier in modern societies but was then not yet scrutinized by sociologists. Should this have been the case, they would have diagnosed, at the time of happening, similar trends and processes as they are now registering in Africa and Asia: how the acids of modernity (as defined at the time) dissolve the authorities of former times; how people are becoming surrounded by necessities that they do not understand and cannot escape; the astounding malleability of culture, its capacity to withstand severe impact and absorb elements that one would have judged indigestible, the problems created by increased welfare, by increased speed of so much that used to go leisurely, the alteration of roles, the creation of areas of social tension and political striving that become nuclei for other political, economic, social problems; and the frequently painful adjustment to change brought about by a new technology, the surrender of old loyalties and ways of living, the difficulty of learning new ways, the changed relations of the sexes and of the generations.

The collision of the old and the new order has been noticeable in the protests and disturbances of one agrarian society after the other when they were first disrupted, by nativistic movements, by many forms of inarticulate protest and rejection. People will be taught that inactivity that used to be taken for granted is now being considered as leisure, if not as idleness. In eastern Bengal, the jute-rice farmer used to work only about four months a year. There was little he could do in the remaining eight months. He may now be informed of his idleness, of which he was never aware and which a new ethos now condemns. The worker in the expanding cities has to accept an environment that is impersonal, an assessment that is based on his performance and which has nothing to do with his status away from his work. The encroachment of the technical-scientific life-style does not make everybody happy. Personality problems become common in such periods of transition. According to some, severe mental disturbance in some of the members of a modern society is the inevitable concomitant of a truly and fully rationalized way of life.

Social and cultural change in itself creates difficulties for many reasons. The mere fact of having to change long-established ways of doing things, can be a source of difficulty because feelings of futility, instability and inferiority may occur. Traditional accomplishments worked out with some intricacy and devotion, sources of personal pride and badges of distinction, may become outmoded and may have to be discarded for new ways. The resulting frustration has been noted in all societies subject to rapid change, from the inhabitants of Papua to the old workers in Yankeeville's shoe factory. Older people may be unable to deal with new situations because of an insufficient capacity to anticipate unaccustomed unpleasantness; they may fail to set the correct goals, unable to discard traditional patterns of response after they are no longer useful, unable to adjust to the changed status and role of their womenfolk, unable to discontinue loyalties to groups that are being dissolved as part of the general change that seems to leave nothing untouched. The case histories of hill-billy families that have left their Appalachian homes for the large cities in the north convincingly and tragically demonstrate all these pains of absorption into modern urban society.

Forces bringing cultural and social change are perceived by individuals as overwhelming and impersonal. They are sometimes judged beneficial, sometimes hateful; they are beyond the control of the individual and may implant in him feelings of powerlessness and estrangement from a world he never made and which conditions his life in a hundred ways. He can do nothing about it, nobody can really do much about it even though some may seem to be leaders, taking destiny firmly in hand. They usually mirror change to which they are no less subject than everybody else and are no initiating force. Goethe recognized the illusion for what it was: '*Man meint zu schieben und wird geschoben.*'

In complex urban culture, one should control emotions and postpone gratification. Success in life, advancement, depend on the ability to avoid antagonism in avoiding tense situations. Still, the expression of emotion is a great satisfaction, and so is immediate gratification of desires. The new discipline is often difficult to acquire, interaction with others becomes uninteresting under the new restrictions, impersonality stamps responses and the modern urban dweller may feel himself part of the lonely crowd rather than a full-blooded individual. Little of importance is left to his choice or decision. He is now living in a set-up in which every single social task is taken care of by a large organization, his life has become passive in so many respects that the boredom of so many people fails to surprise.

The supremacy of technical civilization has indeed had a profound impact on the life of man. In former times when man grew up in a much closer relation with nature, he was constantly confronted by the vagaries of his environment. They must have been a tonic factor in his existence, no matter how monotonous this may now seem to us to have been. Living in the controlled, predictable, artificial environment of the city, much of this former stimulation has disappeared in the continuous, monotonous process of making a living in a office, spending the evenings and the nights in suburban surroundings, in a web of social relations which for many has become increasingly opaque and ephemeral. Much nonsense has been written about 'man in mass society'. There are great rewards in city life which no rural existence could ever offer. Yet it seems to me incontestable that there are reasons for the feelings of alienation, of the powerless ennui of man living in a technical habitat.

Could it be that one of the reasons for the striking need of large numbers of people for travel—an entirely new phenomenon in the history of man—originates in his craving for new experience, not provided by his daily existence? It is perhaps no accident that the nation that spearheaded modernity, the United Kingdom, was also the first one where the habit originated of travelling for pleasure, more or less regularly, by numerous people. The names of many hotels scattered over the continent still bear witness to this British proclivity. It is now characteristic of life in all Western countries, as are many other means by which the monotony of one's existence can be broken by controlled adventure. This again: predictable in its course, in its outcome, mildly exciting, but as safe as man can now make even his most exciting escapades.

One can speculate about such matters, one can also deplore the difficulty of making such problems, real or alleged, subject to concrete research up to established scholarly standards. Still, there seems to be no harm in speculations that may stimulate thought. If one pursues one's thought along these lines, various problems that have come to the fore recently, in Western countries, seem to fit well into the more general questions that a thoroughly technical civilization poses.

One instance could be the present concern with euthanasia. Death is one of the few remaining aspects of life in a highly technical environment that still has to be accepted as a natural phenomenon resisting all attempts to master it. It can be postponed by medical know-how, it can be brought about on a mass scale and in a time span as never before in history, but in principle we still have to accept death as one of those aspects of life where nature holds the ultimate decision. The unavoidability of death is, therefore, one of the weakest points in our technical civilization, because its ties with the laws of organic nature have as yet proved unshakable. Our technical civilization has, unavoidably one might almost say, created a new ethical issue in this sphere: the medical possibility of keeping incurable patients alive has increased much more and faster than our willingness to accept prolonged human suffering. This tension between two conflicting trends, together with the force of persisting religious beliefs, has created the problem of euthanasia as a very real issue in modern times.

Another example of the far-reaching implications of the environment towards which modern nations are now tending, is the changed position in total society of the sexes and the generations. I shall resist the temptation to consider here women's changed roles and functions, or to enlarge on the totally altered position and function of the aged. Let us restrict ourselves to a group that is nowadays demanding and attracting so much attention: the adolescents.

Modern man lives longer than his ancestors did, he comes nearer optimal physical development than he used to do, and available evidence tends to show that he also matures, physically and mentally, earlier than his forebears did. According to some, he also ages slower. The working life-span of man has thus increased greatly. In the technically and scientifically most advanced countries today, it is now twice that of a century ago and 50 per cent longer than it was half a century ago. This change has had many consequences. It has, for example, reduced the number of dependents per producer, in spite of the great increase of old people; and also in spite of the longer schooling extended to a constantly larger part of the younger age groups, and their later entrance into the work force.

Young people formerly were gainfully employed when they were 15 years of age, if not younger (and not considering the extensive child labour on parental farms which never appeared in labour statistics). Young people belonging to similar classes are now frequently not employed till they are 18 or 20 years of age. In the more privileged social layers, the education of a professional man may easily last till he is 25 or, in the case of a medical specialist, till he is 33. These extreme cases do not represent a large part of the gainfully employed population, but the fact remains that the increased total wealth and productive capacity of society, the lengthening of the working-life span, social and political democratization, made possible the up-grading of the educational level, which, in its turn made possible the arrival of the knowledge-worker, which ultimately underlies the increased sophistication and technical refinement of present society. It needs the intellectuals, it needs even more the hundreds of thousands who are not intellectuals, who are not culturally or scientifically creative: the well-educated, well-trained, well-paid members of the ever larger middle classes, the technologists, the staff members, the employees, all those well-informed 'white-collar' workers who, in modern society, have become the successors of the skilled artisans of yore.

Now, the crucial point is that the extension of years spent in being educated, meaning a postponed participation in productive life, and postponed granting of social adulthood, goes with the attainment of physical and mental maturity almost ten years before social maturity can be achieved—if we take the large college population in the United States as an example. This gap is typical, or is becoming typical of all advanced societies. The age group concerned is socially defined as 'adolescents'. Unavoidably theirs is a time of conflict between one's physical urges and possibilities, one's mental capabilities and potential, on the one hand, and, on the other hand, one's ascribed social role as a preserved and persistent child, as a non-adult. Certain things these youngsters are expected and permitted to do, others they are not; they constitute an unnatural subsociety, composed exclusively of contemporaries, within the total society, which does not quite know what to do with them, except perhaps to become more and more permissive towards the waiting generation. The transition to social adulthood is not always smoothly achieved. Some youth reject the lengthy preparation for society's highly specialized roles, as their childhood experience did not equip them with the strength and motivation to strive for delayed rewards. Feeling constrained, confused, superfluous and hopeless about future achievement, they often drop out of their educational setting, enlarging modern society's fringe of those who did not make it because they would not meet society's requirements for entering its mainstream.

A common result of failure to meet social expectations is hostility and aggression. The frustrations to which adolescents as an age group are subject may well be related to their penchant for protest, may contribute to juvenile delinquency, hasty marriage, excessive divorce, rejection of the dominant values propagated by adult society—all phenomena that so often seem to be the price societies have to pay for 'modernization'.

Quite apart from the special problems that beset adolescents, rapid social change in societies where 'progress' has become an ideology, generates wide-spread frustration in all layers of society. When change has become normal, continuation of the status quo is defined as stagnation. It becomes a danger to watch out for, old practices become suspect, everybody is supposed to feel committed not only to a new working and social discipline, but to a less immutable view of his world. Not everyone feels up to such expectancies. Many people like the comfort of the familiar, are none too ready to applaud the dynamic perspective and prefer peacefulness in their time. There are other general sources of tension and unhappiness in a society undergoing rapid change. In a complex society any social change is uneven in that it affects some part of the society more than other parts and frequently this lack of synchronization causes social dislocation, dysfunction for many who do not know what hit them and feel themselves ground down by impersonal forces.

Such social dynamics create 'problems' and modern society has adjusted itself to it by educating numerous problem solvers. Partly this reflects the confidence of modern man that any problem can be solved by finding the right answer to it. But quite aside from the fact that many problems cannot be solved—including the problem of life itself—the intuitively obvious 'solutions' to special problems are apt to fall into one of several traps set by the character of complex systems. An attempt to relieve one set of symptoms may only create new behaviour that also has unpleasant consequences. Then, the attempt to produce short-term improvement often sets the stage for a long-term degradation. Then again, the local goals of part of a system often conflict with the objectives of the larger system. Also, people are often led to intervene at points in a system where little leverage exists and where effort and money have slight effect. One always has to take account of the possibility of unintended and unrecognized consequences of social action. Any sociologist who has been called in to help 'solve' problems, in his generally accepted role as trouble-shooter where others have failed, can cite instances of the fallacy that problems can always be 'solved'. Many cannot and attempts to do so anyhow may cause acute unhappiness to those concerned.

Sometimes underlying causes are so general as to escape diagnosis. The growth of 'modernity' in society, with all that goes with it, usually means an increased demand on emotional control. This, and also the suppression of emotion which a complex, highly organized society expects, the postponement of immediate gratification for long-term goals necessary for a rewarding career, the resulting regression of feeling make life within modern social organizations difficult for many, as psychiatric experience and the registered increase of psychosomatic ailments would go to show. The individual in modern society is faced with a wide range of specialist structures and uncoordinated value systems. He has to make razor-sharp distinctions between the expectations posed by his many separate roles, has to obey norms that are often conflicting, has to select between alternative versions of right or wrong supported by different reference groups. There is the strain, imposed by geographical and social mobility, the loss of emotional props when everything from the past is being questioned. The extent to which modern man has learned to accommodate to this complexity is perhaps more amazing than the fact that many do not succeed in reconciling what seems to them irreconcilable and as neurotics become 'the step-children of modern society', to use the phrase of an American psychiatrist.

A generalized wish to challenge authority and to express protests openly seems to have become widespread in modern society even though the means of expression are often irrational. The techniques, so successfully used by youth groups in pressing their demands, have been emulated by others; the results obtained by such action groups have often been remarkable and their activity now seems to form part of modern life. As the behaviour of conflict groups has become increasingly aggressive, violence has become a social problem. No social structure in the present Western world is immune from it and all have experienced some form of it in recent years. It is no easy matter to do something about this. Even without the recent increase of violence, history shows that reduction and control of the human habit of resorting to violence, perhaps the transformation and elimination of this aspect of man's nature, is in actual practice an incredibly difficult task.

Conclusions or hypotheses based on our supposed knowledge of man's 'nature' are notoriously fallacious. Not so those that have to do with the nature within which and by which man lives. He may have emancipated himself from much that kept him subjected; as a living organism that wants to survive he must be able to adjust himself constantly to the changes in his environment, must adapt to his habitat. Each community, human or otherwise, must have regulatory mechanisms that operate to prevent fluctuations in its relationship from reaching an inviable extreme.

Modern man of all living organisms has seen fit to endanger his survival by maltreatment of his natural environment. His aggressive manipulation of nature, the damage which he has done and is doing to the ultimate base of his existence, is too well known to deserve more than passing mention here. But one cannot remain completely silent about it in any discussion of the interrelations between the scientific and technological revolution and the life of society, as is the aim of our present colloquium. The crucial factor in this connection is the technological revolution, not in some vague way any great change in man's mentality. Primitive tribes, peasants and the like are often pointed out as an example of a form of life that knew how to maintain an ecological balance. This is sheer nonsense. There is nothing new about pollution and despoliation. Man has always been a polluter. Neither did primitive man preserve his natural resources. Man's rape of his natural environment is as old as mankind. What is new is the enormous acceleration of the trend, the rapid enlargement of scale. Our power to build and destroy has become almost limitless, and now at last has come the realization that the technically escalated societies on this planet of ours will have to make their peace with nature and must begin to make reparations for the damage they have done to the air, to the land, to the water. Almost unnoticed for a long time, the action of modern technics has exceeded the power of nature to regulate itself. Nature has lost its supreme position, has become the possible victim of man. Thus, a central question now is: how can man continue to control a situation in which his technical proficiency has placed him? Overall trends do not seem to offer much hope for a fundamental change in the near future—to the contrary. Very few parts of the world remain that have not been adversely affected by man's operations. The supply of usable drinking water, good food, clear air, quiet, is becoming smaller and smaller, deleterious refuse is accumulating in the soil, in the air, in the sea. There are many areas of the oceans where the surface life of the water now seldom has a chance to remain free of pollution for more than a few weeks, or even days. That the very seas should be considered a wasting asset must surely be the essential nightmare of the despoliation of this planet which is perpetrated daily before our eyes, about our ears, inside our nostrils and during our sleep.

According to some, the greatest challenge of modern society now is to handle the transition from growth into equilibrium. This will be no mean task. The industrial societies have behind them long traditions that have encouraged and rewarded growth. The folklore and the success story praise growth and expansion. It is well, then, to remember that technology has not been able to solve all of our problems and has aggravated or created others. There is no certainty at all that technology will be able to keep pace with the wide range of problems that lie ahead. Besides, while modern science and contemporary institutions may have eliminated many problems, they also have the potential greatly to magnify man's mistakes as well as his progress.

We know less about the social function of science and scholarship than one might think. There is a sociology of science, some distinguished sociologists have given attention to it, and they have made worthwhile contributions to it. On the whole, however, the study of science as a social institution is an undernourished field. This, in spite of the fact that few will doubt that science is a most vital factor in the modern world. Some people even argue that all else in modern life depends on it.

There are areas of investigation that need to be explored by an invigorated sociology of science and knowledge. What factors influence the rate of change in particular fields of scholarly endeavour? What is the role of theory in trying to grasp a social problem and why is it that experience shows that structural analysis by itself can provide only limited and descriptive answers to such questions? What are the mechanics in academic life that are causing automatic expansion of knowledge? How do they foster specialization and how does the social expectation of success predestine so many gifted people to be narrow specialists? Does specialization lead to fragmentation and is it true that this tends to be self-destructive? Is the extreme specialization of our time tending to reduce the world of learning to a world of barbarians, with tremendous power in their hands? Why has some academic institutional behaviour apparently enjoyed great success, tended to ossify and become absurdly non-functional? Is there any certainty that we shall be able to expand *ad infinitum* our store of knowledge? How valid is the assertion of those who state that the intellectual life of man has now reached a level beyond which further advance is not only not possible but that retrogression and severe stultification are in prospect unless the things that are necessary for raising the present level are done? What are the consequences of the seemingly increasing social isolation of so many scholars? Why is it that a high level of inventiveness can have a high social impact in some cultures, a low one in others? Is the rate of the growth of science in present Western society perhaps too high and leading us to our doom? All these and similar questions can be asked, some of them are being asked. They constitute a legitimate domain for a special branch of sociology which can produce valuable insight into some questions that concern all scholars and are the scholarly concerns of almost no one. In this connection, one may perhaps include the all-important question once asked by T. S. Eliot: 'Where is the wisdom we have lost in knowledge?'

A flourishing sociology of science should also concern itself with anti-science and anti-technology as a cultural phenomenon. Opposition to the machine and reaction against technological supremacy on aesthetic, moral and philosophical grounds, as inimical to the quest for a 'good life', are as old as the advent of the industrialized way of life, as old in fact as the impossible dreams of utopians who see in the machine the key to a golden future. As regards 'progress' in general, there can be no doubt that the optimism of the eighteenth and nineteenth centuries, among intellectuals at least, suffered a setback of such proportions in the twentieth century that belief in progress as a social ideology has not recovered. In view of the enormous abuses of the machine in the Second World War, one would not be surprised to find that the last thirty years have produced an abundance of anti-technological thought. It has not happened; the mechanophobes are not numerous and this in itself is a curious thing. It is worth studying in a time when the dangers that beset *Homo sapiens* by what his rational powers have brought forth, are becoming clear to many thinking people.

In the irrational sphere the attitude of rejection of rational soceity is more than a century old. It has been a traditional element in the anti-Americanism of European intellectuals. As we know only too well, cultural anti-Americanism in the Old World usually rejected and feared as supposedly 'American' traits, what it already feared and condemned in Europe. It has for many decades been the most noticeable reaction against modern machine civilization.

It is a cliche to remark that trying to look into the future is a notoriously tricky thing to do for anyone attempting to understand the world he is living in. Still, if one accepts this challenge, one cannot exclusively move in the present—which actually is always the immediate past. Is there nothing in the way of reasonable expectations one can outline as probable for the near future? Such views are not by necessity sheer speculations, though one has to admit that they form a poor basis for discussion.

It seems difficult to contest that the new society now coming into existence will depend more than ever before on society's capacity to sustain creativity. The continued systematic, organized production and application of knowledge will be vital. Society will increase its providing of services quicker than its production of goods. The increased proportion of white-collar jobs and the higher level of formal education that will be required for the performance of many functions will mean an emphasis on brain, not muscle. Knowledge, capital equipment and management more than hard manual work will increasingly be seen as the key to efficiency.

The craftsman learned in many years of apprenticeship to perform a limited set of specific tasks very well, in one specific way, that would change little during the rest of his working life. Not he, but the technologist, designer, researcher, programmer who have acquired in some years of intensive schooling skills of a very advanced kind which he can apply to a great variety of tasks, will be the typical worker of the future. Work that involves hard physical labour is already becoming exceptional rather than normal and the trend will probably continue. Toil and thrift may well be declining virtues in an economy relying on capital investment and sustained by high consumption. In modern nations increasing emphasis has been placed on the ability to solve intricate problems and make decisions involving extensive information. This need for people who know and can reason gave rise to a noticeable preoccupation with intelligence testing and comparable techniques. The trend to concentrate on the individual's capacity to learn as well as on his educational experience is likely to continue and perhaps be strengthened.

The experience of the most advanced industrial societies indicates that unemployment will be inevitable among the poorly educated. The extension of educational opportunities to everybody will not make people equal. Innate differences in ability will always exist and the problem of inequality resulting from such differences will not disappear. Science can solve many problems but not this one; it is science itself that sets the conditions of this kind of inequality. Among the adequately trained no perfect equality will exist either. There will always be spheres of life and action where the non-expert has to stay out. Future society will probably contain many clusters of established privilege, of small but powerful groups of intellectually participating citizens who manage, direct, know best, decide. In the new social order, talent, ability and informational creativity will form a basis for ordering society. In this way, a functional establishment will come about, while society will cast out—but certainly will keep alive—those who according to new norms, will be superfluous. The trend may be one of those forces that will continue to steer society away from ever being a real and complete democracy. The nature of economic, political and social control will not cease changing. Already it has ceased to be an open subordination of one class by another. Many intermediate groups have expanded. Prestige, wealth and power no longer automatically go together as different interests organize themselves for exerting political and other pressure. Sheer wealth especially will mean less in terms of social influence than is the case already. The future establishment will, even more than it does now, derive its power from the expertise embodied in advisory and appointive posts, membership in overlapping committees, boards, other decision-making organizations. Access to these will become ever more important with the universal trend toward planning, calculation, rationalistic methods and goals.

Knowledge will be the base of the new society and it will have to be not only scientific, technical, economic knowledge. The need for creating knowledge of what makes society tick will be urgent and vital. As yet, we are not too good at it. The need for priority is great in the behavioural and political sciences, in sociology, anthropology and social psychology. Real talent for valuable work, the providing of new insights especially, is less common here than in the natural sciences. Accumulation of knowledge also seems to be a prerequisite in these fields and the development of the individual scholar to the level of optimal creativity is slower than in the natural sciences. This again limits the supply of good minds at any particular time. Yet, even today there is a tremendous need for social thought, knowledge and ideas. Real research techniques do not exist which could immediately tackle questions concerning social implications of technological advances. Social scientists note that despite their desire for scientific status for the social sciences, they do not themselves really believe it exists. This, despite their attempts to move from a more descriptive to a more analytical approach and also despite considerable achievement in attempts to make quantifiable what by its nature cannot be measured or counted. It is also sensible not to lose sight of the continued need for theory. Various trends—among which the ascendance of academic man to spheres of influence—make it likely that theory will be granted a central role as the great innovator, for the articulation of social aims as well as for the instrumentation of innovation. For similar reasons we will perhaps witness a proliferation of professional jargon, the use of which will also serve to indicate one's belonging to the group of inside-dopesters, or, in any case, to the powerful new group of intellectual technicians and consultants.

No one will doubt our urgent need for a development of the social sciences that will really provide us with the means to do something about a particular state of affairs in a given society. How ineffectual have been so many well-meaning efforts to bring about economic development in Third World regions! The contrast between exalted visions of the future and the extremely slow pace imposed by human and social realities has painfully demonstrated how insignificant our capacity is in manipulating social wholes living in the deadly shadows of the past. We simply do not know enough about the mechanism of social change.

I find it difficult to be either optimistic or pessimistic about the prospects of the social sciences in the coming quarter century. There is clearly a great and urgent need for usable social science knowledge, but that does not mean that the need will be fulfilled. Present activity seems to warrant the opinion, or shall we say the hope, that we can expect considerable advancement in understanding the complex dynamics of our social systems, but only with an effort that seems almost disproportionate. The application of valid new insight poses another problem. The spread of it, from the rarefied atmosphere of the scholars' study to a wide public, always takes time. This time lag is considerable and it, therefore, constitutes a problem of its own. The research results of today will only after one or two decades find their ways into the secondary schools, just as concepts of basic physics moved from research to general education over the past three decades. For those whose education has been successful in so far as they have grasped what their teachers tried to teach them, it means that their common-sense assumptions have been changed. Common-sense knowledge is not static, it is constantly being modified by the absorption and useful vulgarization of the knowledge of the academic avant-garde. What to people's common sense is self-evident today is, therefore, different from what it was twenty years ago in many crucial ways. It may be particularly noticeable in the physical science world, it is also to be expected and has, in fact, taken place to a large extent, in the social science world. The trickle-down process is making itself felt in race relations, housing, crime and punishment, leisure and other fields. It is important to watch the process and to understand it, because without a democratic base of common-sense acceptance, no new insights can find wide and easy application. Especially so, now that we have entered a time where science is not only valued as a means of getting to know the world, but that it is also conceived as showing us how to change the world.

The process points to an increased need and revaluation of a step-child of the community of learning. Rapid expansion of the outer fringes of knowlcdge, constantly refined technical aids and instruments, ever higher levels of abstraction are isolating the research scholar more and more in his society. Even in his circle, thought has in certain instances outpaced its means of symbolic expression, so that scholars can no longer communicate. Much less than before is the educated man without special training able to understand the tools and aids with which he is in daily contact. His alienation applies with even more force to the basic principles of top-level research. It seems that today an increasingly important social function can be performed by the scholar who is able and willing to popularize advanced learning. It is an activity that has never been prestigious but there is an increasing need for such creative contact and one can only hope that the adequate performance of the difficult task will receive commensurate social rewards. In the society of the near future, a social structure geared to rapid change under the influence of technology, science and innovation, such mediators will become indispensable.

Man's propensity for taking things for granted is a great saver of thought, it is conducive to peace of mind, but it is not always the wisest attitude. Thinking about the future of Western society is by no means a futile undertaking. There is no reason to assume that Western civilization as we now know it, will last forever. The chances that it will do so are small, in the light of historical evidence. I am not so sure that history has either logic or laws. But I do know that any conviction of one's own uniqueness is fallacious—one's personal uniquencess as well as the unique nature of one's civilization. Great civilizations have come and gone. Why should ours be exempt from this fate? The decline of past high cultures was probably caused by a combination of unfortunate factors, rather than being due to any single cause. If the community that 'is' the culture is in thriving condition, it can overcome widespread damage and adversity. If it has been weakened too much, it will succumb. A decline of culture once initiated in a complex society may be impossible to stop short of complete disaster. Every community is exposed to unfavourable factors that are many and powerful. Contemporary Western culture is not foreign to them. It is good, then, to meet them with awareness.

Social sciences as a channel between science and technology

Salah Kansu

National Centre for Social and Criminological Research, Cairo, Egypt

Science and technology are linked today as if they were situated as two points on the same continuum. In the absence of adequate and favourable conditions, this linkage is harmful for social development. Whereas natural science is neutral and non-ideological, technology cannot be so at present.

Science and technology must be separated now, and we have to declare the ideology by which we are using technology.

This tactical separation paves the way to establishing the strategic connection between science and technology whereby the solid foundation is to be laid.

The social sciences are the cement which binds the joint continuum so as to be directed to the human progress for both. For this, we need no ideology.

Thus, there is an existing pseudo-connection between science and technology without declared ideology (thesis). Such a connection must be negated at the very next stage provided that ideology is declared (antithesis). Then these are to be transcended to reach the real connection through the social sciences without ideology (synthesis).

Certainly, science and technology are related to each other. Without laboratory equipment there is no science. Without scientific knowledge there is no technology. Science and technology are working together within a shared cultural context as if there were an implicit plan for their evolution.

Although science is created and developed by cultural systems and institutions as well as ideological stands, it goes beyond these moments which are not equal nor which correspond to its specific activity. It has its own entity and course which produce a cognitive surplus value.

It would be better if we distinguish between two things: first, the cultural or ideological context through which research processes are developed. Second, the cognitive content of research.

Technology is social investment—achieved in different ways—of the results of natural science. It is quite possible to verify scientific theory but we do not have similar criteria to judge the purposes of using technology. The former criteria are objective, while the latter are ideological.

Obliteration of the difference between the two enterprises or, in other words, hiding the ideological choice at the present stage would lead science to 'scientism' and push technology to 'robotism'. Both 'scientism' and 'robotism' conduce to the alienation of science and man, where science and technology would be exploited for the interests of the enemies of human progress.

An intermediate phase should take its place between science and technology. Ideological commitments must declare the aims which will direct the uses of technology.

Such an intermediate phase opens upon the next phase, in which ideological separation is to be replaced by the social sciences as the connecting medium between objective natural science and purposeful technology. This step is not an idealistic leap, but it is rather a realistic evolution.

Technological revolution on the part of the progressive ideology must not be a spontaneous event which would conduct mankind to either a utopia or to some sort of 'Frankenstein'. Thus, it should be considered as a higher stage of the development of social determination which includes human consciousness and will. Therefore, technological revolution is to be subjected to planning processes: design, prediction and control in addition to learning. These processes are similar to the functions of scientific method: description (which corresponds to learning technology), interpretation (which corresponds to the recognition of technology as a stage of social development), prediction and control. Thus, technology is in need of social science in so far as the consciousness and commitment required for the taming and orientation of technology are the subject-matters of the social sciences.

Social science, therefore, is the only channel that connects science and technology for the following reasons.

First, it helps us to know the explicit or implicit forces which exploit scientific research for their own interests. It clarifies the role of the natural sciences as a principal force for development in such a manner that it reveals the assisting and resisting interests as well. Therefore, we can look at science within the context of the present conditions and problems, and we can anticipate the prospects for its future.

Second, social science studies are what sets society in motion and what drives it to development or deterioration through conflicting interests.

Third, social science follows up the diverse needs and tasks resulting from transformations of social reality. At the same time, it can specify the techniques by which we can achieve the proper tasks.

However, the subject matter of the social sciences is 'man—within society—facing nature'; hence, they are entangled with the involvement of the natural sciences with the concerns of social reality.

Social theory, the culmination of scientific achievements, is expected to be mixed with extraneous elements coming from outside science, specifically from ideology.

It is necessary for the researcher to have his own philosophy and ideology which enable him to produce contributions to science.

Now, how can we judge between conflicting theories? How do we verify them? Is social theory another kind of political party programme which is judged by the number of voters who approve? Social theory must be something else for it must appeal to the scientific model which demands objectivity, the crux of science.

There is no need to insist on the passive significance of objectivity, one which is equal to abstinence from bias and prejudice. The essential significance of objectivity is rather the positive one that means establishment of consensus among researchers. It does not mean abolition of contest; rather it seeks for the application of methods that judge between controversial ideas and theories in order to reach consensus.

The distinctive characteristic of science is its ability to allow the same conclusions to be reached by those who apply the same methods. In this regard, it differs completely from religion, art, philosophy and ideology.

The question which has to be raised concerning the social sciences is: Should we develop such science or not? The entanglement of social science with ideology must push us to fall between two stools.

Then, how can we disengage social science from ideology? Science and ideology do not share the same criteria, yet ideology penetrates into social theories both explicitly and implicitly as well.

There are many approaches by which to handle this question, such as those put forward by Marx, Mannheim, Myrdal and the New Left.

Marx linked objectivity primarily to class position. If a class is rising, then its ideology is objective, and if it is deteriorating, then ideology loses objectivity. This approach may be a good description of the problem but it is not a solution because we are still in need of an objective means to judge whether a class is rising or deteriorating, and to find out the stages of rising and deterioration.

Mannheim suggested that objectivity may possibly be achieved by the 'free-floating intelligentsia' who belong to a relatively independent social stratum which is not deeply rooted in the class conflict. In fact, this situation is interesting because Mannheim himself belonged to this social stratum, and consequently every word he uttered must be objective.

Myrdal discriminated between 'beliefs' and 'evaluations' both of which compose 'opinions' that result in social theories. Accordingly, the ideal solution for maintaining objectivity must be such that the research worker declares his beliefs and evaluations which will be put separately in two logical systems. This prescription may lead to a vicious circle because the researcher cannot, from the outset, sort out his biases and prejudices, factors which he cannot recognize. His inability should not be blamed as bad faith; otherwise it would be very easy to achieve an objective social theory if we would only appeal to good conduct, to standards of morality.

The New Left, or radical social scientists, do not admit this problem. To them, objectivity is not a legitimate quest. It is neither feasible nor desirable because the task of the social sciences must be confined to description and interpretation; rather it must be devoted to criticism and exposure. The social researcher has to choose an ideology upon which he establishes his partial analyses, including his social and political commitments. Only an external or transcendent observer, if any, would be able to achieve objectivity, the alleged demand of the social sciences.

This approach simply puts us outside the problem. Social theories have been doomed to be deadlocked as long as the scientific community cannot agree upon any method to substantiate the radical theory itself.

Denial of the disengagement of ideology from social theory means actually laying down one's arms, abandoning them to the enemy because then every viewpoint would be permitted according to its ideology, since there is no common ground for ascertaining what is right or wrong. We must inevitably look for a way to establish objective social theory.

Such disengagement is not identical to the denial of the relationship between ideology and theory. Also it does not claim neutrality and passivity. It only searches for a common language which is valid for the discussion of theoretical disputes raised by conflicting ideologies, for otherwise we shall confront a vicious circle with no escape except by means of propaganda, corruption and suppression. Preferably, ideologies and philosophies would be sources for social theories, but they are not of the same nature.

It is useless if we preach only abstinence from the inclusion of ideology in social theory. The proper way must be indirect, i.e. an opening or aperture which does not permit anything but objective scientific statements while ideology waits outside until it can be linked to scientific components so that they can penetrate through this aperture. This aperture is the scientific hypothesis. Scientific hypotheses may come from any kind of inspiration: religion, art, philosophy, ideology, etc. But they have to be formulated in such a manner that they are suitable subjects of test and validation in order actually to be scientific facts and theories.

Different ideologies should be considered as 'funds' deposited in banks, yet they are valueless in regard to science unless they are transferable to free currency, namely hypotheses, circulated within the scientific community. As a matter of fact, social scientists differ widely from each other according to their 'wealth'.

Out of an ensemble of verified hypotheses, regardless of their sources, objective social science too can emerge. It may validate a known theory, or initiate a new one.

This suggestion is a mere demarcation of the problem which needs further methodological elaboration through two proposals: first, social scientists should distinguish factual analytical units from total situations; second, they should also separate the level of description and interpretation on the one hand from the level of prediction and control on the other.

At present, the social sciences study total situations, even in psychology, but they deal with them as if they are factual units or real variables.

As natural science succeeded in the discovery of particles, atoms and elements, so social scientists believed that they have done the same. The only social science that did achieve this is 'linguistics' where phonemes and morphemes are the abstracted analytical units of any natural language. Any scientific statement or hypothesis should discover the relationship between such units or variables.

These units might be as strange for the layman as 'H_2O' is a strange description of water for him. There is no actual 'H_2O' (scientific water) in his usual life except distilled water, yet these units are the variables which constitute science.

While analytical units and their relationships are basic requirements for scientific description and interpretation, the total situations are the subject of prediction and control.

Social science failure in prediction, and hence failure in control, is due to the *mélange* of units and situations, because what happens actually is a complex of situations. Social prediction needs another kind of treatment and orientation. It needs some sort of mathematical permutation, and a range of combinations of various units and variables.

In this regard, the social sciences are similar to meteorology. This science depends on several other sciences which deal with many validated statements concerning variables like pressure, temperature, etc. When meteorology starts to predict, it transcends these individual facts, to reach total situations under different conditions. The more comprehensive the combination, the more precise the prediction.

The social sciences confront the same situation: the level of description and interpretation is similar to that of chemistry, physics and astronomy as a base for meteorology, and the level of prediction is similar to meteorological forecasting.

This distinction is helpful for the social sciences, for it makes us aware of what we are talking about.

Technological change and social order

Hideaki Okamoto

Faculty of Business Administration,
Hosei University, Japan

Introduction

This article aims to examine the interrelationship between the scientific and technological revolution, social change, and the social sciences, as they are implied in a specific social development proposal in Japan entitled *Informational Society in Perspective* which had been influencing the opinions of important decision-makers. In so doing, however, the present writer does not intend to focus on any unique aspects of the Japanese situation, but hopefully to extract broader implications for many societies. At the outset, a few preliminary remarks appear to be necessary in order to clarify the meaning of the basic terms now on the agenda.

Scientific and technological revolution

This we take to mean the very dynamic relationship between science and technology *and* production that have become unprecedentedly speedier and interpenetrating since the latter half of the 1960s. The core element in the process may be said to be the development of cybernetics in science. Already, it has been endorsing, together with the computer, the unmistakably profound impacts upon conceptualization and information processing in science, in technology, and in production. Since science thus came to be a part of the productive forces to a much greater extent than before, it came to be identified as a strategic agent of change in the economic and social development plans. The trend is also very clear in Japan. The white paper on science and technology issued every year invariably stresses their bearing on social needs and their relationships with economic and social development plans. And in many societies, one observes the emergence of goal-directed interdisciplinary areas like 'life science', 'atomic technologies', etc. And no doubt cybernetics and computer technologies stimulated development and progress in such fields.

Social change

In this article, the present writer takes 'social change' to denote changes in individuals, in organizations and in social institutions that are significant in their reciprocal relationships, i.e. the social system. In any society, the individuals, groups and social institutions each act as the agent of change of social relations. And naturally, this fact is salient in a society where the power of government is more restricted. But any society would require deeper understanding of emerging tendencies in individuals, in groups and in social institutions in order to obtain support for public policies and to implement them effectively. So one way of looking at the 'adequacy' of the economic and social development plan may be to examine the implied perceptions of the aspects of social changes relative to trends in social realities, as well as to examine the degree of congruence of the proposed programme with trends in real life. And these are what the present writer wishes to do, though inevitably at a higher level of abstraction and with tentative statements.

Social sciences

Here 'social sciences' is understood in the usual usage of the term. Its area customarily has been decided by the area of social action, like economics, political science, education, religion, and also by the factors influencing social action like psychology, sociology and cultural anthropology. The former may be referred to as institutional variables and the latter as behavioural.

At this stage, in order to explore the interlocking relationships between the scientific and technological revolution (STR), social change and social science and the role of social science in this spectrum, an examination is necessary to set a few propositions on the relationship between the STR and social change. And these may be stated as follows:

The STR ultimately originated in, and was stimulated and conditioned by the preceding institutional and behavioural variables, that are in turn the outcome of the antecedent social change.

The STR interacts with various aspects of concomitant social change and is influenced by them.

The STR supports the impact on the various aspects of the social system and may induce a qualitative change in the system. Social scientists can focus upon any of the origins, processes, and impacts, with their respective roles, such as in the applied, academic or critical traditions. But of course, one common social role for social science, is to contribute to deepening insight into social realities on the part of the wider public, by attempting to be as objective as possible under given specified conditions.

In the early phase of the process, naturally, concern as to the ultimate outcome particularly mounts, resulting in various pronouncements and large-scale predictions and plans. It may be that the policy-makers involved usually tend to be optimistic, while the reverse tends to be the case for those who are affected. And at such a stage of social science, the need for empirical and objective detection of the emerging or latent trends may be said to be great, regardless of the researcher's own personal inclinations. The present writer aspires to take this position at this stage to the extent possible.

The soft sciences

The cybernetics and computer technologies have had a great impact on science, both in the natural and the social fields, by generating a set of applied interdisciplinary policy sciences. Those that tend to be more akin to natural science and engineering include environmental engineering and science, systems engineering, human engineering, information engineering and sciences, medical and hygiene engineering, safety engineering, life sciences, etc. And then these are those that are more closely related to the social sciences covering such fields as behavioural sciences, management engineering and sciences, urban engineering, social engineering, transport engineering, policy sciences, educational engineering, etc. Applied economics, sociology and psychology also receive impact from cybernetics and computer technologies in their conceptualization of system and in their statistical methods. One important base for these sciences and engineerings is the 'soft sciences' as they are called in Japan, which do not have any specified object but comprise methods involving operations, i.e. research, systems dynamics, pattern dynamics, multi-variate analysis, simulation, network, PERT, delphi, cluster analysis and so on. And therefore, hereafter in this article, the set of goal-oriented applied sciences and engineerings given above shall be called the 'soft sciences'.

Regarding the implications for the customary social sciences, such soft sciences as the behavioural sciences, management engineering, urban engineering and social engineering, educational engineering, applied economics have the character of applied fields, although the latter group is usually much more interdisciplinary. And it may aptly be said that in the field of the social sciences, dynamic interaction among the pure, applied and practice is becoming more intensive and to that extent, the social sciences are coming to be a part of 'viable' social forces much more directly than in the past.

It is the extent of this development to which the present writer wishes to pay attention in what follows, and later to discuss their implications for social change.

Informational society

In May 1972, the computerization committee of the Japan Information Development Association published a report entitled *Information Society in Perspective*. Briefly, it contains the following major points:

Long-term aim: to develop a society which enables the intellectual creativity of all citizens to flower. The plan envisages completion of the basic computerization condition for such a society by the year 2000. And the intervening thirty years from 1970 was divided into two major stages, based on the perceived evolutionary stages. (See Table 1.)

The first stage (the dual targets): first, to cope with the mounting international and domestic problems resulting from the 'industrialization' of the 1960s. Second, to enable the society to evolve step-by-step effectively into the 'informational society'. And thus to knowledge-intensify the industrial structure through a series of informationalization measures involving computerization.

The third sector (public policy-oriented): the plan stresses the public policy-oriented development through the third sector which is neither private not governmental. The plan explicitly negates the present guideline-oriented approach, not to speak of the *laissez-faire* approach which the committee thinks does not fit in Japan. Some of the contrasts among the three approaches shown are given in Table 2.

The third sector may take the forms of public corporations or boards. But in order to avoid falling into the bureaucratic-

TABLE 1. Stages for computerization

Development stage	First stage, 1970–90	Second stage, 1980–2000
Main area of development	Society	Individual
Goal	GNW	GNS
Societal goal	Social welfare	Self-fulfilment
Main actor	Society	(Conditions of) human beings
Basic sciences	Social sciences	Behavioural sciences
Informational pattern	Problem-solving	Creative-type

administered society, three separate machineries have to be established. The first is the National Council on the Informational Society, with representatives of managers, union leaders, consumers, etc., to deliberate on national policies. The second is the information assessment machinery composed of scientists and third-party experts to review continuously the situation. And the third is the citizens' Policy Formation System by which citizens use the policy development modules to make proposals for the policy decisions at the national and local levels.

Programmes in first stage: the major projects with target year 1985 includes: a broad and long-distance medical system; broad pollution-preventive system; home-terminal experiment; MIS for small business; administration data bank; computopolis; brain group centre; Manpower redevelopment centre; computer-oriented education experiments, computer peace corps. Regarding some of these, a few remarks will be made later in this article. In passing regarding the peace corps, the plan stressed the need for simultaneous development of industrialization and informationalization in the case of developing societies, implying a differing opinion from the Rostow type of staging theory of modernization.

In August 1975, the Ministry of International Trade and Industry (MITI)

TABLE 2. Comparison of the three approaches to informationalization

	Public-policy oriented	Guideline oriented	*Laissez-faire* oriented
Major actor	Government	Mixed	Private business
Steering motor	Plan by policy	Guideline	Free competition
Orientation	Societal	Industry and enterprise	Commercial
Tempo	Speedier	Unbalanced	Spiral
Scale	National and international	Region and industry	Industry and enterprise
Industrial structure	Knowledge-intensive	Assisted industries	Service and leisure
Function	Computer utility	Industrial data bank	Commercial CAIV
Type of culture	Fulfilling	Welfare	Emotion
Merit	Creativity	Welfare	Leisure
Demerit	Bureaucratic control	Privacy (monopoly)	Unemployment and crime
Investment	National network medicine education international assistance	Transport pollution system super-market automation	MIS commercial resistant to informationalization

issued the 'long-range vision' of the Japanese industrial structure. Under this, the relative share of the government capital formation within the total national spending will increase from 8.6 per cent in 1970 to 10.3 per cent in 1985. Adjusted to the price of 1970, the increase means three times the amount, i.e. from 62,000 billion yen to 192,000 billion yen. Just what per cent of this will go for the informationalization is unclear. But in view of the past accelerating ratio, it may be safely concluded that the 'plan' is on the right track. Knowledge-intensification is also clear in the envisioned industrial composition of the working population. The portion of the primary sector will decrease from 17.4 per cent in 1970 to 8.9 per cent in 1985, that of the secondary will increase from 35.2 per cent to 37.5 per cent, that of the tertiary from 47.4 per cent to 53.7 per cent. But within the tertiary, services will increase and not commerce. And within the secondary, the share of heavy-chemical industries like textile, paper and pulp, chemical, steel and non-ferrous metals, etc., will decline. Instead electrical, mechanical, metal fabrication will show some increase. MITI estimates that by 1985, university education will cover 50 per cent of the contemporaries at the entering age.

Regarding the degree of computerization, the total set of all types increased from 1,937 in 1966 to 6,718 in 1970 and 23,443 in 1974 and are estimated to increase to 83,000 in 1980 and 167,000 in 1986. The demand for systems engineers and programmers was 197,470 persons in 1975. It is estimated to increase in 1980 up to half a million.

HEALTH PROBLEMS

The plan anticipates approximately five medical information centres in each of the fifty-two prefectures, each with an emergency medical information subsystem and a periodic health examination subsystem and wherever necessary with the special subsystems for the remote area. Another project involves automated hospital service with a medical engineering centre. Like elsewhere, these are as yet at the experimental stage, but will probably spread rather fast. Although the average longevity reached the comparatively higher level of 70 for the case of males and 75 for females, the number of patients continued to increase. It was 5 million per day in 1962, 6.6 million in 1972 and if the present trend continues, it will be 8.6 million in 1985. Against this background, the medical information systems and automated hospitals will surely have merits, though only potentially thus far.

Although, to be fair, one should note that there has been a somewhat greater readiness among people to go to the hospital, there has been an unmistakable tendency towards an increase in the number of patients. And the important point to be raised here is that the tendency is related to the STR and the concomitant social changes. Progress in medical technology and public hygiene policies reduced tuberculosis and other contagious diseases with a resultant marked decline of the death rate. But on the other hand, the so-called 'civilization diseases' like high blood pressure, apoplexy, diabetes and neurosis increased very sharply. And it is said that there are considerably many 'half-diseased' people. Some estimate that the number amounts to more than 20 million, and while this latter figure can be disputed, there had been again an unmistakable tendency toward a decline of the health standard. The Ministry of Welfare attributed the causes for the phenomena to four factors: (a) insufficient exercise; (b) overeating; (c) mental stress; and (d) environmental pollution. And even the first three of those are related to the STR and social change.

To make a long story short, such variables as the knowledge intensification of work, individuation in work and leisure, meritocratic tendencies in organization, and changes in the social relations in the community, etc., are of considerable importance. Out of these, the knowledge-intensification of work tends to be the generally important variable. Obviously, the tasks become much less labour-intensive and much more mental-intensive. In the case of typical office work, the required level of energy consumption during working hours was in the mid-1970s 2,500 calories but it declined to 2,300 calories. In the case of rolling-mill workers, it declined from 3,700 calories to 2,500 calories and in the case of farming, it declined from 3,500 to below 3,000 calories. In turn, while the feeling of alienation at work in the traditional sense declines, the knowledge-intensification is accompanied by other problematic mental problems: (a) psychological distinction between work hours and non-work hours becomes blurred; (b) mental fatigue increases but this is cumulative; (c) the discrepancy between ability and requirements becomes more likely, but this is difficult to identify; (d) the expertise comes into greater conflict with promotion and transfer possibilities; and (e) performance is more influenced by organizational variables. That these are important causes for growing mental stress is indisputable in the opinions of many. We will speak about the other variables later, but for now the important point is to be conscious of the relationship between disease and the newer life situations.

To what extent medical informationalization would serve to counter the health problem of knowledge-intensification is as yet difficult to judge. But it is clear that for such phenomena, preventive measures are extremely important, and the involvement of such social scientists as psychologists and sociologists is also necessary.

And it must be mentioned that these are precisely the areas in which Japanese society has shown an unmistakably clear backwardness as compared with other advanced countries. But to turn to the broader implications, one should note the increasing need for the active involvement of the social scientists in this area. The newly developing area of 'life sciences' would be a fruitful one for the behavioural scientists to move into. Fortunately, the white paper of the Bureau of Science and Technology is beginning to call for this, but only recently.

The pollution problems

The 'plan' envisages the completion of the regional pollution-prevention system for major problematic areas and all the large metropolitan areas by 1980 and, by 1985, a National Pollution Information Centre will be founded. The regional system would set various indicators for the factories and other places linked with the system of prediction, warning and control. The metropolitan system will take similar actions in respect to many polluting factors other than, say, the current air pollutants. The national system would deal with the nation-wide trends as well as with prediction and control by the super-computer, with ecology models.

Because of the serious situations caused in the late 1960s in many industrialized and urbanized sectors in Japan, it may fairly be said that the informationalization is relatively advanced. Although pollution problems are serious, studies in both natural and social sciences for this area are at present shifting from pollutant-measurement and short-term action-oriention. In other words, it is recognized that pollution continues to be serious without having proper treatment and actions on the eco-systems in localities. Naturally, the pollution arises as a result of the interaction between natural conditions; soil, air, botany, animals and social system with both of the institutional and behavioural variables, the economy, and other social-ecological relations. Measurement and analysis of this latter aspect are extremely important. Although this latter aspect appears to have been overlooked by the 'plan', any effective large-scale pollution-preventive system will probably soon demand the active involvement of economists, human ecologists, urban sociologists and other social scientists.

Work and leisure

The 'plan' envisages that about 1984, the home terminal will become marketable, in view of the long-range income trend and the declining tendency of the cost of information relative to its performance. But given the present conditions, my own calculations tend to conclude that the appearance of the home terminal on the mass market will be much later (by about ten years). But surely by 2000 it will become widespread among ordinary homes. The implications of home terminals for individual and social life is yet unclear, but some of its significance could be speculated in respect to growing trends in the individuals' lives.

An important contingency of the STR is the marked reduction of working hours. In 1965, the average Japanese worker worked 233 days a year. In 1975, it was reduced to 229 days. But it is expected to decline to 227 days in 1985. In other words, the average worker will have one non-work day in every three days. And according to my own estimate, the free days will probably be more than this, due to the probable need for work sharing. The threat of unemployment will keep reducing the working hours. Hereafter the process may be quicker in Japanese society, because the employment needs have historically been respected by every citizen, including the policy-makers. So we will probably have significantly more free days than one in every three days. Together with this, the Maslow-type hypothesis became persuasive in a Japanese setting, as elsewhere. Increasingly, work and leisure come to be seen as the occasion for existential fulfilment; the social meaning of work and leisure has been changing, both are valued positively and are coming to be valued in both positive involvement and autonomy. And along with this trend, individuation, and, therefore, the diversification of interest and concerns, have been the growing trends. Of course, there have been the differences among socio-economic groups and generations. The tendencies may be said to have been most conspicuous among the 'knowledge-workers' and among the younger generation, but they are also occurring among the rest.

One can be fairly safe in these statements, because these are areas in which behavioural sciences most conspicuously contributed to enable the people at large to see some of the changes occurring inside themselves. And along with these, changes have occurred in affiliated needs and relational values: loyalty to any specific organizations and social groups have tended to decline, while relational values became more universal than particularist values. Broadly, one may also add in this latter respect that achievement became more valued in social relations than ascription and membership. Although there have been attempts to re-enforce the loyalty on the part of many organizations though reparticularizations and reascription with notable results, the latent tendencies towards individuation, universalist and achievement orientation appear to have grown and are likely to continue. It must be noted in this connection that the trends in needs and values have the potential to generate social alienation on the part of less adaptive personalities in respect to habit, intelligence and other characteristics.

Social alienation in this context tended to be significantly greater among the middle-aged and elderly. This is in part due to the changes in the family life cycle. The prolonged longevity creates old-age problems both for individuals and for society. And this became one of the most important current social problems. This is one of the least studied and least understood areas in the social sciences.

The midde-aged have a set of serious problems. While they are naturally to share the responsibility for the aged, they are uncertain whether to expect their sons and daughters to have their share for their parents. In addition, the middle-aged have the burden of prolonged schooling of younger generations, while continuing to pay for their mortgaged houses. At the place of work, they tend to be less adaptive compared with the young who are flexible and better educated, so that wage differentials among the age groups diminished considerably in the last decade and the trend is likely to continue.

In retrospect, the STR historically created similar problems. It changed the family life cycle and it created the declining status problems among the middle-aged. Each time, some collectivist measures were needed either in forms of charities, or poor laws or welfare policies. So, among the middle-aged particularly, there tends to be the quest for recollectivization, reaffiliation, re-ascription and more re-particularization. And this appears to be one major cause for the turn of national policies from 'industrialization' to emphasis upon 'welfare'.

Just what the home terminal will do to the individual is as yet difficult to answer, because experiments are being conducted among knowledge-worker types of families with as yet no involvement of social scientists. It appears to be congruent with the major latent values in society. On the whole the reactions from the tested families are reported to be good. But there is a possibility of intensifying the value conflict of the kind mentioned above that may counteract the welfare demands.

Organizational life

The 'plan' envisages widespread application of the so-called MIS (management information system). The degree of its application to the Japanese factory may be said to have been comparatively high. The plan sees its extensive application to the middle management level by 1980 and to the top management for decision-making by 1985 and to the multi-national firms' policy decisions by 1990. To see the plan's relevance and probable consequences, one must see the trends in organization.

Particularly within the private sector of the economy, the principles of organization have shown marked changes. It may be described as 'de-bureaucratization' or more specifically in staffing organizations. Stated in reference to the Weberian formulation of bureaucracy, the new trends have approximately the following characteristics:

From job to personal ability: the tasks are being assigned not in the form of job, but as a form of project designed to meet with one's ability.

From supervision to autonomy: the process of execution is being left to the individuals, but with accountability; so-called management by objectives quickly became widespread.

From feedback to feedforward: the decisions are not based on the precedent of pre-established manuals, but rather to be made by systems conccpts, utilizing information supplied by data processing.

From hard science to soft science: established traditional administrative disciplines like law, accounting, industrial engineering are declining in importance—instead, a series of soft sciences (systems engineering, material engineering, environment sciences, behavioural sciences, etc.) are increasing in importance.

From qualification to performance: educational and vocational certificates are declining in importance and, instead, the record of on-the-job performance and of experience in the firms' career development are becoming much more important. Higher education is losing its significance in providing occupational preparatory training; and basic vocational training institutions for the youth (as a public activity) became virtually extinct in Japanese society.

From job evaluation to merit rating: job evaluation lost its significance, but merit ratings become extremely important in running the organization as the tasks are coming to be assigned on the ability of the staff, but not as assignment to the bureaux (desks).

Again these trends are more conspicuous among the work places of the knowledge-workers, but trends are observable throughout the organizations. These factors deal with what is happening relative to the role system. Regarding the authority system, the policy decisions tend to be more centralized, but the executive decisions tend to be more decentralized. Regarding the reward system, while the Japanese pay structure is still isocratic by comparison (equality oriented), it is increasingly tending towards a more forward meritocratic direction.

The MIS would probably facilitate these trends. So far, the changes are most conspicuous in the role system. But if meritocracy continues to be strong, it would certainly aggravate the problems of the middle-aged and other maladjusted population. Ability retraining centres can be in part the answer for this, but only in part.

Urban life

The plan envisages the computopolis as a future city in that there will be: home terminals; computerized vehicle service; automated supermarkets; servo-ventilation; and local health administration system.

As for other urban problems, the plan proposes transport systems, material goods-flow systems, and local government administrative systems.

Urban life in Japan has attracted the interest of overseer social scientists in that the urbanization did not have close correlation with the suicide rate and that it also contained the communal social relations inside the city in the neighbourhoods. But this is no longer the case in the 1970s. And there have been attempts to create new towns modelled after the British concept; it is towards this new town that the plan is directing its attention, making it a 'computopolis'.

But, seen from the sociological perspective, the new town has not been a success. People become the movers, they change their place of living and also they travel farther in a day. What has been happening is conurbation. The regional life-span has expanded and it satisfies different needs in different communities. Communal relationships have emerged among the movers based on various moments of life; occupations, schools, hobbies, generations. And this calls for a new conceptualization of urban life.

The significance of small neighbourhood groups declined, and the life-span of larger-than-apartment blocks, or even beyond the territory of present administrative units, is becoming important; and as for community relations, recommunization of groups in associations has become rather marked. In these respects, the computopolis will show inherent limitations.

Other concerns

First of all, stratification; the plan conceives of the giant brain centre, with facilities also for civic participation. However, the needs appearing from the trends in social stratification appear to be of a different kind. Although the majority of population now come to have middle-class identification, as the basis of stratification shifts markedly from inherited wealth or property possession to merits in the broad sense, and as the average consumption levels rise, the organizational and occupational affiliations, as the symbols of merit, continue to be significant. Also the meritocratic tendencies in the organization tend to develop a sort of invisible stratification; the division into people who are fulfilling their lives and who are active and associative; and those who are in social alienation and passive and less associative. This would call for a sort of re-decentralization of brains and talents and re-decentralization of the opportunities for participation.

Second, education; the plan envisages computer-aided problem-solving, self-learning type of education, spread by 1985. Indeed computer installation is spreading fast in schools. In a few years, it will probably become a compulsory subject in senior high schools, and at the junior high-school level it will become an important area of people for extra-curricular activities. There are, of course, plenty of foreseeable merits. But the already emerging educational reform proposals, like a non-year grading system, dual progress plans, team-teaching, tend to imply a direction towards the separation of talented from those who are not. And if individual ability-based learning systems as valued by the plan joins with the proposals of reforms, the chances are that meritocracy will dominate the educational process and aggravate the very problem of social alienation in school that has been one of the most serious social problems.

Third, social administration; the plan envisages a government data bank and also data bank centres for local governments. The former was scheduled to be realized in 1976, and the latter by 1985. Each of the centres is to have policy scientists and participation systems. Also the data would be disclosed to universities, enterprises, unions, consumers, etc. If it could be realized, it would certainly contribute to democratization. However, the trend so far appears to run counter to the proposal. Each section of government separately establishes its data banks and the disclosure systems still have not been installed. And given historical centralism and sectionalisms in the governmental process, the proposed vision will meet with difficulties.

Implied perceptions

It seems plausible to say that the 'informational society in perspective' prepared by the leading managers, scientists and engineers in the soft sciences reflects perceptions of the social impact of the STR in the 1960s as held by a good many of the soft scientists and engineers; the STR under the 'market economy', however guided and whatever good it produced, was problematic in looking back at the 1960s. It has led to a life situation undesirable for physical and mental health. It produced by-products or even main-products, the polluting effect of which was not provable by established knowledge, but the ecology was effected. Therefore, the long-term health information system and the extensive and comprehensive pollution preventive system was proposed.

The STR tended to proceed along with the emergence of the monopolistic big businesses, competing with each other and with the gaints overseas. These firms prefer exclusive possession of inventions to the acquisition of patents, and, therefore, the factories began to move out of the national boundaries. Thus the STR indirectly stimulated the multinationalization of firms. The administration of businesses inside and outside became complex, and long-range planning became essential for investment in science, technology and production. Therefore, the MIS was strongly recommended.

The STR stimulated excessive urbanization by enlarging the size of headquarters, sales and research staffs and by reducing production workers in the local area where excessive dispersion continued. So the transport system and computopolis came to the fore. The STR exerted pressure against seniority-based stratification in society, and age-categorization of education institutions, and, therefore, the training centre with civic participation systems and self-development, with problem-solving types of learning, was suggested.

To say all this is surely a sort of story-making and is over-simplification, but it should show some aspects of the perceptions congruent with the good intentions relative to their positions in society.

The counter force

In the summer of 1972, the Society to Oppose National Back-numbering and to Protect Privacy was formed. The National Council of Japanese Trade Unions, All-Japan Telecommunication Workers Union and other large unions and groups of social scientists, lawyers and other citizens joined to protest against the potential moves to the back-numbering, to protect privacy, and to extend the right to know of the public. 'To protect privacy and the right to know' is the leading principle for the campaign as well as 'for peace and for democracy'. This body, which organized mass meetings in the Osaka area, is engaged not only in a protest movement on general and specific issues, but is also increasingly involved in the information process by calling attention to the computerization, now largely at the legal government level, particularly where progressive parties are in power. In addition, the society is being pressed to formulate an alternative proposal to the Information Foundation Bill soon to be proposed by the government. What happens in the future will largely depend on the final bill and its related legal framework.

Computerization involving economic and social planning at various levels of social life, however, has a historical momentum. The soft sciences are increasingly being used by the unions and other progressive forces too. It will no longer suffice, in formulating wage demands, to base arguments on trends in prices and family budgets or on the relative share of labour and capital. The white papers for wage offensives increasingly use data from the econometrical analyses and sociological insights as to the conditions and consequences of wage increase targets on members and on their populations, even basing their assumptions on alternatives in the economic and social development plans. The unions' counter-industrial policies, say, for securing employment and for protecting the interests of the middle-aged and the elderly, the victims of the changing life cycle, involve analysis of trends in marketing, investment, production planning and even trends in research and development, covering the wider areas of the management sciences. The organizers for unions and other civic bodies concerned with safety, health and civil rights are coming to have varying degrees of expertise in safety engineering, in ecosystem sciences and in urban planning. Regarding the behavioural sciences, the major unions under Sahyo (the General Council of Japanese Unions) grouped themselves together to found the institute known as the Labour Research Council in 1970, so as to reflect the changing needs, values, aspirations and social situations of their members. To date, the leaders in unions and mass-based political parties would say that the present effort lags behind in this respect, to counteract effectively the government and management, and therefore, there is the urgent need to strengthen these trends.

Soft sciences and social sciences

While expectations of the soft sciences are increasing from the various sectors of society due to these various reasons, there have been, it appears, renewed disappointment in, and attacks upon the social sciences from various sectors and for various reasons. And there tend to be somewhat uneasy relations between the soft sciences and the social sciences. Indeed, there has been a sort of mutual sniping between the two.

In part, this could be understood by reference to the differences in methodology. The soft sciences, when they are in action, have the characters of problem-solving application, futuristic types of orientation. In short, epistemologically, they are teleological, i.e. purposive and value-committing. But the social sciences tend to have differing traditions, they tended to orient themselves to value-free neutrality or to be critical, objective or provocative, empirical or historical. But given the trends of the soft sciences, the importance of the ontological traditions of being either value-free or critical, is in the position of assuming far greater social responsibility.

Research and development needs for solving global problems

John Platt
University of Michigan, United States of America

If we could first know where we are, and whither we are tending, we could better judge what to do, and how to do it.
Abraham Lincoln

Our order-of-magnitude technological changes since the Second World War are forcing all of humanity towards a more strongly interacting global society. If we can control these new powers without destroying ourselves, it will mean an evolutionary jump to a new mode of collective human organization. Already we see the reversals of many old attitudes and laws that had existed for generations. The changes are often initiated by catalytic analyses of our pressing new problems, amplified by crises with the help of television and the new mass media, and bring forth mass protest movements producing major reforms in just three to ten years.

As these pressures for restructuring move to different nations around the world in the next decade, we may expect a turbulent transition period. There are many potential disasters, from nuclear war and terrorism to environmental catastrophe and megafamines, but the mutual necessity for avoiding them or for dealing with them may lead to the creation of new international covenants and management mechanisms.

There is no time to lose in starting the technical and social research and development on new mechanisms and innovative structures for dealing with these crises and creating a less dangerous and more sustainable world system. Several special characteristics of these new social problems are listed here, along with a classification of a hundred areas that need urgent study, from the natural sciences to the social sciences and systems sciences.

Technology and the great world transformation today

The great technological changes and inventions since the Second World

War go far beyond the inventions and advances of all previous centuries. Look at the list: nuclear power, nuclear weapons, jet planes, intercontinental missiles, space travel, electronic computers, television, DDT, penicillin, oral contraceptives, new grains, new biology, new psychological discoveries, satellite communications, new global networks. They are so world-shaking that we can destroy ourselves with them if we do not agree quickly on how to manage them collectively; but if we can manage them, it seems almost certain that they will press us steadily toward a more stable and more interconnected global society.

From the point of view of the human future, all this adds up to a great evolutionary jump, like the coming ashore of the land animals, and it is all happening within a single generation. Our human institutions never had to deal with new powers and problems on such a scale, and the result is that they are all now changing before our eyes—farm, factory, family, school, church, corporation, city, national government, and world organization. The changes, because of their technological base, jump across national boundaries, and I think we can see similar changes around the world, even in countries with very different political and economic systems.

Is this waterfall of global transformation irresistible, or can we still choose among alternative futures? Let us look at some of these recent changes in detail, to see what we can do.

REVERSALS OF ATTITUDES IN THE UNITED STATES SINCE 1968

We can see dozens of reversals of old attitudes and laws in the United States and in several other countries since about 1968. They include:

Détente and trade agreements after 1969 between East and West, representing a reversal of the Cold War attitudes and recognizing the need for political coexistence on a finite planet.

Special Drawing Rights (SDRs), the first international money not based on gold or a national currency, recognizing that we are increasingly in one world economically.

New laws for environmental protection, including limits on some new physical or biological technologies on the grounds of human danger.

Changes in sex laws in the United States, Italy, France, and elsewhere, on homosexuality, pornography, abortion, and contraception, recognizing the individual rights of consenting adults, reversing the attitudes of Western Christianity for 2,000 years.

Reduction of birth rates to replacement level or below in the United States and almost a dozen other countries since 1971, by the free choice of hundreds of millions of families, recognizing a responsibility for population limitation.

Wide-spread legal reform in the United States and other countries, lowering the voting age to 18, decriminalizing divorce, car accidents, and drunkenness, and increasing the rights of prisoners and the mentally ill.

The greatest reform in the universities in a generation, in the United States and other countries, treating students as adults and as equals.

The greatest ferment and reform in Western religion since the Protestant Reformation, according to many observers.

The greatest reform in the rights and opportunities of minorities, blacks and women in several generations.

The greatest reform in American politics, in election laws and the public accountability of elected officials, in this century.

A turnaround in attitudes, as shown by surveys of public opinion, toward the 'Puritan' ethic of work and property, the benefits of science and technology, energy consumption, space exploration, the global future, the limits to growth, and systems analysis of global problems and alternative futures.

There are other areas, of course, where the changes are less rapid and dramatic. Some of these may be more closely bound to deep-rooted political or economic ideas, while others may change later, as this great historical evolutionary jump continues to unfold. But the list of major changes just given may represent the fastest and most extensive reversal of attitudes in peacetime in the whole history of humanity. It is worth examining in detail how some of these changes have taken place.

Catalytic→ Analysis	*Catalytic→ Crisis*	*Attitude→ change Key groups*	*Political→ Action*	*Administrative→ Legal Change*	*Follow-through→ New Pattern*
Environmental movement					
R. Carson, 1962	Oil spills, DDT	Sierra Club	1970 elections	Pollution laws 1972, etc.	Environmental Protection Agency
Auto-safety consumer movement					
R. Nader 1965	Corvair danger, etc.	Consumers Union	1970 elections	Safety and consumer laws	State and federal agencies
Birth-rate reduction					
P. Ehrlich, 1968	Pill and abortion contro-versies	Zero popu-lation growth	Birth-control clinics	Supreme Court decisions	Births below replacement level
Limits-to-growth ideas					
D. Meadows, 1972	City costs, smog, energy crisis	Ecologists	City councils	City-state laws and policies	Limits in many cities

CATALYTIC ANALYSES INITIATING A HISTORICAL SEQUENCE OF REFORM

There are undoubtedly many precursor conditions—technical, social, educational and communicational—which must be met before such rapid changes can take place. But once these conditions are met, a number of these reforms show a pattern (see table on page 343), of an initial 'catalytic analysis' followed by a rapid step-wise evolution of public support and change.

In the creation of the environmental movement, for example, Rachel Carson in 1962 dramatized the problem of environmental pollution with her book, *Silent Spring*. The title refers to the mass killing of birds by DDT and other pesticides. This was a 'catalytic analysis' which assembled the evidence and clarified the problem and suggested persuasively what could be done about it.

But such analyses, even if sound, are likely to be neglected by the public and by officials unless they are ratified by events—by the 'reality principle', or by what would be called 'catalytic crises'. These are crises like the oil spills off Santa Barbara, California, that are serious enough to mobilize the time and money and organizational efforts of thousands or millions of people. Such crises, especially when they reach public emotional awareness through mass media like television, lead to rapid growth of protest groups or constituencies. If the problems continue, this then leads to political mobilization, election of new legislators, passage of new laws to deal with the problem, and often to ongoing administrative agencies or other institutions. The Environmental Protection Agency and other agencies, which were set up within ten to twelve years after the publication of Carson's book, represent an ongoing change of governmental policy and a multi-billion dollar commitment to the new concern.

All this is, of course, a much oversimplified account of a complex process of mutual education and action with many tangled strands and thousands of heroes. The process is never smooth or easy, and usually involves opposition and backlash from established administrations and organizations. But this general sequence of steps seems to be present in almost all the successful reform movements in the United States in the past decade. It is very much like the traditional sequence of steps in proletarian revolutions, from the leadership of the intellectual vanguard to the workers' study groups, the growth of awareness, confrontations with the Establishment, and the seizure of power. The difference in these recent reforms within the United States system is that they are informational or organizational reforms, less violently resisted; and that the changes of policy can take place in the short time of three to ten years, as shown by the examples above, at least in this kind of high-education television society with wide-spread communications and leadership.

THE ROLE OF TELEVISION IN PARTICIPATORY CHANGE

The role of television in this kind of rapid wide-spread change should not be underemphasized. It is only half true that it leads to conformity of ideas, consumer purchases and acceptance of government policy. It is also emotional adrenaline that can lead to instant imitation, instant outrage, and instant demands. People take to politics or to the streets to protest, and parallel participatory protests spring up in a thousand cities at once.

This gives us a new kind of change movement, distributed everywhere with each local group initiating action. It differs from the hierarchical pyramid structure of the Army or the Church or the older revolutionary movements where only the leader or a central directorate had the brains and skill to direct the others. It is more like the decentralized feedback loops of a living organism such as the body—like the nervous system or the blood system. The pervasiveness and power of these new movements is shown by the fact that they have thrown out of office in the last ten years so many of the old type of hierarchical managers and manipulators in New York and Washington—'the best and the brightest'—who supposedly had all the advertising and propaganda and conformity on their side, pro-white-supermacy, pro-Vietnam-War, pro-Nixon, pro-pollution, pro-births, and pro-male.

These new movements have gone particularly fast because television reaches all ages and sectors simultaneously. New images and ideas can spread rapidly without having to wait for the older generation to be replaced by the young. The old idea of a 'social lag' of twenty-five years in adapting to technical change has therefore become obsolete. Today, it is social ideas and goals that can turn around in three to five years, as we have seen, and it is now technology that demands a long lag time for research and development to catch up with the new social demands, for example, in fusion power, solar energy, mass transit, and new food sources.

Similar social movements with rapid changes of attitudes and new group consciousness are now appearing in France, Italy, Japan and other countries with television. Television is spreading rapidly because it is the cheapest leisure and information source we have, and it may therefore bring wide-spread changes in attitudes toward international problems and towards improved social goals in the next few years.

The transition period ahead

CRISES OF THE NEXT DECADE

The crises of the next decade that need to be avoided or somehow controlled may be even more numerous and dangerous than the confrontations of the last thirty years. The reason is the increased intensity of global interactions and racial and religious conflicts, the increased level of arms, and the increased number of nuclear powers and of new nations demanding sovereignty and rights that were denied them for so long.

The result is that today we urgently need new catalytic analyses to interpret these major crises of the next decade, and to show how to shape new global institutions for their prevention or management. The crises may include:

Ecological disasters such as chemical spills from big tankers that may wipe out part of the food chain, or global atmospheric changes that affect the climate.

All-out nuclear war, or even terrorist nuclear destruction of major cities, producing savage reprisals or dictatorships.

Megafamines, with 10 to 50 million people starving in some areas after one or two bad crop years, without adequate supply or storage or early warning and management to get food to them in time.

New local wars or dictatorships that could trigger larger wars.

New oil and energy crises or world economic crises.

The collapse of development hopes in the poorest countries.

The above examples could be enlarged for many other scenarios and combinations that can be imagined.

OR MUTUALLY BENEFICIAL COVENANTS FOR WORLD MANAGEMENT?

On the other hand, these crises, fearful as they are, may offer the possibility of being catalytic if they are not too catastrophic—of being the contact with the reality principle that will finally awaken all of us to the urgent necessity of designing and building a more satisfactory global structure, arranging mutually beneficial covenants of such value to all nations that they are not likely to be broken. They might include:

An effective peace-keeping system, emphasizing the positive control of nuclear weapons and the early settlement of causes of dispute and injustice.

A world food stockpile and distribution system, with storage in many places and an early-warning network.

A world population planning covenant, to assure the co-operating nations that resources will be used for improving the quality of life rather than for increasing the number of mouths to be fed.

A world resources board, for improving the planning of resource production and use, and for improving the justice of the system, so that deprived nations may begin to get a fair share of the world's goods.

An ocean management board, for monitoring and controlling pollution and increasing the food and mineral harvest of the oceans on a sustainable basis.

Checks and balances on multinational corporations and other trading entities, with multinational environmental controls, multinational consumer protection, multinational labour organizations, and multinational anti-cartel acts, to enhance the human benefits and the distribution of resources through these powerful networks; and so on.

Special research and development needs for the evolutionary jump

Society is at least as complicated as an automobile. We need research and development teams to design and redesign and try out 10,000 new social arrangements and fit them together to give the best possible alternative future.

SPECIAL CHARACTERISTICS OF SOCIAL PROBLEMS TODAY

Many of our social problems today have special characteristics that we are just beginning to recognize. They are non-linear and aggregative and hierarchical, with feedback loops that may produce either rigidity or instability and escalation, and 'counter-intuitive' results, as Jay Forrester has emphasized. This makes them strikingly different from physics and chemistry problems, and any effective analysis must take these special features into account. These features include:

Hierarchical restructuring to new levels of organization and global management. As in a metamorphosis from caterpillar to butterfly, problems must be solved in an ongoing way, on a transition-path making use of the old systems while creating the new.

Rapid time-constants of change. Changes in existing societies are neither so strongly resisted as some claim (conservatives, radicals), nor so instantaneous as others suppose they can achieve (Utopians). It is essential to know the time-scales for effective change in different areas, whether they are short, as for recent attitude reversals in the United States (three to ten years), or long, as for basic technological research and development (ten to twenty-five years).

Locked-in patterns. Collective behaviour always gets locked in to self-maintaining patterns. Many are necessary and good, like food and trade networks, but others are damaging, like drug networks or governmental corruption. It is essential to understand how to interrupt such lock-ins and establish better patterns, and how to change them from the inside when necessary.

Catalytic analyses and catalytic crises. When conditions are ripe, a catalytic analysis may show millions how to change a bad situation. But it must be ratified by reality, by catalytic crises such as lake pollution, oil spills, or social crises, that show the accuracy of the analysis and mobilize us to effective action. Successful problem-solving will require an understanding of the forces at work and the points of effective intervention and the mechanisms of self-amplifying change.

Social traps. Individuals and societies are frequently trapped by immediate feedbacks or 'reinforcers' into doing things that are damaging in the long run to the group (competitive exploitation of resources, escalation of hostilities); or similarly barred by immediate disincentives from doing things that would be advantageous (saving for old age). We cannot 'do what we want' because there may be opposite short-run and long-run consequences. But analysis of such traps usually shows many individual and social ways to prevent entrapment or to get out, often very easily by changing the immediate reinforcers.

'Green-thumb' methods of inference for analysis and empirical cure of malfunctions in complex living systems. You cannot pull up a tree by its roots every day to see how it is doing. Small natural fluctuations must be used for evidence, and subtle cues must be looked for, and a good theory of subtle cues must be developed, such as seeing the effect of more water or shade on an ailing plant. We must be cautious and patient in our interventions, in the light of the thousands of feedbacks in the tree, or in the human body, or in society, that we do not yet understand.

Cybernetic or goal-directed stabilization mechanisms. The best international interventions may lead us astray because of unforeseen factors or changed conditions. So an ongoing lookout and anticipatory feedback must be designed into successful systems to keep them from going off the track or passing a point of no return. Internal feedback, or checks and balances, are needed for continual monitoring and correction of directing groups so that they do not go astray or misuse their power.

Justice. Justice as 'fairness' (Rawls), in which the lucky and affluent give their maximum help to the worst-off, is probably a necessary requirement of all viable problem-solutions, both on the transition-path and in every future society in a high-information, high-interaction world. This principle alone indicates the only politically viable solutions to many problems, from ghetto education to guerrilla terrorism and multinational resource pricing.

Decentralization of decisions to the lowest level possible. National legislators and administrators often try to make choices that could be made better and more cheaply and more enthusiastically at lower levels, by individuals or families or city or regional groups. On the other hand, some decisions are now too big for any one nation-state, and need to be pushed to a higher level, a super-ordinate authority, as in the case of ocean pollution or world food reserves.

Orchestration. Some must play one tune on the horns while others play a complementary tune on the violins; and the tunes change with time. Developing countries may need growth, while affluent countries are moving to a simpler life style. It is important to encourage diversity and complementarity, as with a good educational system or a good economic system, if various countries and groups are to preserve their cultural heritages with co-existence. We will need diverse experiments, not only nuclear plants but ecology farms—to get the data to make better judgements and combinations.

Match of individual and global solutions. For successful social solutions, the internalized attitudes and behaviour of individuals must support the new global structures or they will collapse. But correspondingly, the global structures must provide as much tolerance and freedom as possible for different cultures and nations so that individuals and groups and regions can develop as far as possible in their own way without damaging others.

Long-range cultural maintenance. Any viable social system or subsystem must teach its children, and must build in by practice, the attitudes and behaviours necessary to maintain the system and necessary to teach their own children in turn.

CLASSIFICATION OF NEEDED STUDIES

The following is a suggested classification of needed studies in technical and social research and development related to solving our world-wide problems of the next twenty years. It is a much shortened version of the list given by R. A. Cellarius and J. Platt in 'Councils of Urgent Studies', *Science*, Vol. 177, 1972, p. 670–6:

I. Physical technology and engineering related to crisis needs.

1. *Energy sources*
improved nuclear power, fossil fuels, non-fossil energy, direct solar power, wood and biofuels, increased efficiency of energy use, possible satellite solar power stations.

2. *Material resources*
water supply, minerals, improved land use, ocean, atmosphere management.

3. *Structures and replacement*
housing, public ways, city and regional planning, faster and lower-cost building and replacement.

4. *Transportation*
auto, rail, air, urban mass transit, marine transport.

5. *Electronics and communication*
improved communications and television for developing countries, satellite education and world communication, new printing methods, large-scale data handling, knowledge storage, indexing and retrieval.

6. *General physical and engineering problems*
technology assessment and monitoring, ocean use, pollution watch, disaster research, systems approaches, special technical problems and needs in developing countries.

II. Biotechnology

7. *Population problems*
better contraceptive methods, demography, genetic surveys, mobilities, forecasting.

8. *Food and famine*
new grains, new fertilizers, microbiological sources, food from oceans, novel sources, forecasting.

9. *Environmental problems*
monitoring, biological handling of pollution,

10. *Health—basic research*
microbiology, genetics, development, reproduction, ageing, neurosciences, world health education and services.

11. *Health—therapy*
disease research, low-level diagnosis, psychopharmacology, psychiatry and mental health, emergency care, nutrition, public health, clean water for everyone, food and drug monitoring, special problems of developing countries.

III. Behaviour and personal relations

12. *Behavioural research*
behaviour modification research, responsive-environment studies, child development, psychotherapy, learning with interpersonal games.

13. *Education*
classroom teaching, programmed and computer-assisted learning, language teaching, vocation skills and retraining, university structures and communities, adult education, teacher education, special problems of developing countries.

14. *Small groups*
methods of responsive living, family and neighbour relations, new housing and group-living experiences, child-care communities, group interactions in school and business and political groups, special new roles of confidant, therapist and ombudsman.

IV. National social structures

15. *Economics*
inflation, unemployment, relation to credit, economic aids to urban restructuring, problem-solving economic aids and incentives, reduced consumption of non-renewable resources, economic methods and payoffs for needed governmental restructuring, long-range systems analysis and normative theory.

16. *Organizations*
new management methods, improved information handling, computerization, systems analysis, small neighbourhood organizations, special problems of developing countries.

17. *Mass communications*
press, radio-television, networks, privacy problems, public feedback, news handling, new media, records and tapes, mass communications for community and world education, role in change, amplification of crises, systems analysis of effects.

18. *Politics*
improvement in public administration and management, responsiveness of élites and checks on them, science-technology inputs, party structures and instabilities, reduction of community hostilities, social indicators of choice and stress, opinion polls, mechanisms of stability and change, redesign of government and constitutional structures for improved effectiveness in a nuclear and TV age, systems analysis, theory, and long-range planning.

19. *Urban and rural problems*
structures, inflow-outflow, new élites and regional planning, political and community structures, law and justice, rural problems, family and community welfare, aesthetics and culture, national-local relations and co-ordination of overlapping governments.

20. *Large-scale change*
population pressures, quality of life, social indicators, theory of change and forecasting, philosophy of individual-group relationships, law and justice and non-punitive social reshaping, special problems in developing countries, reward systems for social inventions.

V. World structure

21. *Peace-keeping and peace-creating structures.*
United Nations revisions, crisis damping, new feedback-stabilization mechanisms, non-zero-sum game theory research, local war mediation, arms control and disarmament, military-industrial conversion, systems analysis of alternative world structures.

22. *Economic development*
investment and growth, monetary stabilization, large-scale housing, managerial education and feedbacks for competence and honesty, special problems of developing countries.

23. *Developing countries*
population, food, health, energy, education, television, preservation of cultural values, damping of hostilities, governmental restructuring, systems analysis, forecasting and planning.

VI. Channels of effectiveness

24. *Political and economic support of urgent research studies*
case studies of innovation, interdisciplinary centres for urgent studies, co-ordination of studies, political and economic channels for application and action, technical advisory services, self-supporting research organizations, public support.

25. *Systems analysis*
mapping of problem areas and progress, long-range world dynamics and systems analysis, transition-paths with matching of new innovations and education to desired long-range directions, democratic theory of group and social choices, and of checks and balances in the process of complex change, long-range evolutionary and normative philosophical and value structure.

Underlying all these areas of study and innovation must be the systems concept of optimizing the performance of the total world system, in human terms. We need to use better information and participatory design to eliminate blocks and malfunctions, anticipating and damping clashes and disasters, or finding anticipatory pay-offs to avert these or to compensate for them. The goal and the necessity is to raise hope and to improve visibly the quality of personal and neighbourhood and national life, improving especially the lot of the worst-off so as to share the benefits of an affluent society with equity and justice for all.

We are one human family, and we will not be able to make this great evolutionary jump successfully or even to survive, unless we begin to apply at least the same standards of competence and satisfaction in running the world that we apply in running a business, and at least the same standards of equity and love that we apply in our own families.

The contribution that science can make towards a creative society

M. W. Thring

Queen Mary College, University of London, United Kingdom

Statement of the problem

At the present time the world's population is increasing rapidly at about 3 per cent per annum in the countries which have not benefited from the industrial revolution and less than 1 per cent in the countries that have. As two-thirds of the world's population is in the former countries, we shall certainly have twice the world's present population by the year 2010 unless we have had a world-wide disaster killing hundreds of millions of people by weapons of war, or famine and pestilence. Thus, the basic problem for any far-sighted scientist concerned about the life of humanity in the twenty-first century is, how 7,000 million people can live within the limited resources of the earth. What are the necessary and sufficient conditions that must be fulfilled so that all men can live in a humanly satisfying equilibrium with the environment?

If we look at the world's problems from where we are now they seem insoluble, as in the case of the man who asked the way to Dublin and was told 'don't start from here'. The purpose of this article is to see where we must be and what conditions must be like in the next century if our grandchildren are to have a decent life. Then having decided what these conditions are, each of us will be more prepared to make the sacrifices of short-term interests and local objectives which are shown to be essential when one considers the condition the world will be in after our personal death. This must take full account of the earth resources available and the number of people using them, because one certain consequence of the technology we already have is that, for good or evil, no single country or even single continent is independent of every other one. Humanity has been shrunk to a single 'mankind' by the rapid transmission of goods, food, people, ideas and weapons.

It is becoming clear to all thoughtful people that decisions taken on a short-term basis, on the basis of the sectional interests of one group of people or on the basis of a rigid framework of preconceived theory, rendered irrelevant by changes in technology, can never enable humanity to avert disaster. Even the descendants of the group who are trying to look after their own interests will inevitably suffer the disaster within thirty to forty years. Four types of world-wide disasters are already threatening us and have been seen and written about by far-sighted people in many countries.

First, famine and pestilence caused by overpopulation; unstable cropping systems; pollution; dependence on exhaustible fossil fuels and phosphate deposits; climatic changes; drought, floods. It is useless to adopt a Micawber-like attitude and assume technology can solve all these problems when they occur.

Second, nuclear war. It has frequently been pointed out that the use of even a fraction of the nuclear weapons in their escalating stockpiles can only lead to both sides having their civilizations as well as those of the rest of the world completely destroyed. Yet looked at from the present short-term position, the politicians cannot find any alternative to spending an increasingly unaffordable fraction of their nations' resources on 'defence'. Ever since the atom bomb was used on Japan, the world has gone in fear of such a holocaust and, as the armouries build up, the fear increases.

Third, we see all over the world increasing local violence and vandalism, ranging from muggings and knifings to hijacking, kidnapping and random bombing. The applied scientist is responsible for the development of hand guns and convenient explosives which have facilitated many of these activities. Many such activities are largely due to a lack of opportunities for a life of creative self-fulfilment, and are increasing because such frustrated lives result from the short-term decisions of expediency. There is a real danger that in large parts of the world we shall revert to a stage in which everyone carries murderous weapons because they can no longer rely on protection by the forces of the law. This means that constant fear will ruin the lives of ordinary people.

Finally, there is the terrible situation called the 'Slough of Despond', a civilization in which the freedom of the individual to develop his own potential is so hedged in with restrictions and the surveillance of 'big brother' that life is scarcely worth living (zero quality of life). Again, there is a real danger in many parts of the world of this disaster reaching global proportions.

The present situation in the richer countries is very closely analogous to the final period of the Roman Empire; we have a breakdown of the classical, moral and ethical belief; penal taxation of those who work hard; bread and circuses for the lazy; vigour-sapping pollution, for example due to lead; imported labour to do the unpopular jobs;

corruption of various kinds and an erosion of the excitement and joy of life. However, we have one great advantage and that is that thousands of thoughtful people are beginning to see the disasters looming ahead and to study the action necessary to avoid them. It is quite clear that only by seeing the problem on a long-term, world-wide basis can we hope to find the way out.

The present human situation

Table 1 summarizes the harmful effects of technology. We are lost in the middle of a forest of trees, every one of which is so important that it has had a world conference about it, in spite of which, it grows worse. Short-

TABLE 1. The harmful effects of technology

	Cause	Cure
Group 1: Damage to the environment		
Air pollution	Cheap engineering	Good engineering
Water pollution	Cheap engineering	Good engineering
Land pollution, litter, dereliction, chemical disposal, fallout	Cheap engineering	Good engineering
Noise and vibration	Cheap engineering	Good engineering
Group 2: Damage to present humanity		
War machines	Careless engineering and sectional politics	Humanitarian politics
Poverty: underdevelopment countries	Careless engineering and sectional politics	Humanitarian politics and special engineering
Accidents	Cheap engineering	Good engineering
Unnatural life especially in cities	Careless engineering	Humane economics and special engineering
Group 3: Damage to future generations		
Fossil fuel depletion		Fuel economy
Radioactive waste from nuclear power stations	Cheap engineering	Good engineering
Metal ore depletion		Good engineering
Misuse of land: monocropping, hedges, burning stubble	Cheap engineering	Good husbandry
Misuse of fresh water	Cheap engineering	Water economy
Group 4: Human paid work		
Underemployment of talents		Good engineering
Unemployment		Humanitarian politics

term decisions avoid one tree and promptly lead us up against the next one. The result is a series of random paths which can only lead downhill to one or another of the world-scale disasters at the bottom.

The only hope for mankind is to use the compass of the conscience of the individual to find our way out of the wood to a decent world at the top.

There are seven basic groups of professional people who each have to climb their own professional ladder from the divisive, short-sighted, self-interest that is creating so much harm at the moment to a long-term service of humanity as a whole. In this article I am concerned primarily with three ladders, those of health, food and technology but these three cannot save mankind from disaster unless the other four also climb their ladders. I have found that all thinkers and writers on the subject of a decent future for mankind seem to converge on certain conclusions. Some of these are:

1. In the long term, the gap between the standards of living in different parts of the world must completely disappear. This does not mean that the differences of traditional life-style must in any way be destroyed, only that it is not a stable situation where some groups are thoroughly wasteful of the earth's limited resources, while others go short.
2. Within one generation, we must find a humane way of levelling off the world's population, since twice the present number will already tax the earth's resources severely.
3. Since a no-growth, unchanging equilibrium economic system is essential, people must measure success in life not by status symbols and wealth, but by creative self-fulfilment.
4. Since it is a basic human instinct to want to feel that society needs one, an interesting, worthwhile job must be available for all. This certainly implies a much higher proportion of people growing food than occurs at present in the rich countries.

The five principles for humane progress

The industrial revolution could have taken place with different financial, political, legal or philosophical systems, but not with a different technology. Perhaps, therefore, a technologist can be permitted to give his conclusions on some principles by which we can avoid the disaster to which the industrial and technological revolution is steadily leading us.

All dogma is disastrous. People matter infinitely more than either ideas or words, just as they are infinitely more important than money, possessions or machines. 'Thou shalt believe this because *X* has written it or said it' leads inevitably to people stifling their own consciences and thus to a gradual movement away from humanitarian actions. The only right principle is that of Buddha: 'Believe nothing anybody tells you, not even what I tell you.' People must be completely free inwardly to have their own thoughts and their own beliefs and all that any education or other outside influence can do is to give them material to help them to work out their own beliefs. The developed conscience of the individual is the only self-correcting basis for society. There are two main reasons why the blind acceptance of dogma is so dangerous. In the first place, *Y* will certainly have laid down dogma which is the exact opposite of *X*'s, so that adherents of dogma always split up into opposing groups and sects. Secondly, the more subtle the problem (and all matters of belief are necessarily subtle), the less accurately it can be conveyed in words, because everyone puts a slightly different interpretation on the words, and the use of words changes with time. Thus, any verbal formulation must be tested by one's own conscience before it is accepted.

Any society in which the selfish motive predominates in the average individual over the motive of a social ethic will necessarily destroy itself. In such a society, the centrifugal forces of crime, violence, corruption and graft will be too strong for the centripetal glue of public morality.

The consequence of this axiom is that since religious morality has lost its hold over so many people, we have to provide and develop a new ethos of public morality suited to an educated people used to technology. I call it new, but of course its basis must be the same as it has always been: the long-term, world-wide love of humanity that the Greeks called *agape*. The enemy of society is not the trade union or the right-winger, but the selfishness within each of us.

An equilibrium society must be based on voluntary public acceptance of the laws impinging on individual freedom. Unless the ordinary person understands and feels that they are for his or her direct or indirect benefit, they will be flouted. The classical examples are prohibition in the United States and the laws against drugs.

Violence is always harmful to the ordinary person and always breeds more violence. Hence, a humane person will always strive by his actions to reduce violence, and will only indulge in it when a situation of desperate, uncontrolled violence is already in existence. Examples where violence is necessary are a mad dog running amok and the action of Hitler in power.

No man can make a real contribution to the solution of these problems unless, before reaching a position of power, he has developed himself to the point where he is quite certain he will act according to what he feels, in his conscience, to be the best, long-term interests of humanity, rather than what he regards as most likely to keep himself in power. Lord Acton wrote: 'All power corrupts and absolute power corrupts absolutely.' This is true of all men who reach power without preparing themselves sufficiently against selfishness.

Men who acquire power for its own sake or, even worse, for self-enrichment will always be corrupted by it to the point where they will regard their political opponents as criminals to be restricted and damaged by all possible means. In recent history, we have seen several examples of this even in genuine democracies.

The way out of the forest

It is no use trying to look back in history and draw conclusions from earlier attempts at utopias because two factors have changed irreversibly when compared with all previous situations in history. The first of these is the tremendous growth in world population, which means that, even with all the resources of technology, we cannot provide everyone with an extravagantly high standard of living, such as is enjoyed in the rich countries at the present time.

The second factor is the immense power, for good or evil, given to mankind by technology. We have shrunk the world by our travel and transport machines and by our communication machines so that it is impossible for any one country to live in isolation from the rest of the world, whether in war or peace. We have created weapons of destruction so powerful that if a small fraction of them are released, the radioactive fall-out will cause all the next generation of children and young animals all over the world to be abnormal. With technology we have given mankind all the magic power imagined in the fairy tales of the past: flying carpets, the ability to move mountains, create great lakes, and even control the irrigation of a whole desert.

These two factors mean that the outmoded political, economic and nationalistic ideas of the past do not contain a feasible future for humanity. We have, therefore, to look at the problem afresh, but of course, on the basis of the traditional knowledge of the wishes and needs and possibilities of human beings.

Returning to the analogy of a group of men lost in a forest, the two things they would need to get out of the forest and to avoid falling into the pit in its centre, are a map and a compass. The compass must be the natural conscience of the individual. The fundamental idea which has been at the basis of all religions is that man has some spark of idealism in him which can enable him to make great efforts to use his creative talents to the full to contribute to a better world. In our present civilization he has forgotten how to use this compass because the primary aim which is taught implicitly in all schools is that success is measured by the acquisition of possessions and money and this, of course, destroys man's ability to contribute to a better world, that is, to follow the direction pointed to by the compass. The map must show the sunlit world outside the forest where all mankind can live a decent life in the twenty-first century. I shall try to show that it is possible to envisage a decent world in the next century subject to certain sacrifices of man's immediate short-term aims.

The creative society

It is quite certain that the population of the world will double once more to reach a figure of the order of 7,000 million early in the twenty-first century. It is also a certain fact that the population growth rate in the richer developed countries is less than one-third of that in the poorer developing countries. It follows from this that the only humane way of causing the growth of the world's population to stop at the figure of 7,000 million is to give everyone in the world a decent education and an adequate standard of living, so that each man can decide for himself to limit his family to the figure of two children each. The problem, therefore, is, how can we provide all these 7,000 million people with a decent standard of living? We can start by considering this from the point of view of energy and consumption since, apart from a few corrections, it is possible to regard the energy consumption of an individual as a direct measure of his standard of living—about 12 tons of coal equivalent. While there are certain correction factors for heating in winter and cooling in summer, it is, generally speaking, true that the standard of living of a people may be linearly related to this energy consumption figure. Now if we were to put all the energy consumption of the 7,000 million people with the American figure of 12 tons of coal equivalent, the world would be hopelessly unable to supply the necessary energy requirements. If, on the other hand, we are prepared to find a way of giving everybody a decent standard of living on a figure equal to the present world average of just under 2 tons, then it is possible to provide twice the present world population with that energy requirement. This implies that the rich countries have to be prepared to reduce their standard of living to the equivalent of the present world average. This is an essential requirement of the creative society.

At first sight this is a very depressing conclusion. However, this is only because our civilization has fundamentally mixed up two entirely different concepts: (a) standard of living or material wealth is a limited cake, so that if one group of people has several times its share, another must go short; and (b) quality of life, by which I understand a person's feeling that his life receives meaning and purpose from creative self-fulfilment. This has the property that, the more a person has of it, the more there is likely to be in his neighbour.

It so happens that the relation between these two concepts for an individual is exactly like the relation between the quantity of food intake and health, namely as one increases the quantity variable (food or money), the quality variable (health or quality of life) first increases and then reaches a peak at an optimum value of the input, and then falls with excessive consumption. Moreover, there is little doubt that the rich countries are in the affluent stage corresponding to obesity, while the poor ones are in the underfed stage. Hence the optimum is probably somewhere around the present world average. This means that the reduction of affluence, which world limitations and population increase will necessarily force on the rich countries can, in fact, increase their quality of life if they can learn to see the value of this quality instead of chasing affluence as their life's goal.

Other requirements of the creative society are that all consumer goods be made to last as long as possible and that as little fuel and raw materials as possible be consumed in their production and use. The whole conception of built-in obsolescence and fashion changes is inconsistent with an overpopulated world. It is also necessary to develop systems of transport which combine the very low energy consumption of the most efficient, moderate-speed public transport system with the convenience and safety and comfort of private transport.

The idea of the creative society is developed fully in my two books: *Man, Machines and Tomorrow* and *Machines—Masters or Slaves of Man*. Here I will simply summarize under five headings.

The first and key part is that we have to educate everybody and develop the idealistic part of them so that each person judges his success in life purely by his creative self-fulfilment, his personal contribution to a better or more beautiful world. When someone does a job, he must do it as much for the sake of making a good job of what he is doing as for the money which he earns for doing it. It is this excessive interest in money which has led us so far in the wrong direction. Fortunately, a few people still do their job well for the sake of doing it well, otherwise society would have broken down completely long ago.

Second, we have to have an equilibrium economics, an economic system in which 7,000 million people live in complete, stable equilibrium with the world. Of course, during the period that we are depending on coal and oil, this will not be a true stable equilibrium. It can be a transient equilibrium if we try to make these fuels last as long as possible by extreme economy and by using only the premium fuels for premium purposes, for example oil only for transport, and electricity only for power and lighting. Ultimately, of course, we must come to a stable equilibrium using solar energy alone. Similarly, in terms of food we have to be in stable equilibrium with the world. We have to develop irrigation systems, techniques for farming the sea, and conservation, so that we are not feeding ourselves at the expense of future generations. Other aspects of equilibrium economics are making machines last for ever, having no fashion changes and zero pollution; also recycling and reusing all waste materials, particularly turning all night soil, animal and vegetable refuse into good compost to go back on the land. We have to grow crops which make the maximum use of the solar energy falling on the land and of the limited fresh-water resources.

The third essential feature of the creative society is that a worthwhile and interesting job is provided for everyone, with a choice to suit himself, and that this job enables him to earn enough to have a decent standard of living for himself and his family. This does not mean, of course, that everyone must be paid identically, since those who undergo more training or take more responsibility should be paid more. On the other hand, all jobs involving danger and discomfort can be done by machines and this is one of the most challenging tasks for the engineer.

Fourthly, there will be leisure, education and opportunities for everyone to find a self-fulfilling creative hobby, other than that which he does in his work, which can restore the balance between head, heart and hands, which is so absent in our present society. This could be in the form of gardening, artistic activities of all kinds, craft activities of all kinds. We have to restore the respect for manual skill which existed in Jewish civilization, but which was largely destroyed by the ideas of Plato.

Fifthly, and perhaps most important, is the problem of education. Our present educational system falls down in three respects. It does not teach people to think for themselves and this is why they are prey to all kinds of outmoded ideas and demagogy and this had led to so many of our present troubles. The only hope for the future is that every intelligent person be taught to think for himself right back to first principles, particularly to his own motivation and purpose in life. It is this lack of education in regard to life motivation which is the second great failure of our present educational system. Students should be taught to keep before them constantly the idea, 'What am I going to make of my life?' 'Will I be satisfied with my life when it comes to the end?' 'Will I feel I have done the best I could with the possibilities I had?' 'Have I contributed to a better world?' They should also be taught to think about and understand the answers to these questions, which many great thinkers have posed through the centuries and answered for themselves.

Finally, we have to instruct people in such a way that all three brains are balanced and particularly that those who tend to be over-intellectual have a really good opportunity to acquire a manual skill in at least one field, whether it be music, or a craft or art, so that they learn to be balanced individuals.

How do we get there?

Religious moralities have failed to keep man from destroying himself with hatred and selfishness: one of the main reasons for this is that, generally speaking, the religions themselves have broken up into increasingly struggling sects by this same personal vanity and selfishness of the individuals leading them. It is my opinion that we have to have a secular morality which can appeal to people of all religions and to atheists and agnostics alike. This must be based on the love of humanity and of the environment and creative work. So much of our present educational system is perpetuating past hatreds by teaching people the hatreds and murders done in the past, telling them that they have to continue these hatreds and murders because of the past. We have to look forward to the future and realize that the whole of humanity is in the same boat, and it is no longer possible for one region or one group of people to survive at the expense of another. We have to draw out and educate the natural conscience of the young, their natural idealism which is as strong now as it has ever been in history, but which receives no help or encouragement from the establishment at all stages of education: kindergarten, schools and universities. We have to consider the moral aspects and consequences of what we are teaching and what we are doing. We have to be as responsible for the students becoming psychologically mature, that is, having their own sense of values and living by them as much as they possibly can. It is for this purpose that I have stressed the possibility of a 'Hippocratic Oath' as follows:

An Oath for Applied Scientists and Engineers
I vow to strive to apply my professional skills only to projects which, after conscientious examination, I believe will contribute to the goal of coexistence of all human beings in peace, human dignity and self-fulfilment.

I believe that this goal requires the provision of an adequate supply of the necessities of life (good food, air, water, clothing, and housing, access to man-made beauty), education, and opportunities to enable each person to work out for himself his life objectives and to develop creativeness and skill in the use of the hands as well as the head.

I vow to struggle through my work to minimize danger, noise, strain, or invasion or privacy of the individual as well as pollution of the earth, air, water, destruction of natural beauty, mineral resources and wildlife.

Above all, I think that we applied scientists and technologists have a tremendous responsibility. We have given mankind the magic powers and now we can see what the future possibilities of these magic powers are better than anyone else. We can see how science can be applied for the complete destruction of humanity or to give all the world's people a decent life.

Appendixes

I. Final report of the Unesco Symposium on the Scientific and Technological Revolution and the Social Sciences, Prague, 6–10 September 1976

Introduction

1. The International Unesco Conference on the Scientific and Technological Revolution and the Social Sciences was held in Prague from 6 to 10 September 1976, in accordance with Resolution 3.211 adopted by the General Conference of Unesco at its eighteenth session which, among other things, authorized the Director-General 'to contribute to basic thinking in the social sciences by organizing research projects and international meetings on selected topics' paying special attention to 'the social sciences' response to the social and human problems which derive from the scientific and technological revolution.

2. The conference was jointly organized by Unesco and the Czechoslovak Academy of Sciences in co-operation with the Czechoslovak Commission for Unesco. It was a great honour for the conference to be convened under the auspices of the Federal Prime Minister, Dr L. Strougal.

3. The conference was officially opened by Academician J. Poulik, Vice-President of the Czechoslovak Academy of Sciences. Greetings were also delivered by Academician J. Kozesnik, President of the same Academy. Mrs M. Hildebrandt, Assistant Director-General for Social Sciences of Unesco, addressed the conference on behalf of the Director-General and Mr V. Mshvenieradze, Director of the Division for the International Development of Social Sciences, Unesco, spoke on the social implications of the scientific and technological revolution.

4. Some twenty-five participants and fifteen observers took part in the conference. Twenty-one working papers were submitted and discussed. The official languages were English, French and Russian.

Summary of discussions

5. The social scientists, natural scientists and experts in the areas of technical and medical sciences who took part in the conference examined the current level of the development of scientific knowledge about the nature and

process of the recent advances in science and technology, often referred to as 'the scientific and technological revolution'. They discussed in particular the opportunities and problems that this opens, with special reference to its social and human implications.

6. Some theoretical as well as practical questions raised by the growth of the scientific and technological potential were examined with due regard to the different regional contexts in the world.

It was pointed out that the social sciences had to tackle with a greater vigour the social issues raised from the technological advances.

7. The conference noted the positive results obtained by the United Nations. Some examples are: the Declaration on the Use of Scientific and Technological Progress in the Interests of Peace and for Benefit of Mankind, No. 3384 (XXX); the Declaration and the Programme of Action on the Establishment of a New International Economic Order, Nos. 3201 (S-VI) and 3202 (S-IV); Development and International Economic Co-operation, No. 3362 (S-VII). While the sixth session laid down the notion of a 'New International Economic Order', the seventh session worked it out into a resolution on development and international economic co-operation.

At the regional plane, one may note the ideas included in the Final Act of the Conference on Security and Co-operation in Europe in 1975 and in the documents of the fifth Conference of Heads of State or Government of Non-aligned Countries in Colombo in 1976.

8. It was stressed that only by close collaboration of natural, technical, medical and social-scientific research, will it be possible to contribute effectively to general welfare in the context of the scientific and technological revolution. The emphasis laid on the social aspect of research into the scientific and technological revolution is fundamental to the successful application of interdisciplinary scientific results.

9. The current state of the scientific and technological revolution indicates that it can become one of the key instruments for the global and positive transformation of living conditions but can also pose a threat to the very existence of humanity. The following questions were raised:

Shall we live to see the turn of the millennium in peace, or will the enormous scientific and technological potential manifest its destructive power?

Shall we succeed in using the impetus of science and technology to close the 'civilization scissors', or will the unevenness of technological, economic and social development in the world further increase?

Will it be possible to control the processes of the scientific and technological revolution, and thus determine whether it is to be a great opportunity or a threat to mankind? Success will depend on the nature of societies, on their organization, their capability of managing their own development. The dangers involved are not inherent to science and technology, but are essentially social.

10. Here lies the fundamental task of developing the theoretical foundations for the practical management of social processes determining whether the achievements of the scientific and technological revolution are conducive to the betterment of the lot and life of populations. As technological progress accelerates, efforts towards solving social problems also have to be accelerated.

11. This gives rise to the key task of the social sciences as an irreplaceable instrument in helping to solve the great problems of our time. The social sciences must be developed more intensively to contribute to the solution of this contemporary task by utilizing all their potential and preventing their possible misuse or their avoidance of real current problems. This, of course, is not only a matter of improving theoretical and methodological levels. The whole complexity of the question arises because their development is linked with the general development of society.

An important task for the social sciences is to delineate the implications and consequences of the scientific and technological revolution by expanding public involvement in its understanding and in the solution of problems and the grasping of available opportunities.

12. If the social sciences are to make substantial contributions to overcome the contradiction existing between the different stages of development in science and technology, on the one hand, and in socio-economic development, on the other, they must play a new and increasingly significant role not merely by confining themselves to formulating goals, but by identifying ways to attain these goals. The public at large should be made aware of the implications and consequences of technological change.

13. Problems related to technological advances include, first and foremost, the question of world peace (the prevention of a world nuclear catastrophe and relaxation of international tension), which is a *conditio sine qua non* for furthering individual rights and liberties on a world-wide scale.

14. The elimination of socio-economic inequalities in different parts of the world should also be tackled urgently as a condition for peace. A 'new international economic order' would be designed for this purpose.

While the idea of a 'new international economic order' is widely supported, it has not yet been satisfactorily elaborated, especially in the area of social research. First of all, it is clear that progressive changes in international (external) economic relations cannot be isolated from national (internal) economic changes. It is also evident that such an innovative phenomenon is complex by nature and cannot be limited only to economic aspects.

15. The scientific and technological revolution continuously raises fresh dimensions in the relation between man and nature. The key to the interrelations between man and nature lies essentially in social processes, as is the case for the environmental problems arising from new technology. This only testifies to the crucial role to be performed by the social sciences. It is a task for them to analyse the processes whereby ways of life are subjected to change. It is also a social science domain to examine social changes in the light of various cultural or ethical viewpoints.

16. It is evident that accelerated socialization of economic and social processes has also presented an obstacle to the solution of fundamental problems. Modern science and technology are of concern to all citizens and not only to certain élites. The danger of the monopolization of decision-making power by technocrats can be met only by the broadest popular participation in policy and the full utilization of the potential of science and technology.

17. Global simulation models and world-wide development forecasts are becoming an important instrument of social science analysis. It is possible to

develop—in a measure by far exceeding that applied hitherto—comparative researches on key social phenomena and their mutation in different social systems.

A further step in the development of social sciences concerns the elaboration of indicators of social development. The data so far available on the international level unevenly cover only limited areas of social life without taking sufficiently into account the qualitative aspects of social phenomena and processes.

The efforts by Unesco to work out a system of social indicators, especially those concerning way of life and cultural development, are to be welcomed.

Priority areas for research

18. The conference proceeded to identify priority areas for research. The importance of interdisciplinary and international collaboration in undertaking the research was repeatedly emphasized.

19. On the question of the interaction between scientific, technological and social development, the following subareas were proposed for research:

Elaboration of the methodology for scientific technological and social prognosis.

The setting up of a system of technological, economic and socio-cultural indicators to evaluate the impact of scientific and technological development on people and societies.

The examination by the use of this methodology of trends of social development, by comparing studies and prognoses on scientific and technological developments so far elaborated.

20. The need was emphasized for research into interaction between natural, technical and social sciences both at the cognitive and institutional level, to strengthen the capacity of the social sciences in tackling complex social problems.

21. The importance of international scientific and technological co-operation was stressed. Multidisciplinary research appeared indispensable as an approach to possibilities and methods of applying science to the solution of problems facing the developing countries. The creation of domestic scientific communities capable of understanding and contributing to development in these countries is essential.

22. Another important area is concerned with the function, role and management of the system or research and development in the context of the scientific and technological revolution.

23. Another and more general area for research is on the interaction between man, society and nature under conditions of scientific and technological innovation.

24. It is also a high priority to conduct research on the interrelation between science and technology, on the one hand, and world peace, on the other.

This should include the question of preventing the use of precious resources and of scientific and technological potential for destructive purposes as well as the elimination of global inequalities.

25. The question of public participation was considered important both in learning about and applying the results of scientific activities in the social, political and cultural spheres.

26. Also important are studies in the methodology of social assessment of technology with special attention to: (a) the identification and selection of appropriate methods; (b) the interdisciplinary organization and execution of research; and (c) the integration of research results into the process of decision-making on technical change at various levels concerned with directing and managing such change.

27. The conference proposed a series of international comparative studies of the impact of certain technologies on: (a) working conditions; (b) urban development; (c) health and social welfare; and (d) the environment. These studies need to be conducted both in developed and developing countries in order to learn from different experiences, and to anticipate adverse effects in order to undertake preventive action.

28. A thorough study is necessary of the bases of social cohesion at local, national and international levels and of the forces leading to its erosion under the impact of the scientific and technological revolution.

29. Finally, research is required on the implications of modern technology and science for individuals and groups, including exploration of the possibilities of setting up indices of international welfare and resource evaluation so that development should not be defined in purely economic terms; and secondly, on the effect of social conditions on the structure of human personality, attitudes, motives and prejudices; on emotional spontaneity and control, especially in the earlier years of life; and conversely the effect of such personality patterns in determining the patterns of society. (This may be referred to as the psychosocial nexus.)

Recommendations to Unesco

30. The conference was invited to make specific recommendations for the programme of Unesco under its relevant projects. The following recommendations, based on the priority research areas described in the preceding paragraphs, are for appropriate action by Unesco over the coming years, starting in 1971.

31. Unesco is invited to:

Organize, in co-operation with appropriate United Nations agencies, National Commissions for Unesco, and non-governmental organizations such as the International Council for Science Policy Studies, expert meetings, seminars, conferences, and symposia on the interface between socio-cultural traditions and scientific progress.

Conduct research on the relations between the scientific and technological revolution and the strategy and policy for the development of science and technology.

Examine the state of co-operation among nations in the context of the scientific and technological revolution with Third World countries, paying particular attention to the question of alternative technology.

Study research and development (R&D) with special reference to its management process, policy and public participation.

Undertake research on science and technology as an instrument of economic, social and cultural development.

Study the implications of the imbalance in the developmental stages of the natural sciences on the one hand and those between natural and social sciences on the other.

Examine how underdeveloped and/or idle human potentials can be better utilized.

Study the ethical and legal aspects of the scientific and technological revolution.

Examine the participation and role of youth in the scientific and technological revolution.

Review and define the interaction and balance of human needs, both material and psychological, at different levels of development.

Study the optimum rates of social change and direction in view of the different needs of individuals, groups and society.

32. The conference appreciated the offers made by the Institute for Applied Systems Analysis, of the 'Club of Rome', of the Centre for the Study of Science, Technology and Development, the Indian Council for Science and Industrial Research, and of the Academy of Science of the German Democratic Republic to contribute to the activities mentioned above.

33. Unesco is further invited to:

Publish the proceedings of the conference in book form and to establish for this purpose an editorial board, preferably with one of the conference participants as editor.

Convene in 1978 an international meeting to organize interdisciplinary research on the social implications of the scientific and technological revolution.

Ensure the widest participation of social scientists in the preparation for the United Nations Conference on Science and Technology, to take place in 1979.

Conclusion

34. The conference was officially closed by the addresses delivered by Mr V. V. Mshvenieradze, Director of the Division for International Development of Social Sciences of Unesco, and Academician J. Poulik, Vice-President of the Czechoslovak Academy of Science. The conference adopted two resolutions, one endorsing and supporting Unesco initiatives and the other appreciating the excellent organization of the Conference by the Czechoslovak Academy of Sciences.

Resolution I

The participants at the Conference on the Scientific and Technological Revolution and the Social Sciences organized under the joint auspices of Unesco and the Czechoslovak Academy of Sciences in Prague, 6–10 September 1976, greatly welcome the initiative taken by Unesco in organizing this important conference and thus enabling social and natural scientists to engage in a multidisciplinary dialogue on the implications of the scientific and technological revolution on man, society and nature. They further note with appreciation that Unesco emphasizes the social sciences and tries to establish close collaboration between the social and natural sciences to improve understanding and analysis of the social aspects of the scientific and technological revolution.

The participants attach great value to Unesco's medium-term programme for the years 1977–82. They consider further that Unesco should continue to convene conferences of this kind with a view to increasing the role of social scientific research on the complex problems related to the scientific and technological revolution.

They recommend to Unesco several important tasks to deal with some of the major problems facing mankind today. These tasks are in harmony with various objectives set forth in the Medium-term Plan of Unesco for 1977–82 (quoted at the end of this resolution.)

While the participants are to bring the conclusions of the conference to the attention of their respective National Commissions and governments, Unesco is also invited to bring the deliberations and recommendations to the special attention of various National Commissions.

Objective 2.1 Promotion of peace research, in particular on manifestations of violation of peace, causes preventing its realization, ways and means to eliminate them and proper measures to be taken in order to maintain and reinforce a just, lasting and constructive peace at the level of groups, societies and the world.

Objective 3.1 Promotion of the formulation of a global, multidisciplinary interpretation of development, having regard to the interrelations between the various contributing factors to this and which are, in turn, affected by it.

Objective 3.2 Studies of socio-cultural conditions, systems of values, motivations and procedures for participation by the population likely to foster endogenous, diversified development processes in keeping with the practical conditions and needs of the different societies.

Objective 3.3 Contribution to the development of infrastructures and programmes in the social sciences with a view to increasing the ability of different societies to find ways of solving social and human problems.

Objective 3.4 Development and application of tools and methods of socio-economic analysis and development planning.

Objective 4.1 Investigation of interactions between science, technology and society, as well as of the implications of scientific and technological change for man, within the context of the long-term development of science and technology in line with social progress and changing ways of life.

Objective 4.2 Promotion of the formulation and application of policies and improvement of planning and financing in the fields of science and technology.

Objective 4.4 Development of a better understanding of the nature of science and technology and of their role in a changing society, by improving and extending teaching in school and adult education, and by promoting public information in these fields.

Resolution II

The participants express their appreciation and thanks to the Organizing Committee for its excellent preparation of the conference.

They wish to thank the Czechoslovak Government, the Czechoslovak Academy of Sciences as well as the Czechoslovak National Commission for Unesco for the warm and generous hospitality extended to the participants.

They also wish to record their thanks to all those who helped in various ways to make the conference a success.

Attendance at the conference

PARTICIPANTS

Professor A. Mahamoudou Cissé, École Normale Supérieure, Bamako, Mali.*
Academician P. N. Fedoseyev, Vice-president of the Academy of Sciences of the U.S.S.R., Leninskij prospekt 14, Moskva V–71, USSR.*
Assistant. Professor J. Filipec, Institute for Philosophy and Sociology, Czechoslovak Academy of Sciences, Praha 1, Jilská 1, Czechoslovakia.*
Dr G. Friedrichs, Wilhelm-Leuschner-Strasse 79–85, 6 Frankfurt-am-Main, Federal Republic of Germany.*
Dr D. M. Gatovskiy, Corresponding member of the Academy of Sciences of the USSR, Institut Ekonomii AN SSSR Volchonka 14, Moskva G–19, USSR.*
Dr D. M. Gvishiani, Corresponding member of the Academy of Sciences of the USSR, Deputy Chairman of the State Council for Science and Technology, USSR Council of Ministers, ul. Gorkogo II Moskva, USSR.*
Professor E. Hegazy, Sana'a University, P.O. Box 1273, Sana'a Yemen Arab Republic.*
Professor H. F. Hrubecky, School of Engineering, Tulane University, New Orleans, LA 70118, United States of America.
Dr S. Kansu, Head of the Methodological Research Unit, National Centre for Social and Criminological Research, Gezira P.O., Cairo, Egypt.*
Dr N. M. Karanda, University of Dar es Salaam, P.O. Box 35062, Dar es Salaam, United Republic of Tanzania.
Academician J. Kozesnik, President of the Czechoslovak Academy of Sciences, Narodni 3, Praha 1, Czechoslovakia.
Professor dr G Kröber, Direktor des Instituts für Wissenschaftstheorie und Organisation, Akademie der Wissenschaften der DDR, 108 Berlin, Otto-Nuschke-Strasse 22–23, German Democratic Republic.*
Professor E. Mendelsohn, Department of the History of Science, Harvard University, Cambridge, Massachusetts 02138, United States of America.*
Professor H. Okamoto, Hosei University, Faculty of Business Administration, 2–17–1 Fujimi, Chiyoda-Ky, Tokyo, Japan.*

* Authors of papers which are included in the present publication.

Professor E. Olszewski, Committee for Science of Polish Academy of Sciences, Warsaw Technical University, al. Niepodlegtosci 222–13 a, 00–663 Warszawa, Poland.
Dr A. Peccei, President of Club of Rome, Via Giorgione 00147 Rome, Italy.*
Professor P. Piganiol, 5, Rue le Dantec, 75013 Paris, France.*
Professor J. Platt, Mental Health Research Institute, University of Michigan, Ann Arbor, Ml 48104, United States of America.*
Academician J. Poulik, Vice-President of the Czechoslovak Academy of Sciences, Národní 3, Praha 1, Czechoslovakia.
Dr A. Rahman, Council of Scientific and Industrial Research, Rafi Marg, New Delhi 110001, India.*
Academician R. Richta, Director of the Institute of Philosophy and Sociology, Czechoslovak Academy of Sciences, Jilská 1, Praha 1, Czechoslovakia.*
Dr S. Sigal (Peru), Centre d'Etudes des Mouvements Sociaux, Bureau 814, 54 Boulevard Raspail, 75006 Paris, France
Dr K. Stroetmann, Vice-President and Senior Research Associate, Abt Associates GmbH, Tiergartenstrasse 15, D-6900 Heidelberg 1, Federal Republic of Germany.*
Dr G. R. Taylor, The Hall, Freshford, Bath BA3 6EJ, United Kingdom.*
Professor M. W. Thring, Queen Mary College, University of London, Mile End Road, London E1 4NS, United Kingdom.*

OBSERVERS

Dr F. Adler, European Co-ordination Centre for Research and Documentation in Social Sciences, B.P. 974, Grunangergasse 2, A-1010 Vienne I, Austria.
Academician S. Ganovski, President of World Association of Philosophy Bulgarian Academy of Sciences, ul. '7-noembri' 1, Sofia, Bulgaria.*
Ing. M. Holub, CSc., Institute for Economy of the Czechoslovak Academy of Sciences, Třida plitických věznu 7, Praha 1, Czechoslovakia.
Doc. dr. J. Kunz, CSc., Institute for State and Law of the Czechoslovak Academy of Sciences, Národní 18, Praha 1, Czechoslovakia.
Academician B. Kvasil, Rector of the Czech Technical University, Horská 4, Praha 2, Czechoslovakia.
Dr S. R. Mikulinskiy, Corresponding member of the Academy of Sciences of the USSR, Director Instituta Istorii nauki Techniki, AN SSSR, Staropanskij pereulok 1/5, Moskva K-12, USSR.
Dr S. C. Mills (United Kingdom), European Co-ordination Centre for Research and Documentation in Social Sciences, B.P. 974. Grunangergasse 2, A-1010 Wien I, Austria.
Dr P. Maydl, CSc., Director of European Centre for Education and Leisure, Jílská 1, Praha 1, Czechoslovakia.
Professor dr. T. R. Niederland, CIOMS, 3 interná klinika LF University Komenského, Bratislava-Kramáre, nemocnice, Czechoslovakia.
Dr L. Novy, CSc., IOPHS, Institute for World and Czechoslovak History of the Czechoslovak Academy of Sciences, Vyšehradska 49, Praha 2, Czechoslovakia.

Doc. dr. J. Patek, Secretary of the Czechoslovak Commission for Unesco, Valdštejnské nám. 1, Praha 1, Czechoslovakia.
Professor S. N. Smirnov, Department of Social Sciences, Academy of Sciences of USSR, Leninskij prospekt 14, Moskva V-71, USSR.
Academician A. Szalai, UNITAR, Attila-út 125, H 1012 Budapest, Hungary.
Ing. T. Vasko, Federal Ministry of Technics and Development, Slezská 9, Praha 2, Czechoslovakia.

UNESCO SECRETARIAT

Dr M. Hildebrandt, Assistant Director-General for Social Sciences and their Applications.
Dr V. Mshvenieradze, Director, Division for International Development of Social Sciences.*
Dr K. Delev, Senior Programme Specialist, Division of Philosophy.
Dr T. Uchida, Programme Specialist, Division for International Development of Social Sciences.

EDITOR

Professor Robert S. Cohen, Center for Philosophy and History of Science, Boston University, Boston MA 02215, United States.

MUMFORD, L. *The Myth of the Machine*. Vol. 1: *Technics and Human Development*; Vol. 2: *The Pentagon of Power*. New York, Harcourt, Brace, Jovanovich, 1967 and 1970.
The final and culminating work of Mumford's decades of study of the relations between technology and science on the one hand and human existence on the other, by the author of the fundamental work on the social history of technology, *Technics and Civilization* of 1934. All of Mumford's books contain lengthy and usefully annotated bibliographies.

NOBLE, D. F. *America by Design: Science, Technology, and the Rise of Corporate Capitalism*, New York, Knopf, 1977.
A provocative and useful analysis of science and technology and their planning, management and social functioning in the development and consolidation of American industrial capitalism (1880–1930).

Man-Science-Technology: A Marxist-Leninist Analysis of the Scientific and Technological Revolution. Moscow and Prague, Academy of Sciences of the USSR and the Czechoslovak Academy of Sciences, 1973.

The Social Assessment of Technology', a symposium in Unesco's *International Social Science Journal* (Paris), Vol. XXV, No. 3, 1973.

Author index

Subject index

The social implications of the scientific and technological revolution

a Unesco symposium

The social implications of the scientific and technological revolution

a Unesco symposium

Published in 1981 by the United Nations
Educational, Scientific and Cultural Organization
7 Place de Fontenoy, 75700 Paris
Photoset by Thomson Press, New Delhi, India,
and printed by Offset-Aubin, Poitiers, France

ISBN 92-3-101664-4

French edition: 92-3-201664-8
Spanish edition: 92-3-301664-1